Crop Husbandry Regimes presents a new synthesis of ancient agriculture in northern England. It is based on a large and completely new data set of charred seed assemblages from late prehistoric and Roman-period settlements in the region. The results are set in their British and Western European contexts.

This case-study provides the basis for a detailed discussion of the various analytical methods used by archaeobotanists today. The case for an ecological approach is clearly stated and the use of multivariate statistical techniques to improve understanding of the structure of such data and to advance the level of their interpretation is advocated.

It is demonstrated that the archaeological relevance of archaeobotany need not be confined to conventional statements concerning the presence of certain crops; this study indicates that differences in the *scale* of arable production can now be identified. Both the cereal crops and their associated weed assemblages indicate that two distinct crop husbandry regimes were present in the area of study during the late Iron Age. One represents small-scale, subsistence agriculture, while the other suggests arable expansion.

The book is written in straightforward, clear language and is aimed at both non-specialist archaeologists and botanists. While the case study is concerned with one specific region, the approach and the methods of analysis described are applicable to charred seed assemblages from any region or period in Old World archaeology.

CROP HUSBANDRY REGIMES

CROP HUSBANDRY REGIMES

An Archaeobotanical Study of Farming in northern England

1000 BC – AD 500

Marijke van der Veen

Sheffield Archaeological Monographs 3

J.R. Collis Publications
Department of Archaeology and Prehistory
University of Sheffield
1992

© M. van der Veen 1992
Publisher: J.R. Collis
Editor: R.B. Adams
Cover design: Marijke van der Veen
Cover realization: David Schofield

This publication was made possible by
generous grants from Durham University,
English Heritage and the Stichting Nederlands
Museum voor Anthropologie en Praehistorie.

A catalogue record of this book is available
from the British Library.

ISBN 0 906090 41 5

Copies of this volume and a catalogue of other
publications by the Department of Archaeology & Prehistory,
University of Sheffield can be obtained from:

J.R. Collis Publications
Department of Archaeology & Prehistory
University of Sheffield
Sheffield   S10 2TN

Printed in Great Britain by

The Dorset Press
Dorchester

Aan mijn ouders

Contents

Extended Table of Contents

List of Figures

List of Tables in the Appendix

Acknowledgements

This book is based on my doctoral thesis which was submitted at the University of Sheffield in August 1990. I am grateful to the Science and Engineering Research Council for granting me an Instant Award which allowed me to carry out this research.

I would like to thank the archaeological directors of the excavations, Paul Bidwell, Peter Clack, Leon Fitts, Tim Gates, Colin Haselgrove, David Heslop, Neil Holbrook, Ian and George Jobey, Chris Smith, and Percival Turnbull for their cooperation at all stages of the work.

Thanks are also due to Prof. W. G. Mook and the Centre for Isotope Research of the University of Groningen, The Netherlands, for granting me the radio-carbon dates, and Jan Lanting for his help and advice regarding these dates and their calibration. I am also indebted to the Programme Advisory Panel of the Science Based Archaeology Committee of the SERC for granting me the accelerator dates, Rupert Housley for his advice regarding these dates, and the Oxford University Radiocarbon Accelerator Unit for providing the dates.

The advice offered to me by Nick Fieller, Brian Huntley, Glynis Jones and Roger Richards regarding the statistics and the computer analyses of the data was very much appreciated. I would like to thank Yvonne Beadnell for general advice on the illustrations and for figures 6.1 and 6.2, David Schofield for realising my design for the front cover and for figures 10.1-3, 10.6, 10.8, 10.10, 10.12, 10.14, 10.16-20, Pauline Fenwick for all remaining figures, and Trevor Woods for their photographic reproduction. I am grateful to Eric Cambridge for his general advice on the text and Caroline Mason for proof-reading the original text.

I would like to mention those colleagues who have particularly influenced and stimulated my research: Willem van Zeist who introduced me to the field of archaeobotany, and David Gilbertson, Colin Haselgrove, Gordon Hillman, Glynis Jones, Martin Jones, and Martin Millett whose ideas and discussions over the years have helped me in the formulation of the issues discussed in this book.

This publication was made possible by generous grants from Durham University, English Heritage and the Stichting Nederlands Museum voor Anthropologie en Praehistorie.

Durham, December 1991.

1. Introduction

Until recently it was thought that arable farming played a minor role in the economy of the Iron Age people living in northern Britain. Piggott described the people living north of the Humber at that time as: 'Celtic cow-boys and shepherds, footloose and unpredictable, moving with their animals over rough pasture and moorland' (Piggott 1958, 25). In his study of the Iron Age economy of Britain, Piggott recognized two types of economy. Firstly, the Woodbury-type, prevalent in the south and east of England, consisting of stable farming communities, growing cereal crops, using well ordered field systems and capable of producing an arable surplus. Secondly, the Stanwick-type of economy, characteristic of the north and west of the country, an economy based on pastoralism and with a probable element of limited nomadism. Arable agriculture in this region was thought to be limited to hoe-cultivation in the Bronze Age tradition. The area north of the Tyne-Solway line was assumed to have remained in a retarded Bronze Age cultural tradition, until refugees from southern Britain arrived in the first century BC (Piggott 1958).

Piggott named the economy of northern Britain after the site at Stanwick in the Tees lowlands, where Wheeler had carried out substantial excavations a few years earlier. On the basis of his work at Stanwick, Wheeler had concluded: 'the crude local pottery of Brigantia faithfully reflects a crude pastoral, semi-nomadic economy; good pottery requires the stimulus and opportunity of a settled agriculture, and its absence at Stanwick helps to complete a consistent outline of a social system in which agriculture played a subordinate part' (Wheeler 1954, 30). Regarding Caesar's generalizations about the British people Wheeler felt 'here he seems to emerge with credit: 'of the inlanders, most do not sow corn, but live on milk and flesh, and clothe themselves in skins" (Wheeler 1954, 29, quoting Caesar, *De Bello Gallico* 5, 14).

During most of the Roman period the bulk of the Roman army was stationed in the north of the country. The supposed absence or subordinate role of arable farming in the North meant that the grain supply of the Roman army was thought to have come from some distance, either from the south of the country or from abroad (Piggott 1958, Rivet 1969, Wheeler 1954).

Underlying this viewpoint was the more general assumption that the climatic and soil conditions in the north of the country made the area unsuitable for arable production. Variations in the settlement pattern, farming practices, etc. in the different parts of Britain have often been explained by differences in the environmental conditions in those regions (Evans 1975, Faechem 1973, Fowler 1983, Piggott 1958, Wheeler 1954). Archaeologists have often used a geographical division of Britain into a Highland and a Lowland Zone. The dividing line is drawn from the mouth of the River Exe in the South West to the mouth of the River Tees in the North East, separating the outcrop of the old Palaeozoic rocks from the younger Mesozoic and Tertiary rocks (Figure 1.1; Evans 1975, Fox 1932, Stamp and Beaver 1971). The Highland Zone lies to the west and north of this line and is characterized by upland regions with 'old, hard' types of bedrock, thin and impoverished soils, and high levels of annual rainfall. The Lowland Zone lies to the south and east of this line and is characterized by lower hills and plains, 'younger and softer' bedrock, fertile soils, and lower levels of annual rainfall. Today most of Britain's cereal production takes place in the Lowland Zone. The Highland Zone is largely, though certainly not exclusively, characterized by animal husbandry and extensive hill farming.

Present evidence suggests that the dichotomy between these two zones came into being with the onset of the climatic deterioration in the late second and early first millennium BC (Evans 1975, Faechem 1973), when an increase in rainfall and a drop in annual temperatures of *ca*. 2 °C , combined with increasing human interference with the landscape, caused a deterioration of the soils and vegetation. In the Highland Zone the regeneration of the soils was insufficient, due to the specific characteristics of the rock and soil types, and the high levels of annual rainfall. The consequent development of vast stretches of podzolic soils and peat greatly limited the subsequent development of arable agriculture in the upland regions of the Highland Zone.

The division of the country into just two zones is, of course, an over-generalization. There are areas of good agricultural land within the Highland Zone, and areas unsuitable for arable farming in the Lowland Zone. Stamp and Beaver described the situation as follows: 'In general, the cultivable and habitable tracts of the Highland Zone form tongues of varying size projecting into a great expanse of moorland and hill pasture; in the Lowland Zone, it is the infertile uplands which tend to form islands in a sea of cultivable and habitable land' (Stamp and Beaver 1971, 238). Fox felt that this broad division was justified as he regarded wealth to be 'mainly based on broad tracts of cultivable land', and he suggested that the 'intermont and coastal lowlands of the Highland Zone are for the most part too scattered and too limited in area to provide the necessary economic basis for independent development' (Fox 1932, 28). In contrast, the presence of these stretches of good arable land within the general area of the Highland Zone has been emphasized by Manning (1975) in his discussion of the grain supply of the Roman army. He pointed out that the documentary evidence indicated that the military garrisons were supplied with grain from the closest available source due to the high cost of overland transport, and that the movement of supplies over long distances overland was avoided wherever possible (Manning 1975). He argued that the areas of good arable land within the Highland Zone were sufficiently large to provide at least a considerable part of the army's requirements 'with all that that involves for the expansion of agriculture and its effect on the landscape' (Manning 1975, 116).

Piggott had also partly based his view of the economy of northern Britain on the absence in the North of storage pits for grain, Celtic fields, and finds of carbonized grain of prehistoric date. Recent work has changed this picture quite considerably. While storage pits for grain have still not been found, this is probably due to differences in soil type (Bradley 1978). Field systems have now been recorded for many parts of northern Britain due to extensive fieldwork and aerial photography (for north-east England e.g. Gates 1982a, 1983, Harding 1979, Topping 1989a, 1989b), and the results of recent excavations at Stanwick have called for a re-interpretation of the nature of this important site (Haselgrove 1990, Haselgrove and Turnbull 1983, 1984). Finally, finds of carbonized grain have also been recovered from the Highland Zone since Piggott's study (e.g. Boyd 1988, Caseldine 1989, Hillman 1978, forthcoming a and b, Van der Veen 1985c, 1987a, 1987b, Van der Veen and Haselgrove 1983), although the database is still very small. As a result, Piggott's view of the north of Britain as a sparsely populated region with a subsistence base of semi-nomadic pastoralism no longer finds general favour. Recent surveys of the archaeology of northern Britain (Clack and Haselgrove 1982, Harding 1982, Chapman and Mytum 1983, Miket and Burgess 1984) are all rather cautious in expressing their ideas about the prevailing economy. It is generally accepted that the population was larger than originally postulated and that some form of crop production was present, especially in the lowland coastal areas. However, notions about the role of arable farming within the overall subsistence base remain vague, largely due to a persistent lack of evidence. Cunliffe concluded that 'intensively farmed agricultural landscapes like those in southern Britain were largely unknown in the North. The inescapable implication must be that the food producing strategy was differently structured, with a far greater

emphasis on pastoral activities though not to the exclusion of crop production' (Cunliffe 1983, 88).

In his recent syntheses of the development of crop husbandry in Britain, M. Jones (1981, 1984a) recognized two important periods of change in the history of crop production. During the first half of the first millennium BC a number of new crop plants were introduced, such as spelt wheat, bread/club wheat, oats, rye, and Celtic bean, and emmer wheat was replaced by spelt wheat as the principal wheat crop. A second important change took place during the first millennium AD, when the glume wheats (emmer and spelt) were replaced by free-threshing cereals such as bread/club wheat and rye. These changes have been recorded across large parts of Britain and, to a lesser extent, also across northern Europe, but regional differences in the time when and the degree to which these changes were adopted have been recognized (M. Jones 1984a). Jones suggested, in fact, that the replacement of emmer by spelt wheat did not take place in the Highland Zone where emmer remained the commonest wheat (M. Jones 1981).

Jones postulated that during the first half of the first millennium BC an increase in population and a decline in soil fertility must have triggered an increase in the scale of arable production, involving a diversification of both the crops and the soils cultivated (M. Jones 1981). This phase was followed in the first millennium AD by a specialization in particular crops (bread/club wheat and rye) which later became the major crops of the historical period (M. Jones 1984a). The regional patterning in the adoption of these changes was taken to 'demonstrate the insufficiency of demographic pressure alone to explain these agricultural changes, and the importance of social factors that may introduce variation at both the regional and local level' (M. Jones 1984a, 124). The data base for Britain is still very patchy and for most of the Highland Zone no information is available at all, so that this regional patterning and its underlying causes are still poorly understood.

It is the aim of the present study to analyse and interpret recently collected archaeobotanical data from one part of the Highland Zone, in order to improve our understanding of the role of arable farming in this region and to assess the extent to which the increase in the scale of arable production, as witnessed in parts of Britain and Europe, took place in this region.

The region chosen for this study is north-east England. This area forms the eastern part of the Highland Zone in England (Figure 1.1), and is defined as the area south of the Tweed Valley, north of the Tees Valley, and bounded to the west by the hills of the Pennines and Cheviots. The area roughly

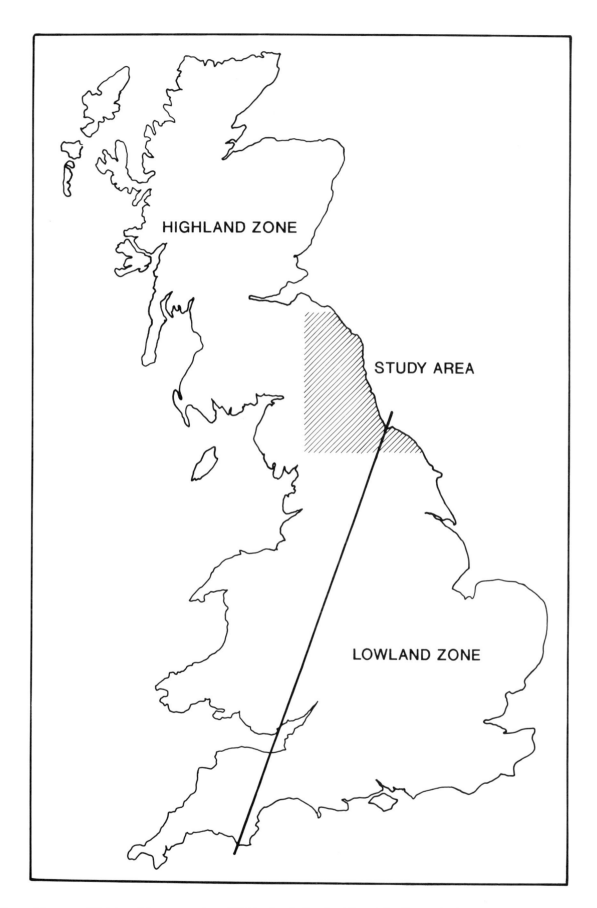

Figure 1.1   Division of the country into a Highland and a Lowland Zone (after Evans 1975, Fox 1932, Stamp and Beaver 1971), and location of the study area.

coincides with the present-day counties of Northumberland, Tyne and Wear, Durham, Cleveland, and a small part of North Yorkshire. The southern part of the study region forms the boundary between the Highland and the Lowland Zones. The region does, in fact, form one of the lowland areas within the Highland Zone, but the site of Stanwick which gave its name to Piggott's Highland Zone economy, is located within the study region.

The data used in this study are carbonized seed assemblages collected by the writer between October 1981 and October 1987 from seven later prehistoric and two Roman period sites in the region. With 325 samples and *ca.* 89,000 seeds this data base represents one of the largest collections of charred seeds from the Highland Zone to date. Carbonized seed assemblages represent, in virtually all cases, the harvested cereal crops and their associated impurities (M. Jones 1984b). As such they form an ideal type of data to study the development of arable farming.

The structure of the present study is as follows. In Chapter 2 the physical and cultural background of the region is described. Chapter 3 gives an account of the methods of analysis used in this study. Chapter 4 presents a catalogue of the sites and plant remains examined. Chapter 5 discusses the results of the radio-carbon dating programme carried out on the plant remains. Chapter 6 gives a general discussion of the results of the archaeobotanical analysis. Chapter 7 analyses the taphonomic role of crop processing. Chapter 8 attempts to make an economic classification of each site. Chapter 9 describes the different approaches to weed ecology and their application to archaeobotanical data. Chapter 10 discusses the results of the multivariate statistical analysis of the data, using the autecological data for the weed species. Chapter 11 puts the archaeobotanical results into their wider archaeological context. In Chapter 12 the results are summarized and some conclusions are drawn.

The convention for referring to radio-carbon dates in this study is as follows. Actual radio-carbon dates are given in uncalibrated radio-carbon years BP, whereby the present is defined as AD 1950. General references to chronological periods, such as 'the first millennium BC', are expressed in uncalibrated radio-carbon years BC or AD. Calibrated radio-carbon dates are referred to as calendar years cal BC or cal AD. The convention of using upper and lower cases, i.e. bc/BC or ad/AD, for uncalibrated and calibrated dates respectively, is not adhered to here.

2. Background Information to the Region

2.1 Physical background

2.1.1 Relief

The north east of England can be broadly divided into two north-south running zones, the lowland coastal plain in the east and the uplands of the Cheviot and Pennine hills to the west (Figure 2.1). The coastal plain is a fairly narrow strip of land less than 200 m O.D., some 20 km wide in the northern part of Northumberland, becoming wider (*ca.* 30–40 km) further south. The upland hills reach altitudes of 500 m O.D. and more. Rain water run-off is drained into the North Sea by a series of west-east running rivers which dissect the region, the most important ones being the Rivers Tweed, Tyne and Tees (Figure 2.1).

2.1.2 Climate

In the atlas of agro-climatic regions of Europe (Thran and Broekhuizen 1965) the north east of England is classified as region 46 (Figure 2.2). This region is characterized as wet and temperate, cool in summer and mild in winter. The average annual temperature is 7.5 $^\circ$C, there are only four months of the year with average temperatures above 10 $^\circ$C, and there is no month with average temperatures above 15 $^\circ$C (Thran and Broekhuizen 1965).

A slightly different characterization is reached when a British classification is used, i.e. a classification looking at the British climate only. Shirlaw (1966) divides Britain into six climatic regions (Figure 2.2). North-east England falls into the region described as having a 'cold dry winter and cool dry summer', but the most western, upland part of the region falls into another zone characterized by 'cold wet winters and cool wet summers' (Shirlaw 1966).

The fact that the most western part of the region falls into a different climatic zone from that of the eastern part is a direct reflection of the influence of relief on the climate. The main relief feature in the north of England is the upland spine running north-south. The prevailing winds are westerly and moist and, since the uplands force the air to rise and lose much of its moisture as precipitation, the area west of and including the uplands is the wettest (Jarvis *et al.* 1984). The coastal plain in the east, lying in the rainshadow of the uplands, is the driest. Consequently, the main climatic variation within the region is east-west, with rainfall increasing from east to west. The annual rainfall lies between 600 and 700 mm on the coastal plain, but is 1000 mm or more in the uplands (Jarvis *et al.* 1984). Agriculture is influenced by temperature as well as rainfall, but instead of the annual average temperature a different measure is normally used, that of the accumulated temperature above 0 $^\circ$C for the early half of the year (grass and cereal leaf extension is maintained, albeit slowly, down to 0 $^\circ$C; Jarvis *et al.* 1984). The median accumulated temperature above 0 $^\circ$C (day degrees) is a measure of the amount of heat energy available to plant growth; it takes into account both the length of the growing season and the range of temperatures above the minimum. The coastal plain has between 1250 and 1400 day degrees centigrade in an average year, while in the uplands above *ca.* 300 m O.D. the totals are less than 1050 day degrees (Jarvis *et al.* 1984).

These figures refer to the present-day situation and their validity for the later prehistoric and Roman period is difficult to assess. Some information is, however, available about the climate during that time. Lamb (1981) has been able to reconstruct the broad pattern of climatic change for the British Isles, using information from peat stratigraphy, pollen analysis, oxygen isotope measurements, tree-rings, varves, ice-sheet stratigraphy, etc. He has established that during the second millennium BC the climate was warmer than that of today, but a gradual decrease in temperature occurred during the millennium. Around the turn of the millennium there was a final warm period after which there was a sharp decline in temperature associated with an increase in rainfall. Lamb estimates a drop of nearly 2 $^\circ$C in overall mean temperatures between 1000 and 750 BC, which may have meant a shortening of the growing season by more than five weeks (Lamb 1981). From about 150 BC the climate did become milder again as well as drier and less stormy. By the time of the Roman conquest of Britain (first century AD) the climate was probably very similar to that of today. The late Roman period (AD 250–400) was slightly warmer again, but soon after AD 400 a sudden turn to wetter summers and colder winters has been suggested (Lamb 1981). Turner (1981b) has suggested that the improvement of the climate at the end of the first millennium BC may have started slightly earlier than Lamb has suggested, i.e. at *ca.* 400 BC instead of 150 BC. Thus, present evidence suggests that the climate during the late Iron Age and Roman period was not dissimilar to that of today.

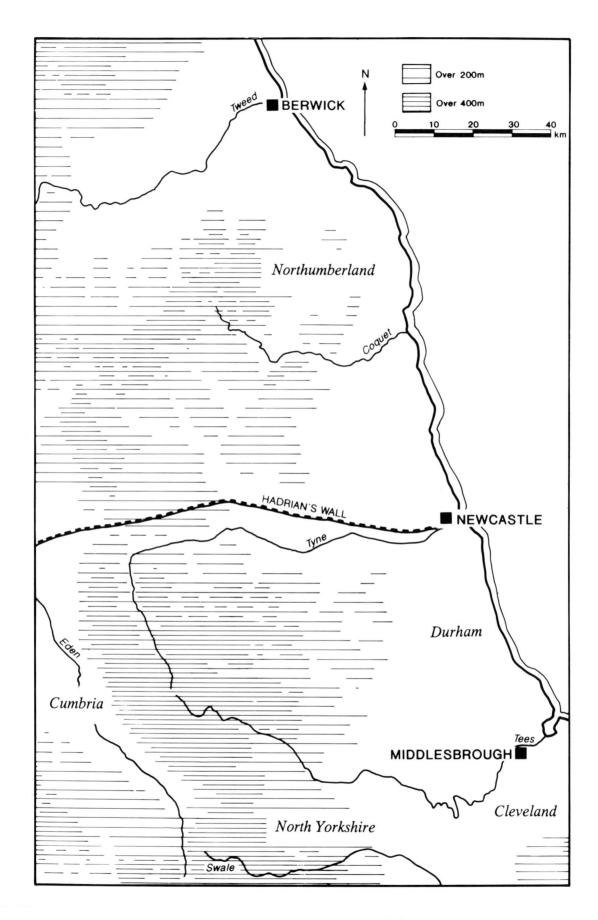

Figure 2.1   Map of the study area showing geographical features mentioned in the text.

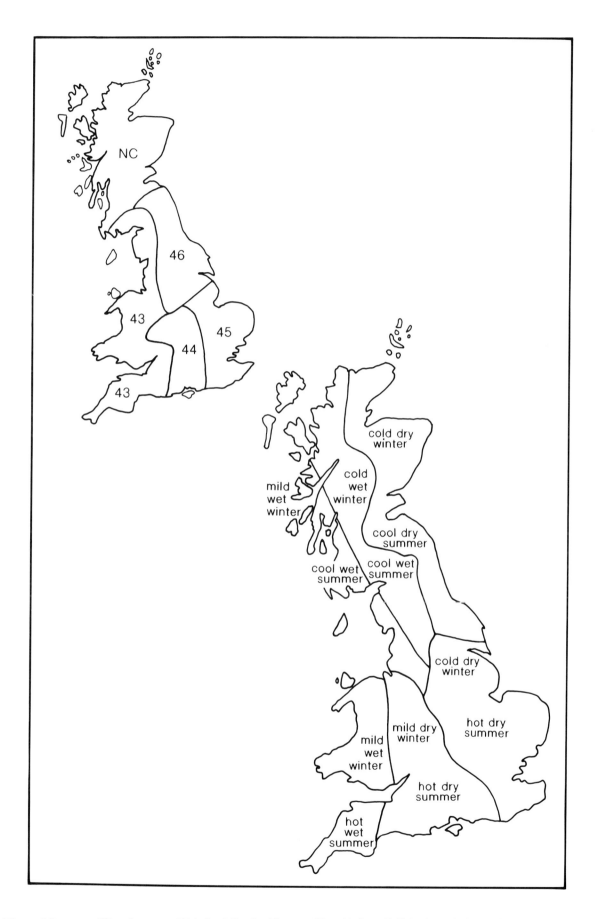

Figure 2.2 Climatic zones of Britain, following Thran and Broekhuizen 1965 (top) and Shirlaw 1966 (below).

2.1.3 Geology and soils

The solid geology of the region consists in very broad terms of a series of bands of rock running north east to south west across the region (Figure 2.3). From north to south these bands consist of Carboniferous Limestone, Carboniferous sandstones and shales, Carboniferous shales and sandstones, Permo-Triassic limestones and finally Permo-Triassic sandstones (Jarvis *et al.* 1984).

Over most of the region the solid geology is obscured by superficial or drift deposits, the most extensive of which is a glacial till (Figure 2.3). The till is very heterogeneous giving rise to considerable local variations in both soil texture and drainage. Surface-Water Gley Soils tend to occur where the drift is thick and overlies impermeable shales, while Stagnogleyic Brown Earths are found where the underlying rocks are more permeable (Jarvis *et al.* 1984). The generalized soil map of the country (Figure 2.4) classifies most of the soils on the coastal plain and in the major river valleys as Stagnogley Soils, i.e. non-calcareous loamy, clayey and loamy and loamy over clayey soils in which drainage is impeded at moderate depth by a relatively impermeable subsurface layer (Mackney 1974). The next most common soil type is that of Brown Earths, i.e. moderately deep and deep, well drained and moderately well drained, loamy non-calcareous soils. The upland areas are covered by Stagnopodzols, Stagnohumic Gley Soils and raw peat (Mackney 1974).

Soils are, however, not just a product of parent material, relief and climate; the activities of people through time have also greatly influenced the present-day soil conditions. Consequently, it is very difficult to assess how these present-day soils relate to those prevailing in the Iron Age.

2.1.4 Vegetation history

The vegetational history of the region during the later prehistoric and Roman periods has been recorded in 25 pollen diagrams (Figure 2.5). Table 2.1 gives the location of each pollen core and a reference to the original publication of the data. The results of this evidence have been discussed in a number of 'overview' papers (Turner 1979, 1983, Van der Veen 1985b, Wilson 1983). Fluctuations in the relative proportions of tree pollen and pollen of certain herbaceous species have been interpreted as reflecting the degree of forest clearance and human interference with the vegetation. The relative proportions of so-called 'arable' and 'pastoral' indicators as well as the occurrence of cereal pollen have been used to make inferences about the type of agricultural economy prevalent at any one time or place. There are a number of reasons why this last aspect of the interpretation of pollen diagrams has met with criticism.

First of all, the confidence attached to the identification of cereal pollen has been questioned (Edwards 1989). The distinction between pollen of cereals and those of wild grasses on the basis of grain size alone is no longer regarded as satisfactory, and additional criteria, such as pore and annulus sizes, and surface sculpturing of the pollen exine, are required (Edwards 1989). It is possible that some early identifications of so-called cereal pollen need to be revised.

Secondly, with the exception of rye, cereals are self-pollinated; they produce low quantities of pollen and these do not travel far. Consequently, both the distance between the pollen core and the arable field and the density of vegetation in between the two have a crucial effect on the likelihood of cereal pollen being detected in a pollen diagram (Behre 1981b, Edwards 1982, 1989, Maguire 1983, Groenman-van Waateringe 1988). Thus, while the presence of cereal pollen can in most, but certainly not all, cases be taken as evidence for arable agriculture, the absence of cereal pollen is not in itself significant and cannot be interpreted as an absence of cereal cultivation.

Thirdly, the use of certain plant species as arable or pastoral indicators relies heavily on the uniformitarian principle that the ecology of these species will not have changed through time (Maguire 1983). In the light of the considerable evidence for changes through time in the composition of especially anthropogenic vegetations such as arable fields and grassland, this assumption is unfounded and has now met with criticism (Behre 1981b, Maguire 1983, Groenman-van Waateringe 1988, see also Chapter 9). The pollen of Gramineae, *Plantago lanceolata*, *Rumex acetosa/acetosella*, and *Ranunculus* spp. are commonly used as pastoral indicators, even though all these species can and do occur in arable fields and have been recorded in past charred weed assemblages, including those analysed here. An example of the confusion is the fact that different authors have used the same genus (*Rumex*) as a different indicator (Behre 1981b, Maguire 1983). Furthermore, past weed assemblages were not uniform, as will be demonstrated in Chapters 9, 10 and 11, but varied in composition depending on climatic region, soil type and cultivation regime. What is more, the Compositae, freqently used as an arable indicator, are a group of weeds absent in one weed assemblage studied here, and only present in low numbers in another. Another complicating factor is the fact that several of the so-called pastoral indicators, such as Gramineae and *Plantago lanceolata*, are high pollen producers in contrast to some of the arable indicators. Thus, the present use of these indicator species tends to result in a general

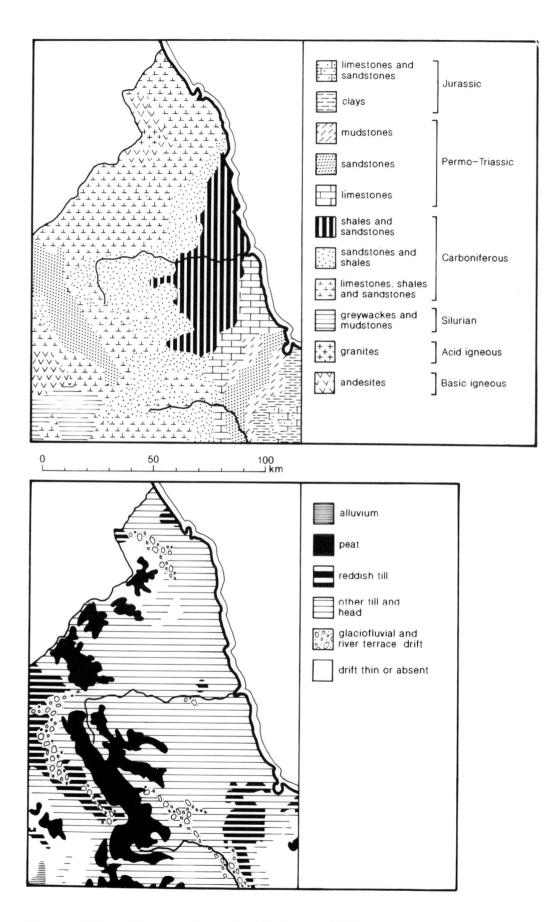

Figure 2.3   Solid and drift geology in the region (after Jarvis *et al.* 1984).

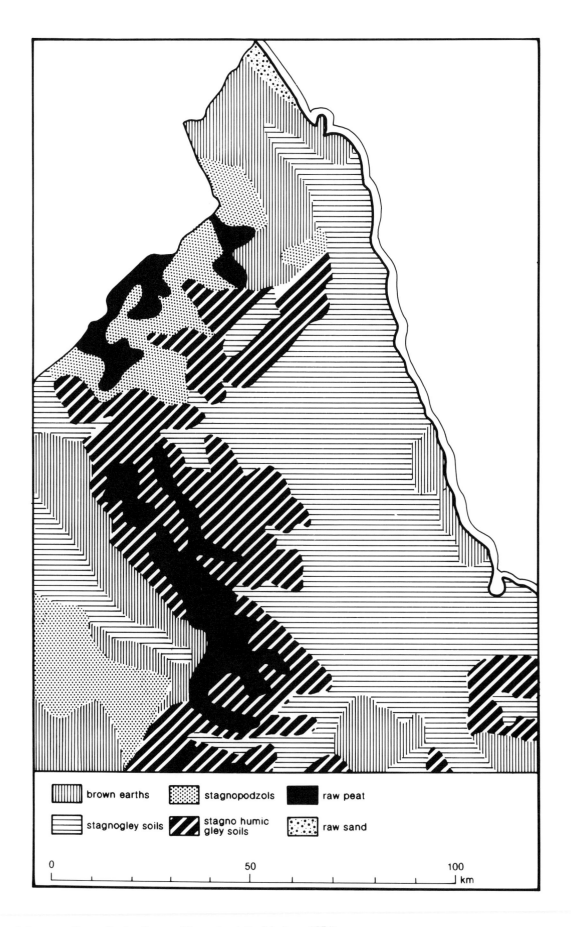

brown earths

stagnopodzols

raw peat

stagnogley soils

stagno humic gley soils

raw sand

0   50   100

km

Figure 2.4   Generalized soil map of the region (after Mackney 1974).

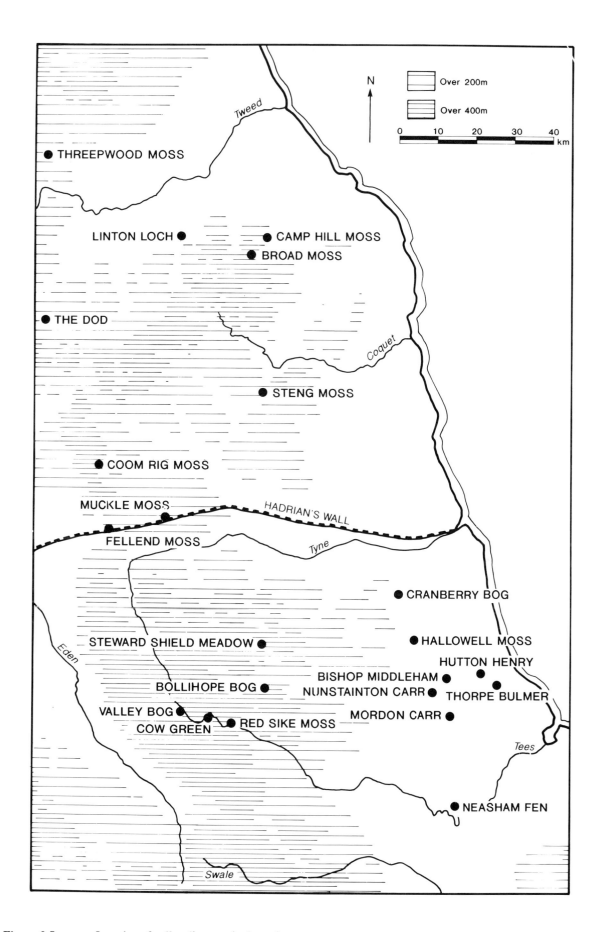

Figure 2.5 Location of pollen diagrams in the region.

underestimation of arable and an overestimation of pastoral agriculture.

Finally, this approach assumes a dichotomy between crop and animal husbandry which is unlikely to have existed in the past (Behre 1981b). Past farming economies are likely to have been mixed to a far greater degree than in the present-day situation with its sharp division between cereal, dairying and beef farms.

Thus, the interpretation of the presence or absence of cereal pollen and arable/pastoral indicator species in pollen diagrams in terms of the presence or absence or the relative importance of crop and animal husbandry can be quite misleading as it suffers from both theoretical and methodological problems, and tends to overestimate the role of animal husbandry. In the following discussion of the vegetational history of the region the emphasis is put on the degree of forest clearance and the time period at which the large scale opening up of the landscape was initiated, which reflects the degree of population increase and agricultural expansion in the region. No attempt is made to try and define the relative importance of crop versus animal husbandry.

The results from *ca.* 25 diagrams in the region can be summarized as follows (Turner 1979, 1983, Van der Veen 1985b, Wilson 1983). During the Bronze Age and early Iron Age a series of clearance phases can be recognized in all diagrams, many of them associated with cereal-type pollen. All these clearances (with the possible exception of the diagram from Bishop Middleham, see below), were followed by forest regeneration and the landscape in general remained largely wooded. Towards the end of the Iron Age or the beginning of the Roman period most diagrams show evidence of a considerable increase in forest clearance compared to the preceding period, with tree pollen percentages dropping to levels not dissimilar to those of today, resulting in a largely open landscape. Unfortunately, the beginning of this clearance phase has only been dated by radio-carbon dates in eight diagrams. These dates do, however, suggest that this phase of forest clearance was initiated prior to the arrival of the Roman army in this region (AD 70), see below. Furthermore, one diagram, that from Bishop Middleham, shows evidence for large-scale forest clearance during the Bronze Age. At this site a dramatic decline in tree pollen starts at *ca.* 3660 ± 80 BP (GaK–2072) (Bartley *et al.* 1976). The intensity of this clearance is not matched anywhere else in the region, although at the nearby site of Hutton Henry the diagram also shows a substantial clearance during this time (at around 3544 ± 80 BP (SRR–601)), but this is followed by forest regeneration. The Bishop Middleham diagram does not cover the first millennium BC, so that it is not clear whether this clearance was temporary or not.

In order to compare the radio-carbon dates for the start of this large-scale clearance phase with those from the seed assemblages discussed in this study (Chapter 5), and in order to compare them with the historical event of the arrival of the Roman army, the dates have all been calibrated (using two standard deviations) against the high-precision calibration curve published by Stuiver and Pearson (1986), using a computer program developed by Van der Plicht and Mook (1987) (see Chapter 5 below). In order to detect any potential differences between the upland and lowland zones of the region the dates have been grouped into those from diagrams located in the uplands, and those located in the lowlands, and within these groups the sites are listed from north to south (Figure 2.6 and Table 2.2). As the sample age of the peat sample used for radio-carbon dating is not usually known, and as the sample age of these samples can be as much as 200 years (Turner 1981a), it is important to treat these calibrations with great care. They tend to give a false sense of accuracy.

The probability of the forest clearance having been initiated prior to the arrival of the Roman army has been calculated using the normalized cumulative probability distribution calculated by the program (Van der Plicht and Mook 1987). The date for the potential impact of the Romans on the vegetation has been set at *ca.* AD 90. This date was chosen to represent the moment in time when most of the forts along the main Roman roads through the region were built, but before the withdrawal from Scotland took place (see also section 2.2.4.1 below).

The probability of the date representing a start of the forest clearance before AD 90:

Upland zone

Steng Moss	1970 ± 60BP (Q–1520)	88%
Fellend Moss	1948 ± 45BP (SRR–876)	86%
Steward Shield	2060 ± 120 BP (GaK–3/033)	90%
Bollihope	1730 ± 100 BP (GaK–3/031)	5%
Valley Bog	2212 ± 55BP (SRR–88)	99%
Valley Bog	2175 ± 45BP (SRR–89)	99%

Lowland zone

Hallowell Moss	1956 ± 70BP (SRR–415)	79%
Hutton Henry	1842 ± 70BP (SRR–600)	18%
Thorpe Bulmer	2064 ± 60BP (SRR–404)	99%

If the samples used for these dates did not have too great a sample age, we may conclude that these dates do not point to any marked intra-regional differences in the start of the clearance phase, and, as expected, confirm the conclusion reached by Turner (1979) that the clearance was, in most localities at least, initiated by the native population, rather than influenced by the arrival of the Romans.

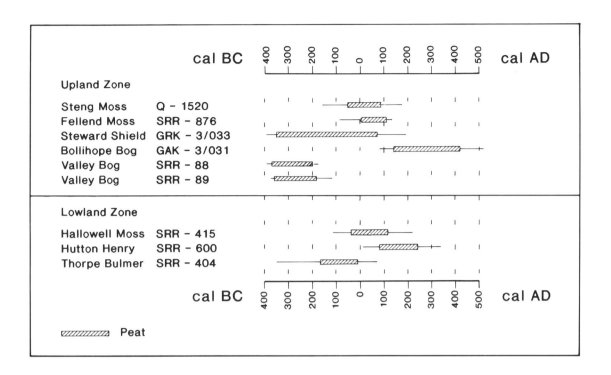

Figure 2.6   Calibrated radio-carbon dates from pollen diagrams in the region dating the start of large-scale deforestation.

2.2. Cultural background

2.2.1. Settlement pattern

In the last 25 years the region has seen a great deal of research on the settlement pattern and economy of the later prehistoric and Roman period. However, it remains a very difficult task to provide a synthesis of the new evidence and even more difficult to evaluate its importance. The absence of a detailed chronology lies at the root of these problems (Young 1987b). There are only a few radio-carbon dates from excavations in the north east of England, and the material culture recovered on excavations is generally poor. The native pottery is insensitive to chronological seriation and several cultural groupings appear to have been aceramic. Consequently, it is very difficult to date the large numbers of sites that have been recovered through aerial photography, field survey, and, to a lesser extent, through excavation. It has been estimated that fewer than five per cent of the known sites in the region are closely datable (i.e. within 200–300 years; Chapman and Mytum 1983). As a result there has been a tendency in the published work to equate morphological similarity with chronological contemporaneity (Young 1987b). While in the absence of new information there is little one can do but make such basic assumptions, care must be taken when interpreting the data, especially when analysing functional differences between sites.

Synthesis of the settlement pattern is further hindered by the uneven distributions of areas of mining, quarrying, afforestation and large conurbations, which affect the potential recovery of new information. The amount of field work carried out in the region is also unevenly distributed. Much work has been done in the Cheviots and their foothills, and also in the Wear and Tees lowlands, but other parts of the region are less well known. The description below of the settlement and subsistence pattern of the region is, therefore, incomplete, and should be seen as a reflection of our present knowledge, rather than as proof that this pattern did actually exist.

2.2.1.1 Tweed-Tyne area

Unenclosed settlements, consisting of one or more hut circles, would appear to form the earliest type of settlement in the area north of the Tyne during the period under study. Radio-carbon dates from eight out of the 90 known sites span the period of 1750–450 BC (Gates 1983). Their concentration in the Cheviots may possibly be explained by the recent intensive fieldwork in that area, while their apparent absence in the lowlands may be a function of the intensity of recent settlement and ploughing (Gates 1983). Many of these settlements have been found associated with cairns and/or small field systems, containing plots of *ca.* 0.2 ha (Gates 1983)

Timber-built palisades commonly occurring as earthworks in the Border Counties, but more widely distributed as chance discoveries in excavations beneath the more substantial defences of Iron Age hillforts (Gates 1983) are the second type of settlement in the region. They appear as early as the 9th century BC and are common during the first half of the millennium, which means that there is a considerable chronological overlap with the unenclosed settlements (Gates 1983).

Soon after 500 BC many palisaded enclosures were superseded by a new type of enclosure, the bank and ditch. These new settlements are usually classified as hillforts; they are often built over palisaded enclosures, but also occur on new sites. In several, but by no means all cases, a sequence from palisade to univallate enclosure, and then to multivallate enclosure has been recognized. Most of these hillforts were quite small, only 0.4 ha or less, but a small number of larger ones exist (e.g. Yeavering Bell (*ca.* 5 ha), and, across the border, Eildon Hill (*ca.* 16 ha) and Traprain Law (*ca.* 20 ha). The distribution of hillforts is largely confined to northern Northumberland.

In the coastal lowlands another type of settlement is present, that of the sub-rectangular enclosure with one or more hut circles inside it. This type of settlement is not closely dated, and examples belonging to both the Iron Age and Roman period are known.

Many of the hillforts were abandoned towards the end of the millennium or the first century AD, though the exact timing is not known due to the lack of excavations and radio-carbon dates. Some sites, however, apparently continued to be occupied. In fact, the largest hillfort, Traprain Law in East Lothian, is thought to have been occupied throughout the Roman period (but see below). It has been suggested that the abandonment of the fortifications was demanded by the Romans (Higham 1986), but there is no actual evidence for this.

The settlement type during the Roman period is characterized by small, stone-built, non-defensive enclosures with one or more hut circles, and often with cobbled inner enclosures on either side of the entrance. In the northern part of Northumberland these enclosures are usually curvilinear in outline, while in the southern part they are rectilinear. On the coastal plain rectilinear enclosures are the most common settlement type. They are all small (less than 0.2 ha), with only one or two hut circles inside the enclosure.

2.2.1.2. Tyne-Tees area

The settlement pattern in this area is much less clear, possibly due to the greater extent of conurbation and damage by the extracting industries. Virtually no unenclosed settlements or palisaded sites are known from this region. There are a few curvilinear enclosures which vary in size from small circular ones to large oval or irregular shaped ones (Haselgrove 1982). None of them have been excavated, so that their dates are unknown. They tend to occur close to other enclosures and may represent earlier occupation, but without excavation this remains speculation (Haselgrove 1982).

The most common settlement type in this part of the region is the rectilinear or sub-rectangular enclosure, usually enclosed by a bank and ditch. Some are enclosed by a palisade, or are preceded by a palisade (Haselgrove 1982). Those that have been excavated date from the mid first millennium BC through to and including the Roman period. They can be divided into three size classes: the largest category (at least 0.7 ha) and the medium one (0.3 – 0.5 ha) are both pre-Roman in origin (Haselgrove 1982). Some of the larger ones did continue in occupation into the Roman period. There are usually several hut circles inside the enclosure. Many of the medium size category settlements are located on or above the 125 m contour, and these appear to have been abandoned by the time of the Roman conquest or even earlier (Haselgrove 1982). Those situated in the lower-lying areas appear to have continued in occupation. The smallest category of these settlements (less than 0.2 ha) is presently dated to the Roman period only. This category and the medium-sized one both only have one hut circle inside the enclosure.

In contrast to Northumberland there are almost no defended sites in this area. It is difficult to explain their absence by the topography alone, as there are enough gently rolling hills to provide suitable locations (Haselgrove 1982). There are only four sites in the entire area which can be called defended, of which Stanwick in North Yorkshire is best known (see below).

The settlement pattern of the uplands of this area is more or less unknown. There are a few possible hillforts in the upland margins and unenclosed hut 'clusters' are known further south in the Cleveland Hills and North Yorkshire Moors (Haselgrove 1982), but apart from that the distribution map for the uplands remains blank. However, see also Coggins 1987, Inman *et al.* 1985, Vyner 1988, and Young 1987a.

2.2.2 Subsistence and land use

2.2.2.1 Location and land use

Changes through time in the distribution and location of settlements and/or their physical structure have sometimes been used to detect changes in the economic base of inhabitants of the settlements. The adoption of palisades in Northumberland has been interpreted as a move away from a mixed economy to a greater emphasis on animal husbandry (Higham 1986). The palisaded enclosures may have been used for coralling livestock or to exclude animals from the living area. Their location on the upland margin is also seen to point to the ranching of extensive herds (Higham 1986). The change-over from palisades to a ditch and bank may reflect the increasing effect of deforestation and the lack of suitable timber to build the palisades (Higham 1986). But they may also reflect the need for more defensive structures. Gates (1982a, 1983) has also pointed to the possible connection between palisades and hillforts and animal husbandry. The choice of location of these sites appears to be independent of the requirements of arable farming, especially at a time when the climate was thought to have been less favourable for upland agriculture, and this, combined with the fact that no field systems have been found associated with these sites (in contrast to earlier and later sites), is suggestive of an emphasis on animal husbandry on these sites (Gates 1982a, 1983).

The information provided by field systems is illustrative in another respect. Gates has found a marked contrast between the size of the fields associated with Bronze Age unenclosed settlements and those associated with Roman period enclosed settlements (Gates 1982a, 1983). The earlier, prehistoric ones are small (often no more than 0.2 ha) and irregular, while the later ones are more regular, larger (0.5 – 0.7 ha), and often integrated with trackways. Gates has tentatively suggested that the earlier, smaller fields reflect the product of hand cultivation, while the Romano-British ones may be the result of the use of the traction plough and better provision for controlling stock (Gates 1982a, 1983). A survey of cord rig cultivation (very narrow ridged cultivation) in Northumberland and the Borders, broadly dated to the Iron Age, has also identified a basic division in plot size (Topping 1989a, 1989b). The presence of ard marks underneath some of these cord rig plots has here, however, been taken as tentative evidence that this type of cultivation was associated with the ard plough rather than the spade (Topping 1989a, 1989b).

In the Tyne-Tees lowlands the location of approximately fifty per cent of the rectilinear settlements on or close to the 125 m contour line has been used to make inferences about the past practice of stock rearing (Haselgrove 1982). Such locations allow easy access to the well-watered lowlands for cattle, and to the uplands for sheep. Most of the remaining sites are located in the lower lying areas. Sites on or above the 125 m contour appear to have been abandoned by the Roman period, or before, which may point to a shift towards more arable agriculture.

The evidence for field systems in the Tyne-Tees area is very fragmentary, and detailed information about size and date is not available.

2.2.2.2 Subsistence

Excavated evidence for subsistence practices in the form of faunal and botanical remains is very rare for the region. The soil conditions, being rather acid, are not favourable for the preservation of bones, and only a very limited sample is available. The small number of bones recovered from most sites prevents a detailed analysis of the nature of the animal husbandry, and it is rarely possible to go beyond a list of species and their relative proportions. Table 2.3 lists the sites from which animal bones were recovered and the relative proportions of the main components. All assemblages have one thing in common, i.e. the predominance of cattle bones over those from other species.

Before the writer started to collect charred seed assemblages such plant material was unknown from sites in the region, with the exception of Doubstead, a small rectangular enclosure on the coastal plain in northern Northumberland (Figure 4.14). Here a small sample was collected by Donaldson, which contained two grains and one rachis internode of barley and a few weed seeds (Donaldson 1982). The material collected by the writer from settlement sites dated to the first millennium BC and the Roman period provide the data for this thesis.

There is very little direct artefactual evidence for arable farming, but this may be a function of the small amount of excavation in the region. Wooden ard fragments have been found to the north west of the region, at Milton Loch and at Lochmaden (both in Dumfries and Galloway) dated to the later Iron Age (Rees 1979). Iron plough shares have been found at the Roman fort at Chesters (Northumberland) and to the north of the region at Blackburn Mill (Borders) and Traprain Law (East Lothian), all dated to the Roman period (Rees 1979). Iron ox goads are known from Traprain Law and the Roman fort at Newsteads, both to the north of the region (Rees 1979). Manual cultivation tools, such as hoes, mattocks, spades, spuds, and turf cutters, have also been found, but only at Roman forts (Rees 1979). All of them could, of course, have been used for non-agricultural purposes. Iron reaping hooks, pruning hooks, scythes and mowers' anvils are also known from Roman forts in the region, as well as from Blackburn Mill and Eckford (Borders) and Traprain Law (East Lothian) (Rees 1979). The predominance of the findspots of these agricultural tools in Roman forts is difficult to interpret. They could point to the fact that the soldiers were actively involved in farming (see also below), but it must be remembered that the amount of excavation carried out in the region is biased in favour of Roman forts and against native settlements.

The presence of quern stones has also been used as evidence for arable agriculture (Haselgrove 1982), although their presence can only be linked with consumption rather than production of grain. Saddle querns, beehive querns and rotary querns are frequently found on excavations. The introduction of beehive querns is thought to have taken place around 200 BC (Heslop 1987), while the rotary querns are dated to the Roman period.

Another type of artefact related to arable production is the plough mark, i.e. the scratches into the subsoil made by the ard tip, which are sometimes preserved beneath later deposits. Ard marks have been found underneath later Roman structures along Hadrian's Wall, at Wallsend, Walker, Newcastle, Denton, Throckley, Rudchester, Wallhouses, Carrawburgh, Dimisdale, Tarraby Lane, and Carlisle (Bennett 1990), as well as on two native sites, Belling Law and Fenton Hill, both in Northumberland (Higham 1986). At most sites the overlying deposits only provide a *terminus ante quem*, but the lack at several sites of a soil horizon between the plough marks and the Roman deposits sealing them might suggest that little time elapsed between the ploughing episode and the construction of the Roman structures (Bennett 1990).

2.2.3 Social structure

The two parts of the region, north and south of the River Tyne, have been identified with two different tribal groupings, the Brigantes and the Votadini, known from classical authors (Figure 2.7). The Votadini are known from Ptolemy's *'Geography'* (*ca.* AD 150), while the Brigantes are mentioned by as many as six Roman authors, all writing during the period AD 60–150, i.e. the period of the Roman conquest of the North (Hartley and Fitts 1988).

The term 'Brigantes' is usually translated as 'the hill people', and their territory is usually referred to as the area from the Rivers Humber and Trent northwards to the Solway and the Tyne (but see also Chapter 11 below), although separate groupings, such as the Parisi in Humberside and the Carvetii in the Eden Valley, are also mentioned (Frere 1978). The Brigantes are usually described as a confederation of tribal septs, without much social cohesion. The large fortified site at Stanwick, North Yorkshire (Figure

2.7), has been identified as the residence of their queen, Queen Cartimandua (Haselgrove 1982, 1990). The main defences at Stanwick enclose an area of *ca.* 300 ha; they are thought to have been built around the mid first century AD (Haselgrove 1982, 1990). Recent excavations have recovered a very rich assemblage of Roman imports, all dated to *ca.* AD 40–70. These are thought to have been the products of diplomatic contact between the Queen and Rome. Tacitus mentions that Queen Cartimandua entered into a treaty with Claudius and that she was 'pro-Rome'. Her loyalty to Rome was apparently not shared by all her people, and some, if not all, of them, and including her husband Venutius, revolted against her. She was rescued by the Romans. Although there is no evidence for any major battles, the site ceases to be occupied in the later part of the first century AD. It is thought that Stanwick was a major centre of the Brigantes, probably the only one in the north-eastern part of their territory, and acted as a focal point for political authority, social status, and economic activity (Haselgrove 1990). The site may have had both diplomatic and commercial relations with Rome in the period after the conquest of southern Britain, but before the military occupation of the North.

Much less information is available about the Votadini. They are mentioned by Ptolemy and are thought to have occupied the Cheviots and the east coast from the River Tyne to the River Forth in southern Scotland. The large hillfort at Traprain Law in East Lothian is often regarded as the capital of this tribe (Figure 2.7). The richness of the Roman imports on this site (including some later first century and many third and fourth century finds), the site's apparent continuation as a defended settlement throughout the Roman period, and the absence of Roman military installations from most of their territory, have been used to credit the tribe with a special relationship with Rome (Breeze 1982, Higham 1986, Maxwell 1980). Higham has suggested that they may have been one of the eleven tribes which entered into treaties with Rome (Higham 1986), although there is no actual evidence for this suggestion. In fact, the absence of Roman finds dated to this early period in contrast to Stanwick, make such a treaty unlikely. The earliest Roman finds at Traprain are dated to the Agricolan period, *ca.* AD 80 (S. Willis, pers. comm.). On present evidence the distribution of early (i.e. AD 40–70) Roman fine wares is restricted to the area south of the Tyne (M. Millett and S. Willis, pers. comm.).

A recent reassessment of the evidence from Traprain Law has produced an alternative interpretation for the site, which sees it primarily as a ceremonial centre rather than the residence of the tribal élite (Hill 1987). Hill suggests that the defences

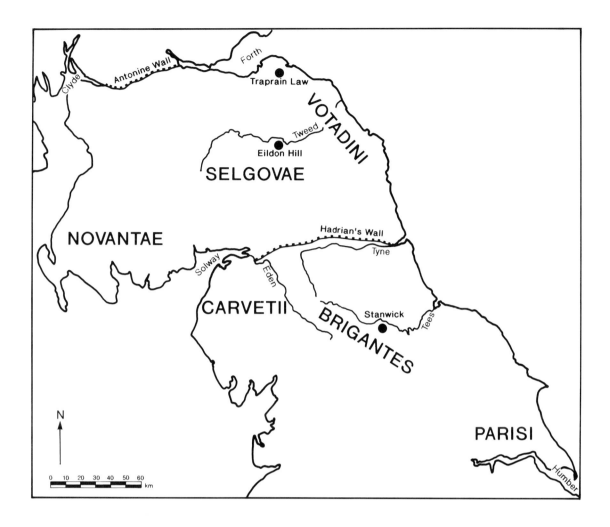

Figure 2.7   The tribes in northern Britain during the early Roman period and the location of the two Roman frontiers (after Breeze 1982 and Frere 1978).

had fallen into decay prior to the arrival of the Romans and that the site was subsequently used as a ritual centre, receiving many votive deposits (Hill 1987). If this view became accepted, it would partly undermine the argument that the Votadini had a special relationship with Rome.

2.2.4   Roman conquest

2.2.4.1   Chronology of occupation

The first contact between the Romans and the native population in the region was in AD 43 or soon afterwards, when the Brigantes became a client state of Rome. During the late 60s AD the loyalties of the Brigantes changed, with Cartimandua losing her throne and the tribe taking on a distinctly anti-Rome attitude. Petillius Cerialis, the new governor of Britain (AD 71–73/74) attacked the Brigantes and during a series of campaigns covered most of their territory, but no military sites were established

outside south-east Yorkshire (Breeze and Dobson 1975). While his successor, Julius Frontinus (AD 73/4–77/78) operated mainly in Wales, the next governor, Julius Agricola (AD 77/78–83/85) concentrated his efforts on the North, marching through the territory of the Brigantes, and campaigning as far north as the River Tay in Scotland (Breeze and Dobson 1975). Two main roads were established, one from York north to Corbridge and Newstead, and further into Scotland; the second along the western side of the Pennines, from Chester to Carlisle and Dalswinton, and further north. East-west connections were established between the two legionary fortresses at York and Chester, between York and Ribchester, via the Stainmore gap, and via the Tyne-Solway gap, linking Corbridge and Carlisle. The exact date of the construction of the forts along these roads is not known, but most were probably built by AD 85 (Breeze and Dobson 1975). Trouble on the Danube and the need for more manpower in

Europe meant the withdrawal of army units from the North. This led to the abandonment of the forts in Scotland during the last few years of the first century and the beginning of the second century, and an increase in the number of forts along the Tyne-Solway isthmus (Breeze and Dobson 1975). The concept of a permanent frontier came under Emperor Hadrian (AD 117–138) and the construction of Hadrian's Wall along the Tyne-Solway gap (Figure 2.7) was started in AD 122/123. The number of soldiers stationed in northern Britain at this time is estimated at *ca.* 30,000 (Breeze 1984).

Many, if not most, of the forts in the Pennines and along Hadrian's Wall were abandoned when the new Emperor, Antonius Pius (AD 138–161) decided to move the army north again, into Scotland. This change in frontier policy is usually explained by the emperor's need for military prestige, rather than by trouble at the frontier itself (Breeze 1980, 1982). A new frontier was established, the Antonine Wall, along the Forth-Clyde isthmus (Figure 2.7), and this frontier was occupied from *ca.* AD 142–163 (Breeze 1982). After the emperor's death, presumably because there was no longer any reason to stay in Scotland, the frontier was moved back to the line of Hadrian's Wall. Problems with the supply of the army so far north, and possibly the need, once again, for manpower in Europe, may have been additional motives (Breeze 1982). While Hadrian's Wall formed the frontier once more, a number of outpost forts north of the wall were still maintained.

The last part of the second century and the first few years of the third were characterized by unrest and warfare. Some forts and part of the Wall are thought to have been destroyed by 'barbarian' troops (Breeze 1982). These problems at the British frontier ultimately led to the expedition in AD 208–11 by Emperor Severus against the Scottish tribes of the Maeatae and Caledonians. The campaign was spearheaded from Corbridge and South Shields, which both saw the construction of new granaries to stock supplies. Peace was made in AD 211 and no further disturbances took place during the remaining part of the century. The fourth century saw a sequence of disturbances again with the Picts and the Scots invading the province on a number of occasions. The end of the Roman occupation of northern Britain is not well known, but it is assumed that Hadrian's Wall ceased to function after *ca.* AD 410. Many soldiers are thought to have stayed in the area, turning their hand to farming or banditry (Breeze 1982). No evidence of them or their activities survives.

2.2.4.2 Army's demand for agricultural products

Throughout the Roman period the bulk of the army was stationed in the north of the country, but the actual strength of the army fluctuated over time, depending largely on political events outside the province which influenced the changes in frontier policy and the location of the frontier. It is impossible to give exact figures for the number of soldiers deployed in northern Britain at any one time, but rough estimates can be provided. Breeze (1984) has estimated that the strength of the army was in the order of 30,000 men around AD 120, 34,000 around AD 210, 11,700 around AD 300, and 18,300 around AD 400. No figure could be provided for the first century AD. An army of this size would require the provision of a wide range of goods (food; clothing; arms; transport animals; tents etc.), which had to be provided by the local population or be brought in from afar.

The diet of the Roman soldier is relatively well known, from documentary, as well as, archaeozoological and archaeobotanical sources (Davies 1971, Dickson 1989, King 1978, 1984). The basic diet consisted of grain (usually wheat), bacon, cheese, vegetables, sour wine, olive oil, and salt (Davies 1971). A much greater variety of food would be available on days of celebration, and the soldiers could also buy extra items out of their spending money (Davies 1971). Plant remains found at Roman forts in Britain include wheat, barley, lentils, Celtic beans, dill, celery, coriander, linseed, poppy seed, figs, strawberries, blackberries, rasberries and hazelnuts (Dickson 1989). Animal bones found at British Roman forts point to the consumption of beef, lamb, mutton, pork, deer, chicken, hare, fish and shellfish (King 1978, 1984).

It has been estimated that each Roman soldier would be given between 2 and 3 pounds of grain each day, which amounts to 330–496 kg of grain per year. Thus, an army of 30,000 soldiers would need some 10–15,000 tonnes of grain each year. In addition to the animals needed for the meat supply, a further 10,000 horses were required at the time of Hadrian for the cavalry units, *ca.* 4000 mules for carrying equipment, and some 2,500 animals for religious sacrifices each year (Breeze 1984). On top of that there was the need for large amounts of leather for tents, shoes, saddles, shields, bags, cases, clothes etc. In addition to the needs of the soldiers, the transport animals would require large amounts of hay and grain as fodder. The overall demand for agricultural produce must have been very considerable.

The food supply of a marching army is largely a problem of logistics and is probably ruled by short term policy, but that of a standing army is influenced by the carrying capacity and economic stability of the region in which it is stationed (Groenman-van Waateringe 1989). During the early years of the conquest of northern Britain most of the supplies were probably brought in by long-distance transport to specially built 'supply bases'. The Agricolan range of warehouses at Red House, Corbridge, may have functioned as such a supply base, while at both Corbridge and South Shields Roman forts, new granaries were built in advance of the Severan campaign into Scotland in the early third century. At South Shields a total of 22 granaries were built at that time. However, we should not rule out the possibility of at least some looting and pillaging of crops. After all, the standard kit of a Roman soldier did include a sickle! (Davies 1971).

When the army was stationed in permanent forts the grain and animals it required would, as far as possible, be obtained through a combination of purchase, requisition and taxation. The extremely high cost of transporting bulk goods such as grain overland meant that the Roman government tried to avoid, as much as possible, having to move supplies over long distances (Manning 1975). It has been suggested that the food consumption of pack animals meant that bulk transport of grain would not be profitable after a distance of *ca.* 100 miles (Groenman-van Waateringe 1980). Transport by water was much cheaper, which is probably why the legionary fortresses of Britain (Caerleon, Chester and York) were located in places which could be reached by water from the sea (Manning 1975). Few of the auxiliary forts in northern Britain can be easily reached by water, however, and local supply sources are likely. There were two possible local sources. First of all, the local farming community, which could have been asked, or forced, to produce a surplus for the army. And secondly, the grain could have been produced on the land around the fort itself.

It is known from other parts of the empire that some fortresses, such as Xanten on the Rhine, had a *territorium*, that is an area of land around the fort, which contained fields, pasture, orchards, woods, etc. (Manning 1975). The land could have been cultivated by the soldiers themselves, or by civilians to which the land was leased (Davies 1971). This could possibly be the explanation for the presence of agricultural tools at Roman forts (see above). Civilian settlements, *vici*, grew up outside the forts, but about their inhabitants little is known (Casey 1982). They could have been the wives and children of soldiers, or retired soldiers, merchants and traders, as well as farmers. Evidence for the existence of *territoria* in Britain is sparse, but there is an inscription referring to a *territorium* for the auxiliary fort at Chester-le-Street in Co. Durham (Manning 1975). There have been very few large-scale excavations of *vici* in northern Britain, and about their nature and function little is known.

Lack of evidence means that we cannot, at the moment, assess how the grain supply of the army in northern Britain was organized, but we should not assume that there was a uniform solution. Not only did the demand vary quite considerably through time, but the ability of the local population to meet the demand may also have varied within the region. Long distance transport did take place, especially of luxury items and items which were not produced by the local community. Figs, lentils, coriander, wine etc. must all have been brought in from the Mediterranean, and the writing tablets found at the Roman fort at Vindolanda, Northumberland, also demonstrate the existence of a considerable postal system, and include letters with requests for items ranging from socks to oysters to be sent to relatives and friends (Bowman and Thomas 1983).

3. Methods

In this chapter the various methods of analysis used in this study are set out. The methods can be divided into four categories: collection, extraction, identification, and quantification of the data. They are discussed below in this order.

3.1 Data collection and sampling

3.1.1 Selection of sites

Samples for the analysis of carbonized plant remains can only be obtained from settlement sites during excavation, which means that the archaeobotanist is dependent for her/his data collection on the 'right' sites being excavated at the 'right' time. This meant that the data for this study had to be collected over a period of six years (1981–87). During that time the directors of all sites excavated in the region and dated to the period under study were asked to collect samples. Several of the sites were excavated for rescue reasons; the collection of charred plant remains did form an important element in the decision to carry out the excavation of two sites (Chester House and Thornbrough, both in Northumberland).

3.1.2 Sampling on site

The aim of the archaeobotanical analysis is to arrive at a characterization of the carbonized seed assemblage from each site, in order to study activities like production and consumption, the relationships between sites, the movement of arable products across the landscape, crop husbandry practices and changes through time. Such an analysis can only be carried out when the plant remains are known to form an assemblage representative of the site as a whole, or at least representative of the area excavated. The strategy for data collection required to obtain such a representative assemblage is one where samples are collected either from all excavated features, or from a random selection of the excavated contexts. On small-scale excavations a strategy of total sampling is usually feasible, but on larger-scale excavations it is rarely possible to process samples from all contexts, and a probabilistic sampling strategy is required. The reasons for choosing a random sampling strategy and the practicalities of such a strategy have been discussed in Van der Veen (1984, 1985a, Van der Veen and Fieller 1982).

On the sites considered in this study the following strategies were chosen: on small sites, or on large sites of which only a small part was excavated, the director was asked to collect a sample from each well-defined, sealed context. This strategy of total or near-total sampling was carried out at eight sites (Hallshill, Murton, Dod Law, Chester House, Stanwick, Rock Castle, Thornbrough, and South Shields). On two of these (Stanwick and Thornbrough), samples were collected from most contexts, but as this resulted in more samples than could be analysed in the time available, only a random selection of the available samples was analysed, using a table of random numbers. At Thorpe Thewles a random sampling strategy was implemented on site, but additional, so-called judgement samples were also collected (Van der Veen 1984, 1985a, 1987a).

Little information is available on the appropriate volume of each sample, as the number of seeds required is dependent on the level of analysis and the number of seeds available varies with the density of seeds in each deposit. It was decided to set the standard sample volume at two buckets of sediment (i.e. 30 litres), as this volume could still be processed relatively quickly and usually produced enough seeds to carry out most analyses (but see section 3.4 below). When a particular context was smaller than 30 litres, a smaller sample was collected, and the actual volume recorded. See Chapter 4 for detailed information on sample strategy and sample volume for each individual site.

3.1.3 Off-site subsampling

Most samples were analysed completely, but in a few cases, when a sample contained more than 500 seeds, only a random subsample was analysed, and the procedure recommended by Van der Veen and Fieller (1982) was followed. The sample was divided into a number of subsamples, using a riffle box, and the analysis was stopped when 500 seeds had been counted. In some cases the entire sample was analysed even though more than 500 seeds were present. This was done either because it was not realized early on during the sorting that the sample was very rich in seeds, or because it was felt useful to analyse a number of very rich samples in case rare species were identified. In the tables the numbers given refer to the actual seeds counted in the sample or subsample. Those samples that were only partially identified are indicated in the tables by an asterisk (*).

In the case of the granary samples from South Shields (i.e. samples from deposit 12236) a slightly different procedure was used. As the number of grains in the samples was extremely large, it was decided to analyse only a proportion of the total. However, to obtain the maximum information from the chaff fragments and weed seeds, which both occurred only in relatively small quantities, the following procedure was followed: the chaff fragments and weed seeds were all picked out and identified, but only the complete grains or large fragments of grain were picked out, and of those only *ca.* 380 grains were identified (selected on a random basis using a table of random numbers). The proportion of the total number of grains represented by these 380 grains was calculated, and the overall total was calculated from this figure. Thus, in the tables for South Shields (deposit 12236), the numbers of grains given for each of the cereal species are estimated on the basis of the information from the 380 grains identified, while the numbers for the chaff fragments and weed seeds are based on actual counts and identifications. The figure of 380 was obtained from Van der Veen and Fieller (1982, Table 4) as the number of seeds required to obtain a 95 per cent chance of estimating the percentage content of a species to within five per cent accuracy (when $N = \infty$, and $P = 50$). Confidence limits were calculated (but are not presented here), and were found to be below the set limit of five per cent. This procedure was not applied to samples 1, 6, 7b, 8, 10b, 10c, 16, 24, 26, and 28, as these samples had cereal grain counts of less than 400.

3.2 Extraction

The carbonized plant remains were extracted from the samples by using manual water flotation and an 0.5 mm mesh sieve. Where necessary the samples were air-dried prior to flotation. Part of the sample was mixed with water in a large bucket, using a hosepipe and tap water. The sediment was stirred thoroughly and lumps of soil were carefully broken up. After the sediment was allowed to settle for a minute, the water with the botanical material (both floating and in suspension) was carefully poured through the sieve, trapping the botanical remains. More water was added to the bucket and the process repeated until no further carbonized material floated to the surface of the water (three times was usually enough). The remaining sediment was quickly searched for small bones, sherds, coins, etc. and then discarded. Once the entire sample was processed in this way, the 'flot', i.e. the material trapped in the sieve, was rinsed and left to dry.

The dry flots were sorted under a Wild M5A stereoscopic microscope using x15 magnification. The charred seeds, fruits, grains, chaff fragments and all other identifiable plant fragments other than wood charcoal were picked out for further identifications. Seeds which were not carbonized were ignored as they are likely to represent modern contamination (Keepax 1977).

3.3 Identification

3.3.1 General

The seeds and other fragments picked out of the flots were identified under the microscope, using up to x60 magnification, by comparing the charred specimen with modern seed reference material, drawings and photographs in seed atlases and other publications, and by consulting colleagues about specific unknown seeds. The seed atlases used were those by Berggren (1969, 1981), Beijerinck (1947), Jacomet (1987a), and Nilsson and Hjelmqvist (1967). Unfortunately, there is no seed atlas specifically dealing with the species of the British flora. The nomenclature used in this thesis is that of Clapham, Tutin and Warburg (1962), *The Flora of the British Isles*, with two exceptions. For the different species of wheat, the nomenclature proposed by Miller (1987) is used, and for the species of *Tripleurospermum* that of Kay (1969) is followed. The general term 'seed' is used throughout, without consideration for the correct botanical terminology. The term is here used to refer to seeds, fruits and false fruits. The full botanical name of each species, common synonyms, and their English names are given in Table 3.1. The abbreviations used in the data tables are given in Table 3.2.

3.3.2 Identification criteria for cereal grain

The identification of charred grains of wheat to species level is very difficult, due to the overlap in grain morphology between species, which is exacerbated by changes in the overall shape of the grains during the charring process. Here the following criteria have been used: grains with a marked dorsal ridge and pointed ends have been identified as *Triticum cf. dicoccum*, emmer wheat. Grains with a low, rounded dorsal profile, rounded ends, more or less straight or parallel sides, and considerably longer than their maximum width have been identified as *Triticum cf. spelta*, spelt wheat. Grains with a slightly more pronounced dorsal curve, a rather steeply placed embryo (compared to spelt), with the greatest width nearest to the embryo, but especially grains which were very compact, i.e. short and fat, not much longer than their greatest width, have been identified as *Triticum cf. aestivo-compactum*, bread/club wheat. From the archaeobotanical remains it is not possible to determine whether we are dealing with *Triticum aestivum*, bread wheat, or *Triticum compactum*, club wheat, as this distinction can only be made when whole ears are available. In club wheat the entire ear is compact, producing only compact grains. In bread wheat some compactness of the ear does occur, but usually only towards the lower and upper end of the ear, producing both compact and non-compact grains. The non-compact grains are virtually indistinguishable from grains of spelt wheat. Here the term *Triticum aestivo-compactum* Schiem. (Renfrew 1973) has been used, as elsewhere in the archaeobotanical literature, to indicate that the grains are very compact. It is not impossible that some non-compact grains of bread wheat have been identified as spelt wheat. Furthermore, a few compact grains were identified as belonging to one of the glume wheats. They had clear markings or grooves at the lower end of the grain (i.e. at the position of the embryo) giving the impression that the grain had been 'pinched'. These markings are interpreted as the impressions left behind by the glumes. While this 'pinching' by the glumes can occasionally be seen very faintly on glume wheat grains, the phenomenon was very marked here and may possibly be explained by assuming that these grains were still enclosed by their glumes when they became carbonized. Grains which could be identified as belonging to wheat, but which could not be identified with some degree of certainty to any of the three categories described above, have been grouped as *Triticum* sp.

The barley grains, i.e. those that were reasonably well preserved, were divided into central (straight) and lateral (twisted) grains and their ratio was calculated. In two-row barley, *Hordeum distichum*, this ratio is 1 : 0, while in six-row barley, *Hordeum vulgare*, it is 1 : 2. The presence or absence of the lemmas was also recorded, as was the presence of ridges or transverse wrinkles on the dorsal surface, and the angularity or smoothness / roundedness of the cross-section of the grains. Grains which show transverse wrinkles on the dorsal surface, have a rounded cross-section, and slightly rounded ends have been identified as belonging to naked barley, var. *nudum*, while grains with longitudinal dorsal ridges, an angular cross-section and/or fragments of the lemma still present, are identified as belonging to hulled barley.

Oat grains are very distinct from those of the other cereal species, being long and thin, and more or less spherical in cross-section. The embryo has a characteristic V-shape and the grains often bear thick hairs. Unfortunately, the grain morphology of the cultivated oat, *Avena sativa*, is very similar to that of wild oat, *Avena fatua*, a common arable weed. While there is some difference in size, the overlap in size range between the two species makes it impossible to identify individual grains to species level on the basis of grain size alone. The floret bases of the two species are different, however, and as the floret bases which were found in the samples all belonged to *Avena fatua*, it is likely that the grains also belong to

this species. All grains of oat have been listed in the tables under the weeds, with the other grasses. It can, of course, not be ruled out that some of them belong to the cultivated variety, but there is no method of testing that at the moment.

Grains of rye, *Secale cereale*, do resemble those of wheat, but can be distinguished from wheat grains by their very blunt upper end and by the large, asymmetrically placed embryo. The grains are usually less rounded than those of wheat, slightly convex ventrally, very ridged, and often look poorly developed.

Grains which were poorly preserved or fragmented could generally not be identified to specific or generic level and are listed in the tables as Cerealia indet.

3.3.3 Identification criteria for cereal chaff

Chaff fragments of cereal grains can often be much more easily identified to species level than the cereal grains themselves, and their identifications are used to check those based on the grains.

The identification of the glume bases of *Triticum dicoccum* and *Triticum spelta* is based on two features, i.e. the venation pattern on the glumes and the angle between the glume faces (G. Hillman, pers. comm. and Jacomet 1987a). On the glumes of emmer wheat both the primary and secondary keels are strongly developed, while the tertiary veins are only faintly visible. On the glumes of spelt wheat the primary keel is prominent, but usually less strongly developed than on emmer. The tertiary veins, in contrast, are so prominent that they can barely be distinguished from the secondary 'keel'. The angle between the glume faces on either side of the primary keel in emmer is equal to or less than 90°, but greater than 90° in spelt. The angle on either side of the secondary 'keel' is distinct, though obtuse, in emmer, while in spelt the angle is hardly present; here, the faces form a more or less smooth curve.

The rachis internodes of both glume wheats, *Triticum dicoccum* and *Triticum spelta*, are listed in the tables as rachis internodes of a brittle rachis wheat. They were generally too badly preserved to allow identification to species level. In a few cases the presence of longitudinal lines near the outer edge of the convex face of the internodes was noted, a feature typical of hexaploid wheats, such as *Triticum spelta* (G. Hillman, pers. comm.). In many cases the surface of the convex face had flaked off, preventing this observation from being made.

The rachis internodes of tough rachis wheats can be distinguished by the shape of the rachis internode, the morphology of the glume base and the presence or absence of longitudinal lines on the convex face

(G. Hillman, pers. comm.). Here only one species was identified, *Triticum aestivum* (or *T. compactum*), bread/club wheat. The rachis internodes were largely identified on the basis of the glume base morphology: immediately below the point of glume insertion there were no lumps, but merely thin, inconspicuous ridges, representing 'inwardly crumpled' tissue attached to the rachis internode. The internodes were very compact, which meant there was no 'room' to express the shape of the internode. Here too, in a very small number of fragments the longitudinal lines near the outer edge of the convex face of the internode were visible, a characteristic of hexaploid species.

The rachis internodes of barley, *Hordeum vulgare*, have a rounded base which leaves a distinct, oval-shaped scar on the lower internode when broken off. The upper part of the internode tapers out and bears the bases of the six glumes. Along the longitudinal margins there are rows of short, densely packed hairs. The rachis fragments in the samples were generally very fragmented, which meant it was not possible to establish the presence of dense- or lax-eared varieties.

The rachis internodes of rye, *Secale cereale*, are similar to those of barley, but they do not taper out at the top, the base of the internode is much less rounded, and the scar left behind is more irregular in shape. They are also generally thinner and smaller than those of barley.

The floret bases of the cultivated oat, *Avena sativa*, can be distinguished from those of wild oat, *Avena fatua*, by the shape of the articulation scar. In cultivated oat the scar is very small, while in wild oat the floret base shows a large, horse-shoe-shaped articulation scar. No floret bases of the cultivated oat were found.

No identification criteria exist for the separation of the culm nodes of cereals from those of large grasses. As the samples consist largely of by-products of the cereal harvests, it has been assumed that the culm nodes present belong to cereals rather than to large grasses.

3.3.4 Identification criteria for wild species

It was not always possible to identify the weed seeds to species, or even to generic level, and in deciding how much time to spend on identifying individual seeds the added information, likely to be gained from the identification, was taken into account. Consequently, many of the grasses and sedges have not been identified in detail, as this was not regarded as cost effective.

The grasses are a particularly difficult group to identify as the seeds of different genera show a considerable degree of morphological overlap. The seeds of *Bromus* were identified down to *B. mollis/secalinus* type. Both are found in cultivated fields, although *B. secalinus* is more common as an arable weed than *B. mollis*. The identification of *Avena* has been discussed above. The seeds of *Sieglingia decumbens* are quite variable, but are generally fairly flat and oval-shaped. The embryo is large in relation to the size of the seed, and the hilum is thin and long, which distinguishes it from seeds of the tribe Paniceae, with which they could otherwise be confused. Most of the grass seeds could not be identified, however, and have been grouped into two size classes. The 'Gramineae' group contains large seeds, similar in size to *Bromus*. The 'small grasses' group contains seeds like *Poa annua*, *Phleum pratense* etc.

A very large number of culm base fragments or rhizome fragments of grasses was present, the majority of which could not be further identified. One type was very distinct, however, the bulbous, slightly pear-shaped tubers. They were identified as belonging to *Arrhenatherum elatius*, var. *bulbosum* (see also Godwin 1975, Plate XIII).

It is notoriously difficult to identify the seeds of the sedge family, the Cyperaceae. Here only two types have been identified to species level, i.e. *Carex pilulifera* and *Carex pulicaris*. It was initially attempted to ascribe some of the remaining seeds to types or groups, but as these groups still contained species of more than one specific habitat, no information was gained by doing so and consequently, the remaining seeds were all grouped under *Carex* spp.

The achenes of *Ranunculus* could only be identified to species when the surface pattern was clearly preserved. Seeds of *R. repens* show a cell pattern of large cells in the centre and smaller ones towards the edge, while those of *R. acris* have only small cells and *R. bulbosus* only large cells (Dickson 1970). The majority of those seeds that could be identified belonged to *R. repens*. In many cases the cell pattern had been eroded away, or the achene itself had completely disappeared, leaving only the internal seed. In these cases the seeds were identified to subgenus only, i.e. *Ranunculus* Subgenus *Ranunculus*, a group which includes *R. acris, R. bulbosus*, and *R. repens*.

The *Rumex* seeds were often poorly preserved, being broken or puffed by carbonization. It is very difficult to make reliable identifications of these seeds, and, with the exception of *Rumex acetosella*, no attempt has been made to identify them to species level. The shape of the better preserved seeds is similar to those of *R. crispus, R. acetosa*, and *R. obtusifolius*.

The seeds of *Chenopodium* could only be identified

to species level when the testa was well preserved. Most seeds could be identified to *Chenopodium album*. In many cases identification was not possible beyond generic level.

The seeds of *Tripleurospermum* were identified to species *inodorum* by the presence of round oil glands (Clapham *et al.* 1962). Note that the nomenclature adhered to here is that of Kay (1969) rather than of Clapham *et al.* (1962).

The category *Vicia/Lathyrus* refers to more or less spherical seeds of either of these species. In most cases the seeds were broken into the two cotyledons and the hilum was broken off, preventing further identification. In a few cases the hilum was, at least partially, preserved, and both *V. hirsuta* and *V. tetrasperma* were identified. Additional species may well be present. The category 'Leguminosae indet. (small)' refers to small leguminous weeds seeds other than the spherical ones. They were generally poorly preserved and could contain a number of different genera, such as *Trifolium, Lotus,* and *Medicago.*

3.4 Quantification

3.4.1 Counting

The numbers in the tables for each sample have been arrived at in the following manner: each grain counts as one; fragments of grains have been combined and the number of complete grains they represent has been estimated. Chaff fragments have all been counted individually, i.e. each glume base, rachis internode or culm node counts as one. In the case of a spikelet fork the count would be: two glume bases and one rachis internode. Glume fragments have been listed and counted separately, as have awn fragments, but they are not included in any quantitative analysis, as they can break into more than one fragment and, in the case of glume fragments, can represent the same part of the spikelet as the glume base. As the barley rachis internodes were often very fragmented, their numbers in the tables always represent the minimum number of internodes present. The weed seeds were all counted as one, even when broken, with the exception of large weed seed fragments when they clearly represented parts of the same seed.

3.4.2 Selection of samples and variables

Before applying the multivariate statistical techniques (i.e. Principal Components Analysis, Cluster Analysis, and Discriminant Analysis), certain samples and species were omitted from the analysis. The number of seeds in the samples varied considerably from sample to sample. The results from the smaller samples, i.e. samples with few identified seeds, are probably not very reliable.

Consequently, samples with fewer than 50 identifications were omitted from the analysis. The figure of 50 is an arbitrary cut-off point and was chosen only because it represents a figure below which the calculation of percentages becomes very problematic. A cut-off point of 100 would have been preferred, but this would have reduced the number of remaining samples by too much. Omitting samples with fewer than 50 seeds meant losing 101 samples out of a total of 325 samples, i.e. 31 per cent (a cut-off point of 100 seeds would have meant losing 145 samples, i.e. 45 per cent of the samples).

Species which occurred only rarely in the samples were also omitted from the data matrix prior to multivariate analysis. The justification for deleting rare species is that their occurrence may be a matter of chance (Gauch 1984). Especially in archaeobotanical samples a certain amount of 'settlement noise' may have entered the samples. Secondly, most multivariate techniques ignore species which carry only a small percentage of the overall information, while other techniques treat rare species as outliers (Gauch 1984). Thirdly, the deletion of rare species reduces the number of zero values in the data matrix (Gauch 1984), and finally, as the number of species in a sample is linked to the sample size, though not in a linear way (Lange 1988), the deletion of rare species may partially overcome the disadvantage of small sample sizes.

As with omitting certain samples from the analysis, the cut-off point for omitting species is an arbitrary one. In ecological studies species are typically defined as rare if they occur in less than 5 per cent of the samples, or in fewer than about 5 to 20 of the samples (Gauch 1984). In other studies a cut-off point of 10 per cent has been used (Dagnalie 1973, G. Jones 1983a, 1984), or species present with less than 11 seeds (Lange 1988). G. Jones did, in fact, test the appropriateness of the 10 per cent figure in analyses similar to those carried out in this study, and found that this cut-off point was 'more than adequate' and that 'the inclusion of rarer species would have been distinctly less cost-effective (G. Jones 1983a, 128). In this study both the 10 per cent and 5 per cent cut-off points have been used, and the 10 per cent figure was found to be very satisfactory. Very little difference could be detected in using the different figures (see Chapter 10). By omitting the weed species which occurred in less than 10 per cent of the samples, as well as species which were identified to insufficient detail, the total number of weed species was reduced from 76 to 32. Most of the species omitted in this way did, in fact, only occur in very small numbers in the samples and the actual number of seeds omitted was only 669, out of a total of 25,020, i.e. 2.7 per cent. A cut-off point of 5 per cent meant reducing the number of weed species from 76 to 36, and omitting 567 seeds, or 2 per cent of the total.

It is important to stress here that the 10 per cent and 5 per cent cut-off points were applied to all species present in the samples, i.e. to all species listed on the data tables, regardless of whether, on these tables, they were grouped under the category 'weeds' or 'other'. However, the species listed in the tables under the category 'other' did always occur in less than 5 per cent of the samples, so that they were excluded from the analysis because they occurred rarely, and not because some judgement was made regarding the likelihood of them originating from arable fields. There are two exceptions to this rule: *Corylus avellana*, hazelnut, and *Calluna vulgaris*, heather. While *Calluna vulgaris* may grow on the edges of arable fields (Hinton 1991), the remains of both hazelnut and heather have here been interpreted as having arrived into the archaeobotanical record by means other than association with the harvested cereal crop. For that reason these two species were excluded from the multivariate analysis of the seed assemblages in Chapter 10.

3.4.3 Standardization of the data

The numbers of seeds in each sample cannot be compared directly between samples, as the actual number of seeds found in any one sample is related to the sample size and the density of seeds in the deposit. While the numbers could be corrected taking into acount the sample volume by expressing the figures as the number of seeds for a set number of litres of sieved sediment, this procedure was not followed as it would mean adding a new, but unknown, variable to the data, i.e. that of the accumulation of deposits (M. Jones 1984b). The factors influencing the deposition of material in archaeological features are poorly understood and are far less predictable than in naturally formed deposits such as peat bogs or lake sediments. Information regarding the amount of carbonized plant remains per volume of sediment can be of interest, however, and is considered as a separate variable in Chapter 8.

Instead of standardizing the data by the volume of sediment sieved per sample, it was decided to use percentage values. There are two possible disadvantages in using percentages. First of all, variation in one variable causes variation in others, but given the relatively large number of species involved (32 or 36), this factor is unlikely to cause serious problems (G. Jones 1983a, Lischka 1975). Secondly, there are differences in the number of seeds produced by each weed species. High seed producers are likely to form a large proportion of the total number of seeds, while not necessarily being very prominent in the arable field. This problem is largely overcome by the fact that we are looking at

changes in the values for each species between samples and sites, rather than comparing the actual proportions between species (M. Jones 1984b).

While standardization by proportional measurement was regarded as preferable to that by volume of sieved soil, the disadvantages of the method were appreciated and a quick test was designed to test for any major differences in results. The values for each species were calculated by both methods for the total figures of each site. The resulting figures are given in Tables 3.3 and 3.4. As will be clear from those two tables, the differences in the values for each species between the sites are very similar for each method. Consequently, it was decided to standardize the data by using percentage values. As the ratios of grains to chaff fragments to weed seeds in each sample is determined by the crop processing stage that each sample represents, it was decided to express the percentage of each species as a percentage of the total for each category of data, rather than as a percentage of the overall total, in order to reduce the effect of the crop processing stages on the figures. Thus, grains have been expressed as a percentage of the total number of grains, chaff fragments as a percentage of total chaff, and weed species as the percentage of total weed seeds.

3.4.4 Transformation of the data

The multivariate statistical techniques used in this study assume that the distribution of the individual variables is not too far from normality (Gauch 1984, Shennan 1988). In order to assess the degree to which this was the case, the coefficient of skewness was calculated for each variable, using the percentage figures as described above, and having omitted samples and weed species as described above (for the weed species only the 10 per cent cut-off point was used here). The values of the coefficient of skewness (Table 3.5) indicate that the distributions in most cases are positively skewed, with a long upper tail. In order to make the data more suitable for the statistical analyses some form of data transformation is required, in order "to 'pull in' the upper tail while leaving the rest of the observations largely unchanged" (Shennan 1988, 110). This can be done by using the square root of each value, rather than the value itself, or by taking logarithms. Both have the effect of compressing the upper end of the measurement scale, thus reducing the importance of large values relative to smaller values (Digby and Kempton 1987). Statisticians tend to favour a logarithmic transformation, as this is thought to provide a better approximation of normality than using the square root (N. Fieller, pers. comm.). A major disadvantage of using a logarithmic transformation is, however, that it cannot be used on data sets which contain many zero values, such as the data set of the present study. This problem is generally overcome by 'unrounding' the zeros, i.e. by replacing the zero values with a small positive value (N. Fieller, pers. comm.), or by adding a constant (usually 1 or 0.5) to all values (Digby and Kempton 1987). The argument in favour of doing this is based on probability theory, i.e. if a larger sample had been analysed, a seed of that species would have been found. The argument against this practice is based on ecological considerations, i.e. this practice would add values for species to samples and sites in which this particular species did not occur. For example, from a probabilistic point of view we might find seeds of *Acacia* trees, if we kept analysing a larger and larger sample, but from an ecological point of view this is an impossibility, because these trees grow in semi-desert environments. As the present study is ecological in nature, it was felt that the logarithmic transformation was inappropriate in this case. Consequently it was decided to use the square root transformation, a transformation used in similar studies to the one presented here (G. Jones 1983a, 1983b, 1984, 1987, forthcoming). Lange (1988) did, in fact, use both the square root and the logarithmic transformation in analyses similar to the ones discussed here, and found that no marked differences could be detected. Here in addition to the square root transformation an alternative method to the logarithmic transformation has been used, i.e. converting the data values into an abundance scale which consists of logarithmic class intervals, the octave scale (Gauch 1984). A further transformation of the data is then unnecessary (Digby and Kempton 1987). This method has the advantage of leaving the zero values as zeros in the data matrix.

Conversion table from data values to octave scale (after Gauch 1984, Table 2.1)

Input			Output
0			0
0	$< x <$	0.5	1
0.5	$\leq x <$	1	2
1	$\leq x <$	2	3
2	$\leq x <$	4	4
4	$\leq x <$	8	5
8	$\leq x <$	16	6
16	$\leq x <$	32	7
32	$\leq x <$	64	8
64	$\leq x \leq$	100	9

The coefficient of skewness for each variable was calculated again, using the square root and the octave scale transformations (Table 3.5). Both methods reduced the skewness for most variables to a satisfactory level (i.e. to within ± 2). In a few cases the direction of the skewness was altered and the magnitude increased. This occurred with those variables which already had a near normal distribution before the transformation, which is the inevitable effect of any transformation. Both methods have been used in the statistical analyses, but no real differences in results were noted (see Chapter 10).

4. General Description of the Sites and their Seed Assemblages

In this chapter the sites from which seed assemblages were collected are described, and information is provided on the type and location of each site, and the sampling strategy adopted. A very brief description is also given of the seed assemblages from each site. Figure 4.1 gives the location of these sites in the study region.

4.1 Hallshill
(Northumberland; NY 906 886)

The site was excavated by Tim Gates of the Archaeological Unit for North East England on behalf of the Department of the Environment and later the Historic Buildings and Monuments Commission. During survey work in 1980, following the notification of the planned afforestation of the area, earthworks were found. These included an inconspicuous embanked circle, *ca.* 15 m in diameter, associated with a low, linear bank enclosing an irregular plot of about 0.6 ha, within which four small cairns were found (Gates 1982b, 1983: Figure 13). Excavations were carried out during 1981 and 1986 to test whether the small circular earthwork could represent the site of an unenclosed round house, and secondly, to test whether the banks and cairns in its vicinity could represent contemporary field clearance for agricultural use.

4.1.1 Location

The site lies at 230 m O. D., 3 km north of the village of East Woodburn, in an area of gently undulating moorland and rough grazing, overlooking the Rede valley to the west (Figure 4.2).

4.1.2 The excavation

The excavations revealed the wall line and post holes of a round timber-built house, *ca.* 9 m in diameter, set within the circuit of a stone bank which was made up of loose stone. In the centre of the house an area of burning was present, while there were also two pits within the area of the house, one of which (context 23) may possibly be stratigraphically earlier than the house (Figure 4.4). No datable finds were recovered, but six samples of charcoal were submitted for radio-carbon dating, two from the hearth, two from the pits, and two from the post holes:

Hearth - context 8 - HAR-4788 2520 ± 70 BP
Hearth - context 8 - HAR-4789 2560 ± 60 BP
Pit - context 23 - HAR-8183 2960 ± 60 BP
Pit - context 25 - HAR-8185 2710 ± 70 BP
Posth. - context 10 - HAR-4800 2780 ± 80 BP
Posth. - context 21 - HAR-8184 3130 ± 70 BP

These dates indicate that the house belongs to the later Bronze Age. The dates from the hearth (context 8), however, are considerably younger than the other dates, which may mean that the remains of the fire are unrelated to the building. Consequently, the description of context 8 was changed from 'hearth' to 'central area of burning' (T. Gates, pers. comm.).

4.1.3 Sampling strategy

Samples were collected from the postholes, pits and central area of burning. The samples collected during the 1981 season were rather small (only *ca.* 2 litres of sediment in volume), with the exception of the sample from context 8, which was *ca.* 70 litres in volume. The samples collected during 1986 were larger, producing an appreciable increase in the number of seeds and consequently in the amount of information available. The volume and context description for each sample is given in Table 4.1.

4.1.4 The plant remains (Table 4.2)

With the exception of the sample from context 8, all samples were dominated by cereal chaff. Only small quantities of cereal grains and weed seeds were found. The sample from context 8 (the central area of burning), in contrast, consisted of more or less clean grain. Very few barley grains were found and these were rather poorly preserved, but could be identified as six-row barley, *Hordeum vulgare*. Both hulled and naked grains appeared to be present, though the results are unreliable due to the very small number of identifiable grains. Both emmer and spelt wheat, *Triticum dicoccum* and *Triticum spelta*, were present, with emmer wheat by far the dominant wheat species. Two seeds of flax were found, probably belonging to the cultivated species, *Linum usitatissimum*. Other possible food plants in the samples were hazelnut, *Corylus avellana*, and blackberry/raspberry, *Rubus* sp. A small number of weeds were also found. The present analysis supersedes earlier reports on this assemblage (Van der Veen 1982, 1985b; see also Van der Veen forthcoming a).

4.2 Murton
(Northumberland; NT 965 496)

The site was excavated by Ian Jobey on behalf of the Department of the Environment and later the Archaeological Unit for North East England, during 1979 – 1981, with a final season in 1983 (Jobey 1987). Aerial photographs show cropmarks of two enclosures, one oval and one circular, on top of the

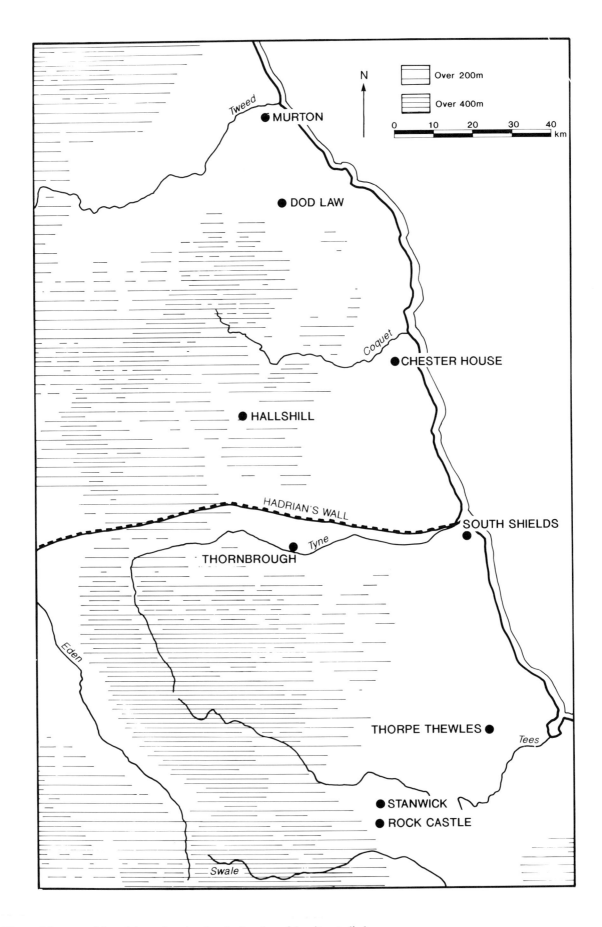

Figure 4.1   Map of the region showing the location of the sites studied.

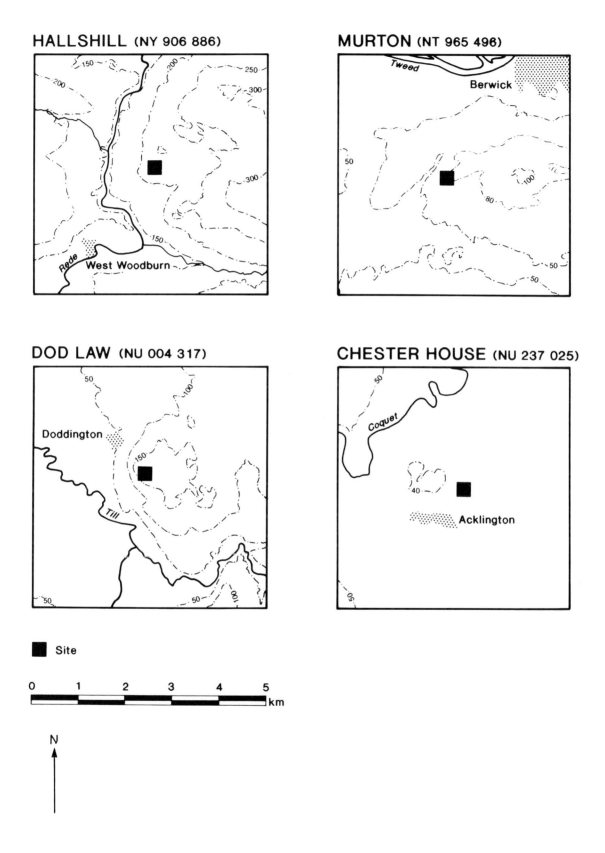

Figure 4.2   Detailed location of the sites studied.

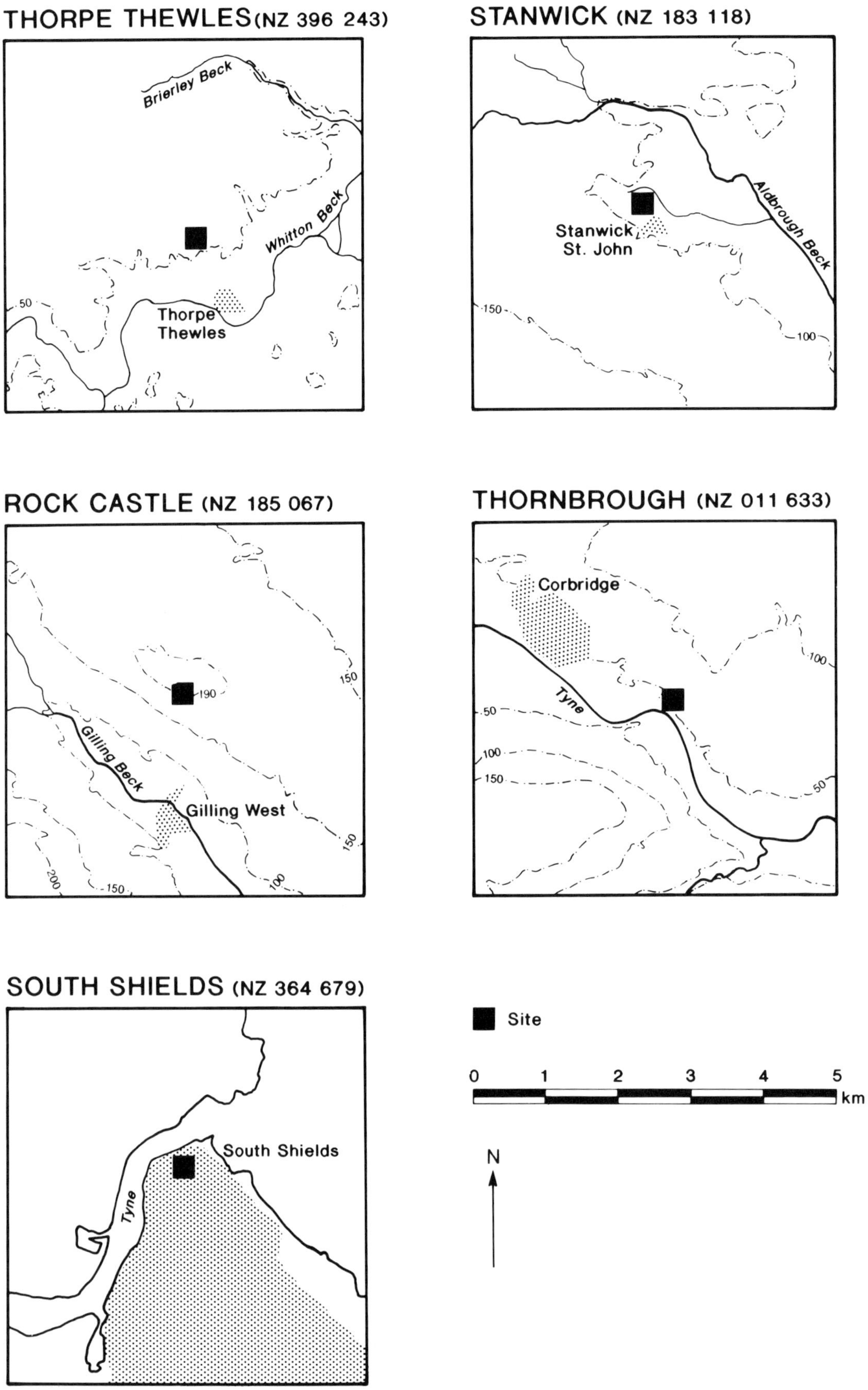
THORPE THEWLES (NZ 396 243)
Brierley Beck
Whitton Beck
50
Thorpe
Thewles
STANWICK (NZ 183 118)
Aldbrough Beck
Stanwick
St. John
150
100
ROCK CASTLE (NZ 185 067)
190
150
Gilling Beck
Gilling West
150
200
150
100
THORNBROUGH (NZ 011 633)
Corbridge
100
Tyne
50
100
150
50
SOUTH SHIELDS (NZ 364 679)
South Shields
Tyne
Site
0
1
2
3
4
5
km
N

Figure 4.3 Detailed location of the sites studied.

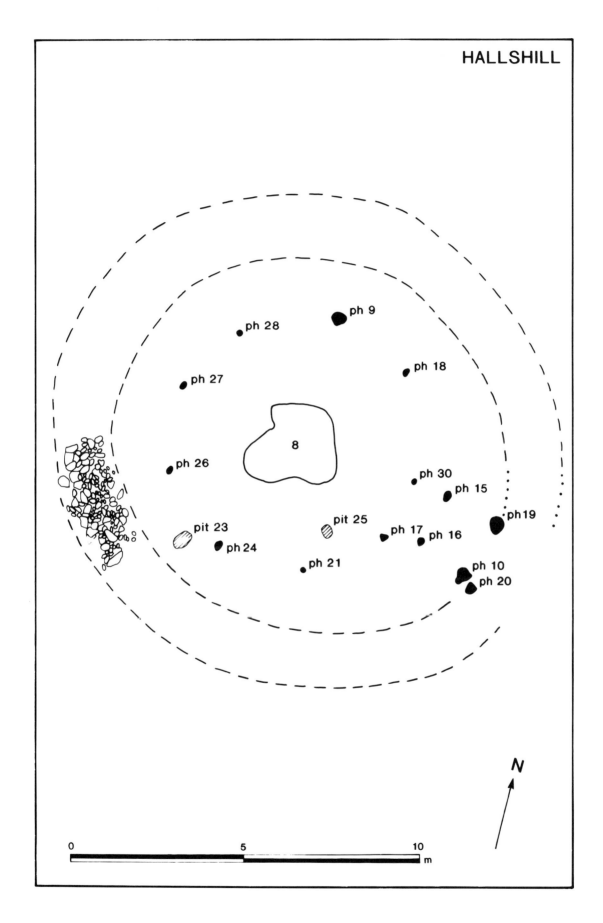

Figure 4.4   Hallshill, site plan (after Gates, forthcoming).

Murton High Crags. The southern, oval, enclosure was partially excavated, as ploughing had started to cause extensive damage and quarrying had also damaged the site.

4.2.1 Location

The site is situated on an outcrop of Fell Sandstone with a maximum altitude of 90 m O. D., taking a typical hillfort position. The site lies *ca.* 5 km south west of Berwick on Tweed (Figure 4.2).

4.2.2 The excavation

Four different phases of activity were recognized:

First of all, there was probably a Bronze Age burial on the site, already disturbed in antiquity. Part of a 'pygmy cup' and a perforated whetstone were recovered in the plough soil, both dated to the period of *ca.* 1500–1200 BC. Other Bronze Age burials are known from the immediate vicinity of the site. Some worked flints also pointed to Bronze Age activity.

Secondly, there is the possibility of an unenclosed settlement of Bronze Age type preceding the main settlement. The evidence consists of two timber-built round houses outside the defensive perimeter of the palisaded settlement. A radio-carbon date from an area of burning earlier than the construction of the first enclosed settlement (2960 ± 80 BP, HAR–6201) may possibly be related to this phase of occupation.

Thirdly, there is a timber-built, enclosed settlement, consisting of three lines of free-standing and embanked palisaded perimeters, defensive in nature, representing two or more phases, enclosing an area of *ca.* 0.7 ha with nine or ten round, timber-built houses (Figure 4.5). The pottery and other finds from this phase are of little value in dating the occupation, but two radio-carbon dates are available, one from a post-slot of the earliest palisade (2130 ± 80 BP, HAR–6202) and one from house T10 (2060 ± 100 BP, HAR–6200), dating this phase to the later Iron Age. All samples come from features associated with this settlement.

Finally, this settlement is superseded by a stone-built one, consisting of nine or ten round, stone-built houses enclosed within a substantial stone wall, sealing the timber built structures (Figure 4.5). The hand-made pottery from this phase probably dates to the Romano-British period. A small number of Roman fine wares suggest a late first century AD date for the start of the stone-built phase, with the occupation continuing until the late second or early third century AD.

4.2.3 Sampling strategy

The collection of samples for this study started after the main excavation of the site had already been completed, but during the short, final season of digging in 1983 it was possible to collect samples from twelve contexts. Three more or less standard size samples were recovered, from contexts 623, 624, and 625. The remaining contexts yielded only very small quantities of sediment. The volume and context of the samples are given in Table 4.3. The samples are all thought to pre-date the stone-built settlement. No samples from this later settlement could be collected because of a lack of sealed deposits from this later phase within the excavated area of 1983.

4.2.4 The plant remains (Table 4.4)

A relatively large number of plant remains were recovered, although the volume of many of the samples was very small. The cereal grains were dominated by those of six-row barley, *Hordeum vulgare*. Most of these could be ascribed to the hulled variety, but a few grains might have belonged to the naked variety. Both the wheat grains and the wheat chaff point to the presence of two species, emmer and spelt wheat, *Triticum dicoccum* and *Triticum spelta*. Emmer appears to be the more abundant of the two species. Only one other food plant was present in the samples, that is, hazelnut, *Corylus avellana*, represented by two small shell fragments. All the other species in the samples are wild plants, many of them common weeds in arable fields and waste places (see also Van der Veen 1987b).

4.3 Dod Law
(Northumberland; NU 004 317)

During the summers of 1984 and 1985 excavations were carried out at Dod Law West, a small hillfort. The excavations were directed by Chris Smith of the Department of Archaeology, University of Newcastle upon Tyne (Smith 1985, 1986, 1990). The hillfort is heavily defended with two substantial, concentric ramparts (Figure 4.6). On the northern side is situated a smaller enclosure or annexe. The interior of the main enclosure is just under 0.3 ha in area. The remains of round, stone-built houses are visible in the north-western part of the enclosure. The aim of the excavations was to establish the sequence of the rampart construction and to establish the chronological range of the site.

4.3.1 Location

The site is located on Doddington Moor, 4 km north east of Wooler (Figure 4.2). At 182 m O. D. it occupies a commanding position overlooking the Milfield Plain to the west and the Cheviots beyond (Smith 1990).

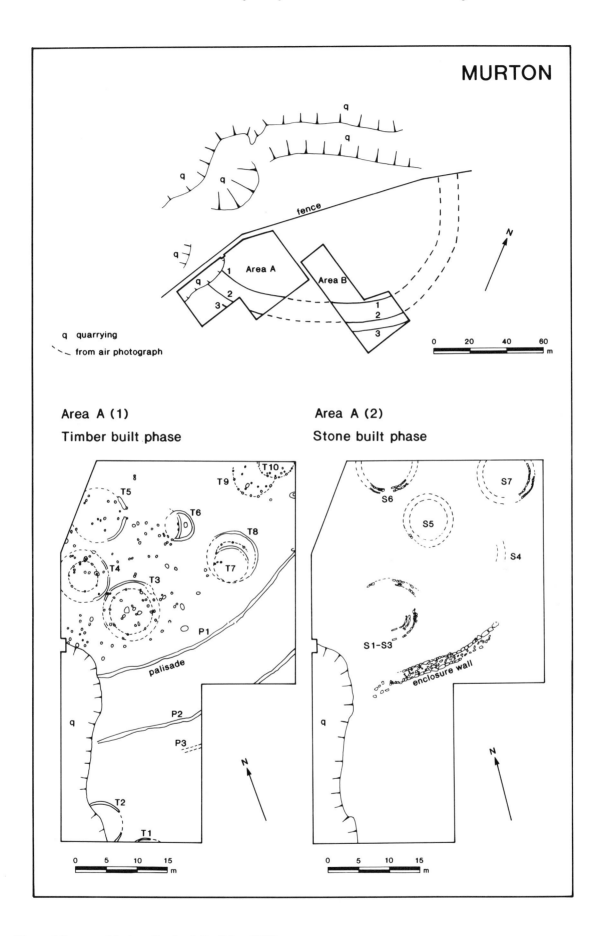

Figure 4.5 Murton, site plan (after Jobey 1987).

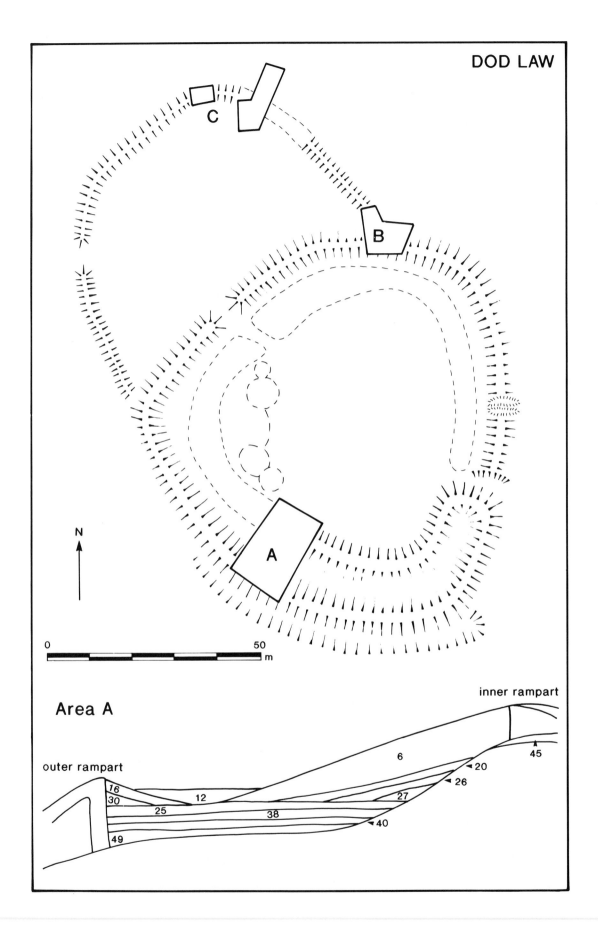

Figure 4.6   Dod Law, site plan (after Smith 1990).

4.3.2 The excavation

Excavations took place in three different areas of the fort (Figure 4.6). Area A provided a section through the inner and outer ramparts. In between these two ramparts a sequence of rich rubbish deposits was recovered, containing pottery, quernstones, metalwork, some animal bones etc. These deposits were sealed by the collapsed inner rampart (Figure 4.6). Two bronze fibulae found within the collapsed rampart deposits give a *terminus ante quem* of *ca*. AD 100 for the rubbish deposits (Smith 1990). Context 30 is a rubbish deposit accumulated against the outer rampart. Most of the samples came from these rubbish deposits.

Area B forms a section through the ramparts and the annexe bank on the eastern side of the fort. Excavations in this area have concentrated on the structural sequence of the defences. Area C lies within the annexe and cuts across the annexe bank. The objective was to look for occupation deposits within the annexe, but lack of time prevented any large-scale work. On the basis of the excavation results the following defensive sequence can be suggested (Smith 1986):

Phase I	outer rampart with palisade
Phase II	stone inner rampart
Phase III	collapse and partial destruction of inner rampart and refurbishment of the outer rampart with an inner stone revetment.

The occupation of the site is thought to have spanned the period from *ca*. 300 BC to *ca*. 200 AD.

4.3.3 Sampling strategy

During the excavation samples were collected from all well sealed deposits. Two samples came from the ground surface underneath the inner rampart (context 45), five came from the rubbish deposits in between the two ramparts in Area A (contexts 25, 38 and 40), one came from a rubbish deposit accumulated against the outer rampart (context 30), three came from Area B, from the rampart deposits (contexts 8 and 51), and finally, one came from Area C, from a deposit rich in charcoal (context 24). The sample volume and context description of the samples are given in Table 4.5.

4.3.4 The plant remains (Table 4.6)

The cereal grains were dominated by those of six-row barley, *Hordeum vulgare*. The majority of the grains belonged to the hulled variety, but 13 grains possessed some transverse wrinkles on the dorsal surface, suggesting that the naked variety was also present, as a minor component. Both the wheat grains and the wheat chaff fragments point to the presence of two species, emmer and spelt wheat, *Triticum dicoccum* and *Triticum spelta*. Emmer was the

dominant wheat species. The other food plants were hazelnut, *Corylus avellana*, rosehip, *Rosa* sp., and blackberry/raspberry, *Rubus* sp. Most of the other species in the samples are common weeds of arable fields and waste places (see also Van der Veen 1990).

4.4 Chester House
(Northumberland; NU 237 025)

The site was excavated by Neil Holbrook of the Archaeological Unit for North East England during 1985 in advance of open-cast mining (Holbrook 1988). The site was known as a crop mark on aerial photographs, which showed a small rectangular enclosure with a single round house visible within the interior (Figure 4.7).

4.4.1 Location

The site is situated on the summit of a slight rise at 41 m O. D., on the Northumbrian coastal plain, 1 km north east of the village of Acklington (Figure 4.2).

4.4.2 The excavation

Only a small part of the site was excavated, recovering part of the enclosure ditch, a palisade and parts of three houses, all of them round and timber-built (Figure 4.7). The ditch enclosed an area of *ca*. 0.2 ha. The fill of the ditch was thought to be derived from both an external and an internal bank, which had either eroded or been deliberately dumped into the ditch. No actual evidence of either bank was, however, recorded. On the southern side of the entrance a palisade trench was identified, running more or less north-south inside the ditch. This palisade is thought to have surmounted the internal bank. House 1 is stratigraphically earlier than House 2, while the stratigraphical relationship of House 3 is unknown, but it cannot have coexisted with House 1.

If an internal bank did exist, and if *ca*. 2–3 m was allowed for this, the bank would overlie the remains of House 2, implying that House 2 (and therefore House 1) predated the enclosure (Holbrook 1988). On the basis of this assumption two, or possibly three, phases of unenclosed settlement are postulated, succeeded by the construction of a rectangular enclosure with which no houses could be associated. No finds were recovered (Holbrook 1988). However, as no actual stratigraphical relationship between the circular houses and the rectangular ditch and bank was recorded, the sequence could equally have been one where the enclosed settlement was superseded by an open settlement.

4.4.3 Sampling strategy

Due to severe plough damage most of the house gullies and post holes had been only partially

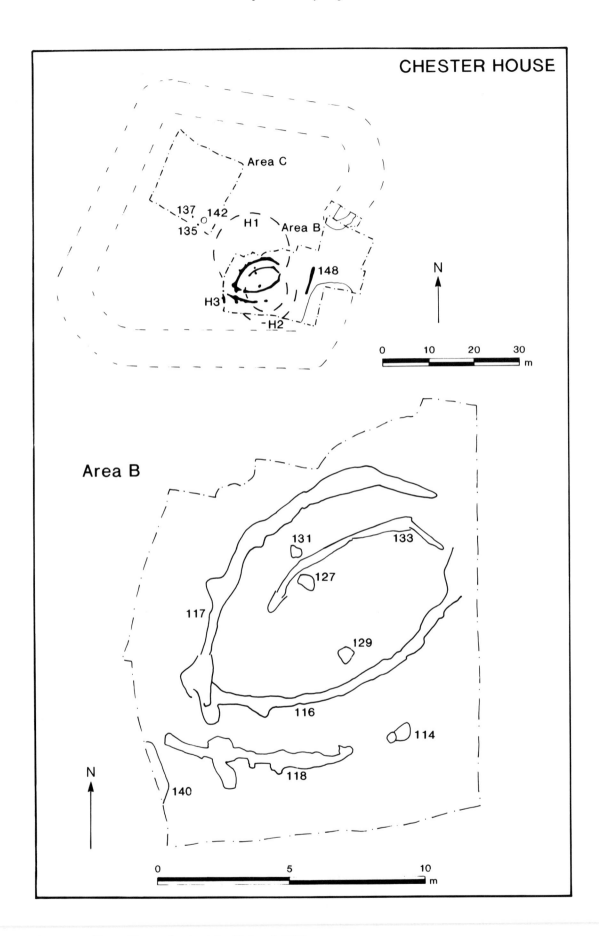

Figure 4.7   Chester House, site plan (after Holbrook 1988).

preserved. In order to maximize the available evidence, it was decided to collect and sieve the entire fill of all features. A total of 31 samples from 14 different contexts was collected. A list of the sample volume and context description is given in Table 4.7.

4.4.4 The plant remains (Table 4.8)

Despite the large volume of sieved sediment the number of seeds in the samples was small. The barley grains belong to six-row barley, *Hordeum vulgare*. Most of them belong to the hulled variety, but some may possibly be ascribed to the naked variety. Both emmer and spelt wheat, *Triticum dicoccum* and *Triticum spelta*, were present, with emmer wheat probably the more abundant of the two. Hazelnut, *Corylus avellana*, was the only other food plant present. The samples also contained a range of weed species (see also Van der Veen forthcoming b).

4.5 Thorpe Thewles
(Cleveland; NZ 396 243)

The site was excavated by David Heslop of the Cleveland Archaeology Unit on behalf of the Historic Buildings and Monuments Commission, during 1980 – 1982 (Heslop 1987). The site was known from aerial photographs, with crop marks showing a large rectangular enclosure and a circular house (Figure 4.8). Continuous ploughing had started to damage the archaeological features.

4.5.1 Location

The site is situated in the foothills of the Durham Plateau, overlooking the River Tees to the south, occupying the summit of a gentle hill, at 60 m O. D., *ca.* 5 km north west of Stockton on Tees (Figure 4.3).

4.5.2 The excavation

On excavation the site was found to be much more complex than had been envisaged on the basis of the crop marks. Four phases of occupation were recognized (Figure 4.8)

Phase I	pre-settlement field boundaries
Phase II	rectangular enclosure with central house and other round, timber-built houses
Phase III	open, nucleated settlement with round, timber-built houses
Phase IV	rectangular enclosure, but without any houses recognized within the excavated area

No dating evidence was available for Phase I, other than that the features pre-dated the Phase II settlement. The finds (pottery, metal work, rotary querns etc.) suggest a starting date of Phase II around 300–200 BC. Phase III started around the turn of the

millennium and finished in the later first century AD, while Phase IV ended around the middle of the second century AD. The results from the thermoluminescence dates (Bailiff 1987) suggest an earlier starting date and a longer timespan for Phase II with a mean date of 485 BC. The mean date for Phase III, 135 BC, is also earlier than the other dating evidence (see also Chapter 5).

4.5.3 Sampling strategy

During the first digging season no samples were collected, but during the second season the collection of flotation samples became an integrated part of the excavation. Because of the large scale of the excavation it was not possible to collect samples from every feature and a program of random sampling was applied (Van der Veen 1984, 1985a). The first season of excavations had shown that the features on the site fell into two categories: linear features (ditches and gullies) and point features (pits and postholes). A random sample of ten per cent was taken from both categories of features, using a table of random numbers (Van der Veen 1984, 1985a, 1987a).

In addition to this random sampling strategy a complementary, subjective strategy was carried out as well, in which the excavator chose additional samples on the basis of subjective criteria such as the occurrence of rich, ashy deposits, or the apparent gaps left by the random sampling strategy. These samples were labelled 'judgement' samples.

A third category of samples was derived from the so-called 'masking layers', which are levels of extant stratigraphy overlying the subsoil cut features. These layers were sampled spatially, by dividing them into grids of one meter squares and collecting samples from one out of every fifteen squares. The deposits on the periphery of these layers were occasionally so thin that the entire square was needed to fill two buckets (the standard sample volume).

Seventy-three random samples, 28 judgement samples and 28 masking layer samples were collected (Figure 4.9). A discussion of the optimal sample size in relation to the Thorpe Thewles sampling strategy is given in Van der Veen (1985b). The number of samples for each phase is as follows:

Phase I	2 samples (1 LS, 1 PF)
Phase I/II	1 sample (1 LS)
Phase II	44 samples (27 LS, 4 PF, 13 JS)
Phase II/III	10 samples (8 LS, 2 JS)
Phase III	29 samples (16 LS, 3 PF, 10 JS)
Phase III/IV	29 samples (1 LS, 28 MF)
Phase IV	12 samples (10 LS, 2 JS)
Unphased	2 samples (1 PF, 1 JS)
TOTAL	129 samples (64 LS, 9 PF, 28 JS, 28 MF)

(Abbreviations: LS = linear sample, PF = point feature, JS = judgement sample, and ML = masking layer).

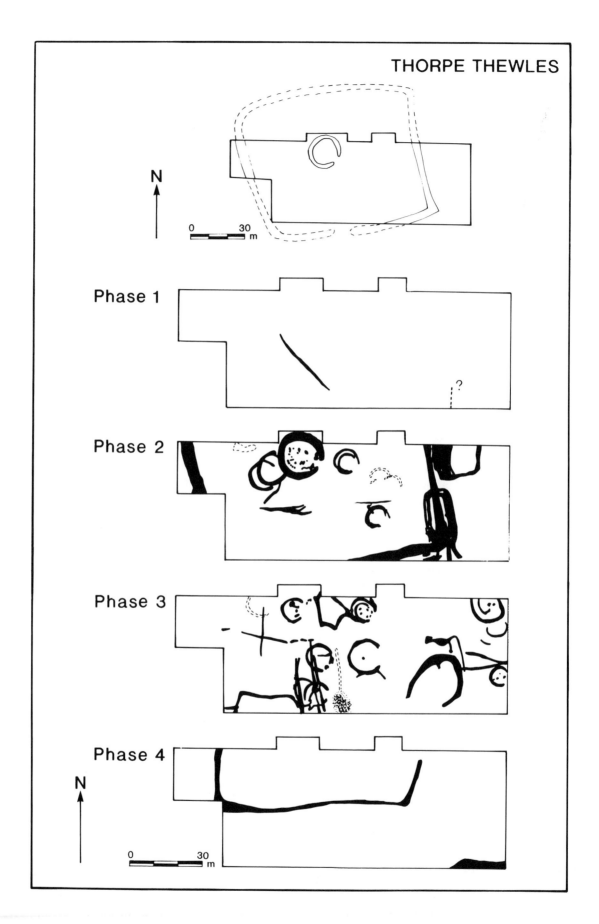

Figure 4.8   Thorpe Thewles, phase plan (after Heslop 1987).

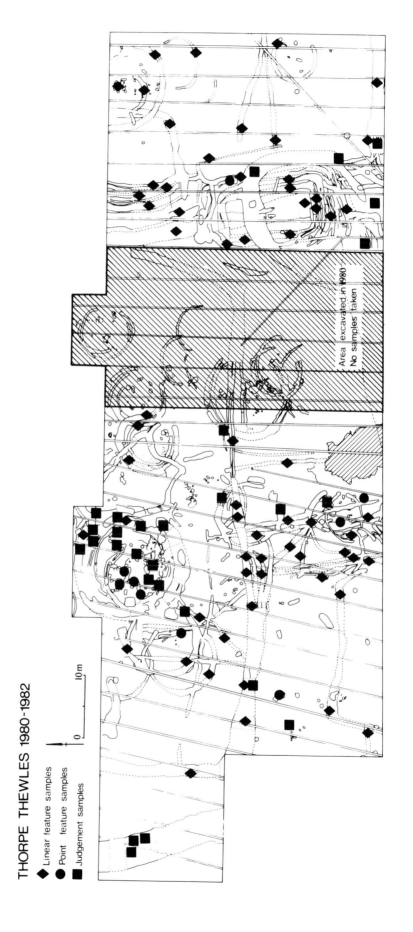

THORPE THEWLES 1980-1982

◆ Linear feature samples
● Point feature samples
■ Judgement samples

Area excavated in 1980
No samples taken

10m

0

Figure 4.9   Thorpe Thewles, site plan (after Heslop 1987).

The sample volume and context description is given in Table 4.9.

4.5.4 The plant remains (Table 4.10)

The assemblage consisted of small amounts of cereal grain and a large amount of wheat chaff, but was dominated by weed seeds. The barley grains belonged to six-row, hulled barley, *Hordeum vulgare*. Spelt wheat, *Triticum spelta*, was the only wheat crop present. There was a very minor trace of emmer wheat, *Triticum dicoccum* in the samples, but this species is here interpreted as a contaminant. Hazelnut, *Corylus avellana*, and the fruit of hawthorn, *Crataegus monogyna*, are the only other food plants present. The weed seeds are dominated by very large numbers of seeds of heath grass, *Sieglingia decumbens*, blinks, *Montia fontana*, spp. *chondrosperma*, and bromegrass, *Bromus mollis/secalinus* (see also Van der Veen 1987a).

4.6 Stanwick
(North Yorkshire; NZ 183 118)

Excavations were carried out at Stanwick, The Tofts, during the summers of 1984, 1985, 1988 and 1989, directed by Colin Haselgrove and Percival Turnbull of the Department of Archaeology, University of Durham, and, from 1988, Leon Fitts of Dickinson College, Carlisle, Pennsylvania. Stanwick represents the largest and most elaborate pre-Roman settlement in the north east of England, and is now regarded as the residence of the ruling élite of the region during the late Iron Age (Haselgrove 1982, 1984, 1990). The excavations are part of a larger project of re-evaluation and re-interpretation of this important site, once interpreted by Wheeler as the centre of indigenous resistance to the advance of Rome (Haselgrove 1982, 1984, 1990, Wheeler 1954).

4.6.1 Location

The site of Stanwick is situated in the Tees lowlands, and the area of The Tofts is located at *ca.* 100 m O. D., 4 km south of the River Tees and 5 km south west of the village of Piercebridge (Figure 4.3).

4.6.2 The excavation

The samples examined from this site derive from the excavations in the north-western part of The Tofts, which occupies a near promontory location at the heart of the earthwork complex, defined by a meander in the Mary Wild Beck and defended by a massive bank and ditch commanding the terrain to the west and south (Figure 4.10). The area of The Tofts is thought to represent the core of the settlement. Occupation here started with a number of drainage ditches and circular house gullies, probably representing an open, unenclosed settlement. During the mid first century AD large earthworks were erected, enclosing an area of *ca.* 300 ha, and a series of enclosures appear to have been constructed within the occupied area (Figure 4.11). During the same period large numbers of Roman imports (pottery and glass) reached the site, corroborating its special nature (Haselgrove and Turnbull 1983, 1984, Haselgrove *et al.* 1988, 1989). The site is thought to have been the seat of Cartimandua, the queen of the Brigantes, who, before the Roman conquest of the North, is thought to have been loyal to Rome and may have entered into a treaty with Rome (see also Chapter 2). The occupation of the site is thought to have started in the first century BC or slightly earlier. The later occupation is dated by the exceptionally rich collection of Roman pottery and glass to *ca.* AD 40 – 70, although occupation may have continued after that.

4.6.3 Sampling strategy

Only a small number of samples have been analysed so far, all collected during the 1984 and 1985 excavations. It was not possible to include samples collected after October 1987; those collected during 1988 and 1989 will be analysed by the writer in the next few years. During 1984 and 1985 samples were collected from all well-defined contexts, but features which were either extremely small (such as stakeholes), ill-defined, or contaminated, were omitted. A total of 112 samples were available for analysis. For the purpose of this study 32 samples were selected on a random basis. As the post-excavation work on the stratigraphy is not yet complete, the samples are treated here as one group, representing occupation deposits dating from *ca.* 100 BC to *ca.* AD 100. The volume and description of each context is given in Table 4.11.

4.6.4 The plant remains (Table 4.12)

The seed assemblage is dominated by weed seeds, with smaller quantities of cereal grains and chaff. The barley grains in the samples belong to six-row, hulled barley, *Hordeum vulgare*. Spelt wheat, *Triticum spelta*, is the only wheat species present. Some wild food plants were also present: hazelnut, *Corylus avellana*, elderberry, *Sambucus nigra*, and rosehip, *Rosa* sp. The weed seeds were dominated by seeds of heath grass, *Sieglingia decumbens*, and bromegrass, *Bromus mollis/secalinus*.

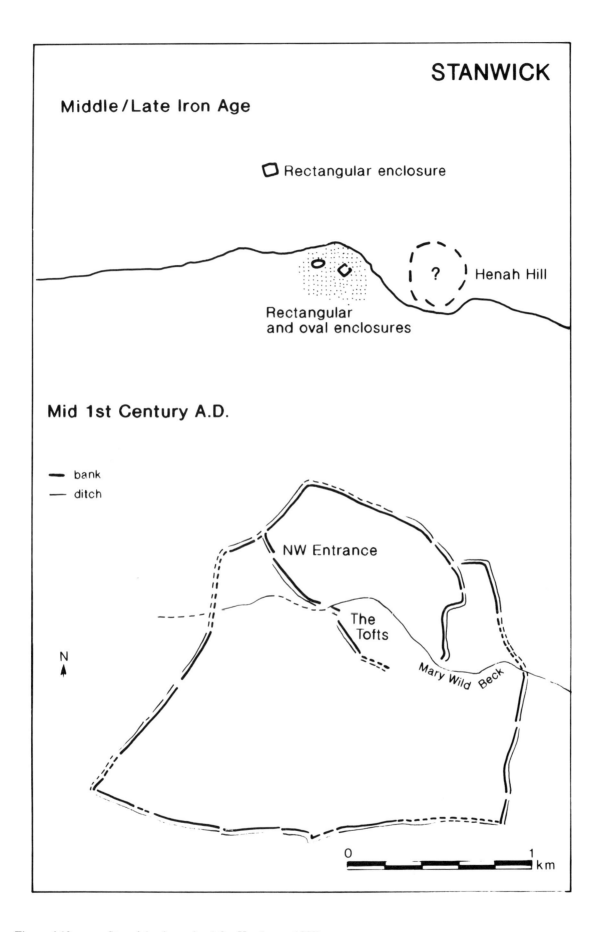

Figure 4.10   Stanwick, phase plan (after Haselgrove 1982).

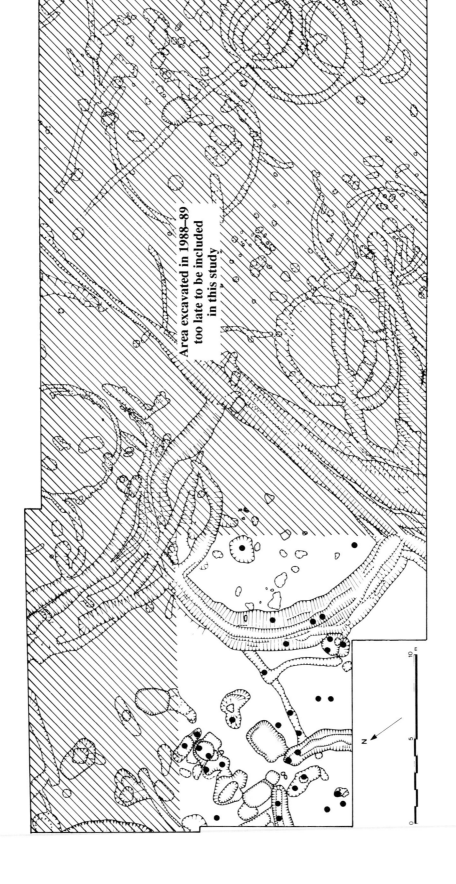

Area excavated in 1988–89 too late to be included in this study

Figure 4.11 Stanwick, site plan (after Haselgrove 1990).

4.7 Rock Castle
(North Yorkshire; NZ 185 067)

The site was excavated in 1987 by Percival Turnbull of the Department of Archaeology, University of Durham, and Leon Fitts of Dickinson College, Carlisle, Pennsylvania (Turnbull and Fitts forthcoming). The excavation forms part of a larger research project focussing on the late Iron Age settlement at Stanwick. The site at Rock Castle is located only 5 km south of Stanwick and is regarded to have been one of the settlements within its sphere of influence. The site showed up with a cropmark on aerial photographs as a sub-rectangular enclosure and central house (Figure 4.12). Small-scale excavation of the site took place in order to study the possible links between this small farmstead and Stanwick, thought to have been the central place in the region.

4.7.1 Location

The site is located in an arable field just south of the A66 trunk road, 6 km north of Richmond, on the summit (190 m O. D.) of a fairly steep, south-facing slope looking into the valley of the Gilling Beck (Figure 4.3).

4.7.2 The excavation

Only a small part of the site was excavated. A series of concentric ditches, postholes, and pits were recovered, probably representing the remains of more than one round, timber-built house (Figure 4.12). The remains of a large ditch were also found, apparently associated with a trackway visible on the aerial photographs. The lay-out of the settlement, i.e. the sub-rectangular enclosure with central house, points to a late Iron Age or Romano-British period of occupation, but while some late Iron Age pottery was found, earlier fabrics were also present. Sherds from two vessels (from contexts 12 and 60) are thought to date to the early Iron Age or even to the late second millennium BC (G. Ferrell and P. Turnbull, pers. comm.). Some Neolithic or Bronze Age worked flints were found in the plough soil.

4.7.3 Sampling strategy

Samples were collected from all well-stratified features. Two samples were later found to come from recent features and were discarded. A list of the volume and context of each sample is given in Table 4.13.

4.7.4 The plant remains (Table 4.14)

The samples contained a large amount of cereal chaff, and a small number of cereal grains, but were dominated by weed seeds. The barley grains in the samples belonged to six-row, hulled barley, *Hordeum vulgare*. Spelt wheat, *Triticum spelta*, was the dominant wheat species, but one context, context 50, contained a large number of rachis internodes of bread/club wheat, *Triticum aestivo-compactum*. Wild food plants were present in the form of hazelnut, *Corylus avellana*, and sloe plums, *Prunus spinosa*. The weed seeds were dominated by seeds of heath grass, *Sieglingia decumbens*, and, to a lesser extent, small grasses such as *Poa annua* (see also Van der Veen forthcoming c).

4.8 Thornbrough
(Northumberland; NZ 011 633)

Excavations took place at Thornbrough during 1983 and 1984 under the direction of Peter Clack of the Archaeological Unit for North East England, on behalf of the Historic Buildings and Monuments Commission (Clack 1984). The site was known as a cropmark on aerial photographs, which showed a rectangular enclosure with an entrance on the eastern side. Excavations took place in advance of gravel extraction.

4.8.1 Location

The site is situated on the northern bank of the River Tyne, at 55 m O. D., 2 km south east of the town of Corbridge, 3 km south east of the Roman fort Corstopitum and 5 km south of Hadrian's Wall (Figure 4.3).

4.8.2 The excavation

During 1983 a small trial excavation took place to assess the survival of archaeological features and environmental data, while in 1984 a larger-scale excavation was carried out. Unfortunately, it proved difficult to distinguish archaeological features from natural ones (the subsoil being very stony). While many postholes and some gullies and ditches were identified, no house plans or other structures were recognized within the excavated area (unfortunately, no site plan is yet available). There appeared to be large areas of cobbling or paving, but it was never clearly established whether these were natural or man-made. The pottery sherds and other finds suggest a Romano-British date for the settlement, with possibly some earlier, late Iron Age occupation as well (M. Millett, pers. comm.; see also chapter 5). The post-excavation work has, unfortunately, not yet been carried out. Despite this lack of reliable archaeological information it was decided to carry out the archaeobotanical analysis of a small selection of the samples, as, with the exception of a small seed assemblage from Catcote, Cleveland (Huntley 1989), no other archaeobotanical material was available from native sites of this period.

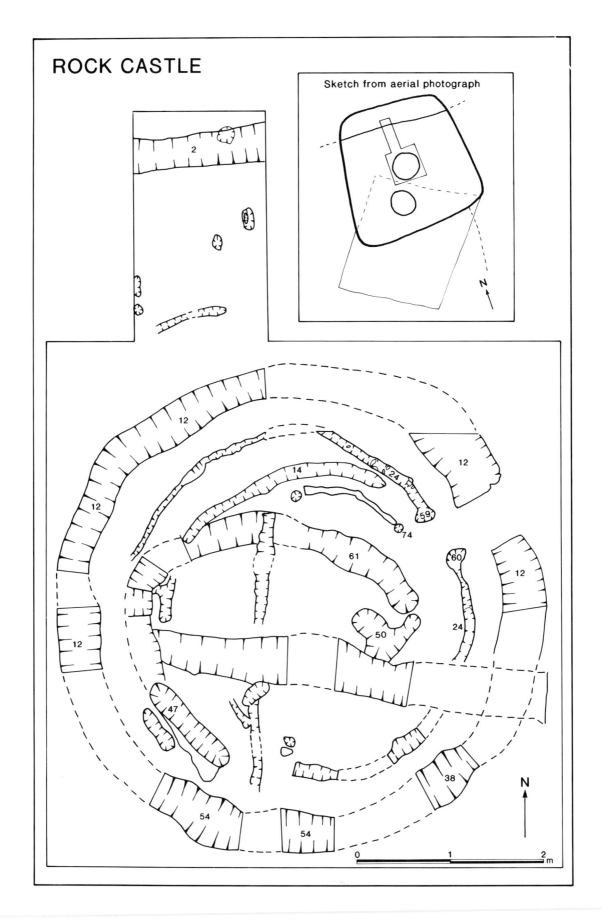

Figure 4.12   Rock Castle, site plan (after Turnbull and Fitts, forthcoming).

4.8.3 Sampling strategy

The aim was to collect samples from all well-stratified contexts. In the event samples were collected from both well- and poorly-stratified deposits, while some contexts were not sampled at all. In total 134 samples from 95 contexts were available. It was decided to analyse *ca.* 20 contexts, as any further work would probably not have been justified in the light of the lack of satisfactory archaeological information about the samples. Nineteen contexts were selected randomly from the subsoil cut features (having first omitted any poorly stratified contexts). In addition to these, two contexts were selected for analysis, because during flotation they were found to be very rich in cereal grains (1983: 10 and 39), and two more contexts were selected because they contained pottery (1984: 46 and 54b) (most of the pottery was found in the plough soil). These last four samples all came from sediment in between areas of cobbling which were only sealed by the plough soil. Thus, a total of 23 contexts was analysed (24 per cent of the total), representing 31 samples (in eight cases two samples were analysed from the same context in order to increase the number of identifications from that context). The volume and context description of each sample is given in Table 4.15.

4.8.4 The plant remains (Table 4.16)

The plant assemblage was dominated by cereal grains; cereal chaff was also common, but weed seeds occurred in low numbers only. The barley grains belonged to six-row, hulled barley, *Hordeum vulgare*. Spelt wheat, *Triticum spelta*, was the only wheat crop present. A very small number of glume bases of emmer, *Triticum cf. dicoccum*, were present, but this species is here regarded as a contaminant, not a crop. A third cereal crop was rye, *Secale cereale*, represented both by grains and rachis internodes. One seed of flax, *Linum cf. usitatissimum*, represents a possible fourth crop species. A few wild food plants were present: hazelnut, *Corylus avellana*, blackberry, *Rubus fruticosus*, and sloe berry, *Prunus spinosa*. The weed assemblage was dominated by seeds of bromegrass, *Bromus mollis/secalinus*. Corncockle, *Agrostemma githago*, is present in two samples.

4.9 South Shields
(Tyne and Wear; NZ 365 679)

The Roman fort at South Shields was built around AD 128, to guard the mouth of the River Tyne, as part of the frontier system of Hadrian's Wall. In AD 208, when Emperor Severus came to Britain to conduct the re-conquest of Scotland, the fort was turned into a supply base for the army operating

there. A total of twenty-two granaries was built for this purpose. By AD 220 South Shields' role as a supply base had been reduced, and several of the granaries were converted into living quarters (Bidwell 1989). Excavations have been carried out in the fort since 1983 under the direction of Paul Bidwell, on behalf of Tyne and Wear County Council. During 1984 a granary was discovered in the forecourt of the original Headquarters building (Figure 4.13). It was probably built in AD 208 along with all the other granaries, but remained in use after AD 220. Most of the other granaries in the fort were excavated a long time ago and no plant remains were collected then. While granaries are a common feature of Roman forts, very little is known about their actual contents. Systematic sampling to recover carbonized grain has rarely been carried out on granaries in Britain. Thus, the excavations of this granary at South Shields offered a unique opportunity to carry out such a sampling exercise.

4.9.1 Location

The fort is situated within the present town of South Shields, on the southern bank of the Tyne, at the mouth of the river (Figure 4.3).

4.9.2 The excavation

As already mentioned above, the excavations in 1984 recovered the remains of a granary in the forecourt of the Headquarters building. Only the sleeper walls of the granary survived, and during the excavation cereal grains were found mixed in with the deposits in between these sleeper walls. Two deposits could be recognized. The lower one (12236) was very thin, with a maximum depth of only 50 mm, consisting of clay and clay-silt, flakes of sandstone and mortar, charcoal and grain. The deposit was sealed when the flagstone floor of the granary was coated with a layer of *opus signinum*, presumably when the flags were worn and beginning to break up (P. Bidwell, pers. comm.). The grain from this deposit probably represents spillage through cracks in the floor. The deposit is dated to the last quarter of the third century AD, through its association with a fire which destroyed much of the fort. The second deposit (12176) represents a layer of debris which had accumulated over the floor of the granary. This deposit contained coins and pottery of mid fourth century date (P. Bidwell, pers. comm.).

4.9.3 Sampling strategy

Samples were taken at regular intervals from in between the sleeper walls, in order to get a total coverage across the whole floor area of the granary. The locations of the samples are indicated in Figure 4.13. In total 33 samples were collected from deposit

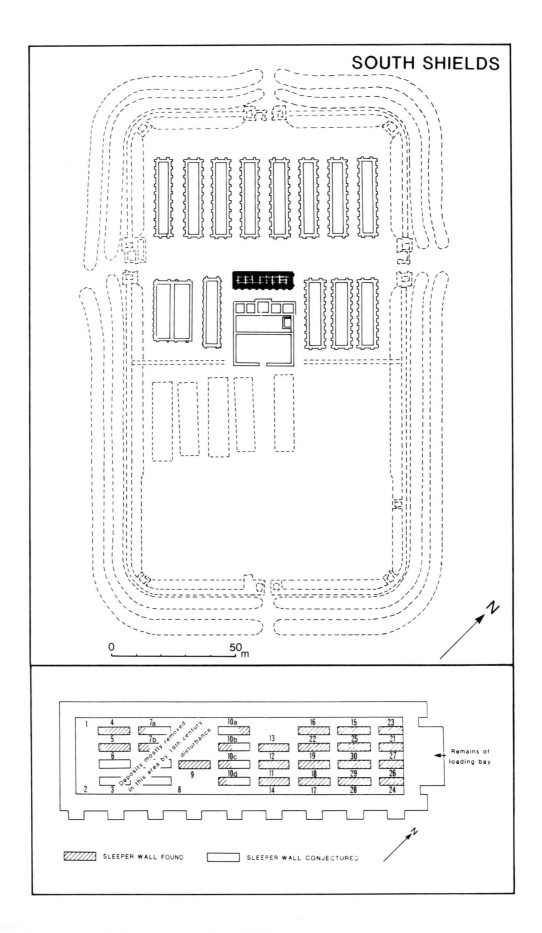

Figure 4.13   South Shields, site plan (after Bidwell 1989).

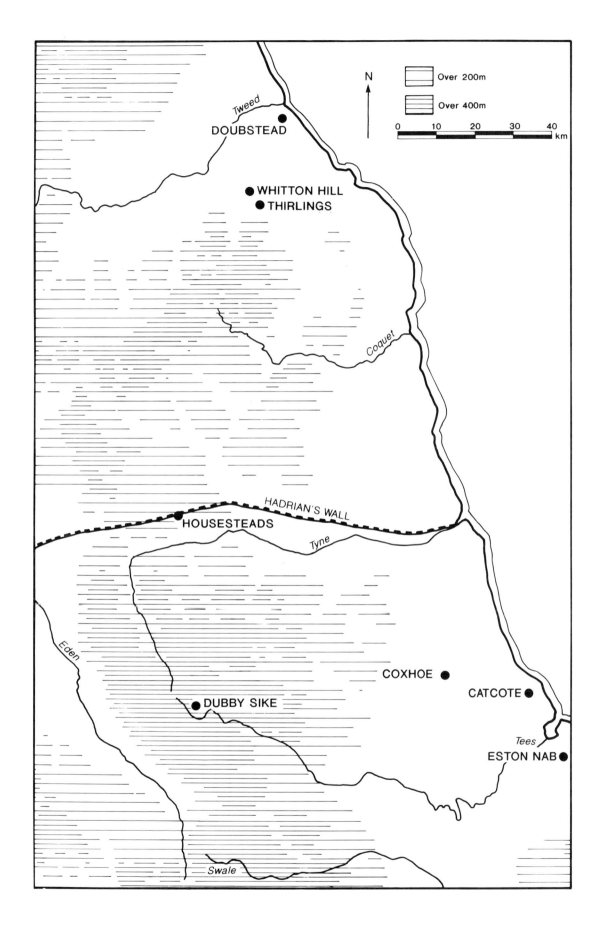

Figure 4.14   Map of the region showing the other sites from which plant remains are available.

12236, and 30 from deposit 12176. The sample volumes are given in Table 4.17.

4.9.4 The plant remains (Table 4.18)

The plant assemblage was dominated by cereal grains with only very few chaff fragments and weed seeds present. Two cereal crops were present: spelt wheat, *Triticum spelta*, and bread/club wheat, *Triticum aestivo-compactum*. A few grains of six-row, hulled barley, *Hordeum vulgare*, were found, but these are here interpreted as contaminants in the wheat crop. The weed seeds are dominated by those of bromegrass, *Bromus mollis/secalinus* and oats, *Avena cf. fatua*, but in deposit 12176 those of heath grass, *Sieglingia decumbens*, are most common (see also Van der Veen 1988a and forthcoming d).

4.10 Other evidence

The only other carbonized seed assemblages available from sites in north-east England are those from:

Neo./BA	*Thirlings* (Northumberland) (Van der Veen 1985b) *Whitton Hill* (Northumberland) (Van der Veen 1985d)
BA/IA	*Eston Nab* (Cleveland) (Van der Veen 1988b)
Late IA	*Coxhoe* (Durham) (Van der Veen and Haselgrove 1983) *Dubby Syke* (Durham) (Van der Veen 1988c)
RB	*Catcote* (Durham) (Huntley 1989) *Doubstead* (Northumberland) (Donaldson 1982)

The assemblages from these sites are not included in the analysis here, because they were too small (Thirlings, Whitton Hill, Coxhoe, and Doubstead), contained little or no remains of cereal crops (Eston Nab and Dubby Sike), or became available too late (Catcote). For a description of these sites and their assemblages the reader is referred to the relevant publications, but for convenience the location of these sites is given in Figure 4.14.

5. Dating Evidence

5.1 Introduction

We have seen in the previous chapter that the available dating evidence for the sites under study suggests the following timespan for the occupation of the individual sites:

Timespan of occupation before dating programme

	late BA	early IA	late IA	early RP	late RP
Hallshill	———				
Murton			———		
Dod Law			———		
Chester House	———— ?		————————		
Thorpe Thewles		?	————————		
Stanwick			——		
Rock Castle	———— ?		————————		
Thornbrough				———————— ?	——
South Shields					——

(Abbreviations: BA = Bronze Age; IA = Iron Age; RP = Roman Period)

However, for a number of reasons, the exact timespan of occupation is not always very closely defined. The coarse pottery found in the region is not very sensitive to chronological seriation, few objects which could be dated independently were found, and the type of settlement (hillfort, sub-rectangular enclosure, open settlement, etc.) is not necessarily restricted to a particular chronological period (see Chapter 2). Only three sites were dated by absolute dating methods (Hallshill and Murton by radio-carbon dates, and Thorpe Thewles by thermoluminescence dates).

As was mentioned in Chapter 1, the first millennium BC saw many important changes in the agricultural economy of Britain. The replacement of emmer wheat by spelt wheat, and later the replacement of the glume wheats by free-threshing species such as bread/club wheat and rye, are seen as particularly crucial developments. The results from the settlements studied here (Chapter 4) indicate that some sites contained both emmer and spelt wheat, but that at others only spelt wheat was present. Bread/club wheat and rye were present at only three sites. The precise dating of the seed assemblages is crucial to our understanding of the development of arable farming in the region. Consequently, it was necessary to improve the dating evidence for the sites and their plant assemblages, before any further analysis could take place.

Four objectives were formulated for the dating programme:

(1) To try and improve the dating evidence for the period of occupation of each settlement in order to allow a better comparison between the assemblages from each site. For this purpose samples were selected from contexts which were either early or late in the stratigraphical sequence, to date the beginning and end of the occupation.

(2) As there was evidence at some of the sites of earlier or later occupation in the vicinity of the settlement, it was necessary to check whether admixture of older or younger material had taken place (material displaced by animal activity, material remaining on the site from previous occupation at the site, unrecognized stratigraphical complications; Waterbolk 1983). For this purpose samples of spelt wheat were selected from early contexts to test that the spelt did not derive from later features, and samples of emmer wheat were selected from late contexts to test it was not derived from earlier occupation. Samples of bread/club wheat were selected as they were thought to represent modern, intrusive material (in two cases), or medieval contamination (one case).

(3) Because the exact date of the introduction of spelt wheat was regarded as important, early records of this species were selected for dating, as well as samples which represented the period when spelt wheat had become the dominant wheat crop. As both bread/club wheat and rye occurred only very rarely in the samples, and as their history is still poorly known in Britain, these species were also dated.

(4) Finally, as the Roman occupation of the region could have had a major impact on the native economy, it was important to date the plant assemblages by absolute methods in order to allow comparison with historical events. For this purpose the results of the radio-carbon dates have been calibrated into calendar years, using the high-precision calibration data recommended by the 12th Radio-carbon Conference, held at Trondheim in 1985, and published in a special issue of *Radiocarbon* (Stuiver and Kra 1986).

In order to meet these objectives, two types of samples were selected and submitted for dating: (a) charcoal samples for 'beta-decay counting' or conventional radio-carbon dating and (b) samples of charred grains or chaff fragments for 'accelerator mass spectrometry' dating or accelerator dating. The first method requires *ca.* 1–5 grammes of pure carbon. It measures the radioactive decay rate of a known amount of carbon, using either gas proportional or liquid scintillation counters (Gillespie 1984, Mook and Waterbolk 1985). This method of dating is usually referred to as 'conventional' radio-carbon dating in contrast to accelerator mass spectrometry dating (Gillespie 1984, Hedges and Gowlett 1986, Mook and Waterbolk 1985), even though this term is slightly confusing as both methods produce a so-called 'conventional radio-carbon age' (i.e. an age expressed in radio-carbon years BP, whereby the present is defined as AD 1950; see also section 5.4 below). The term 'conventional radio-carbon dating' has been maintained here, in line with the publications mentioned above.

The second method requires only 0.5–5 milligrammes of pure carbon, hence the possibility of submitting individual grains or small numbers of chaff fragments. This method measures the carbon-14 concentration in a sample, by using a particle accelerator in conjunction with a mass spectrometer, detecting the atoms without having to wait for them to decay (Gillespie 1984, Hedges and Gowlett 1986, Mook and Waterbolk 1985). This method of dating was of crucial importance in meeting objectives two and three, as they required the dating of actual grains or chaff fragments (these remains are rarely available in quantities large enough for conventional radio-carbon dating). The conventional radio-carbon dates were carried out at the Centre for Isotope Research of the University of Groningen, The Netherlands, while the accelerator dates were provided by the Oxford University Radiocarbon Accelerator Unit (see the Acknowledgements above). A total of 50 samples were submitted to these two laboratories in order to meet the objectives of this dating programme. The nine conventional radio-carbon dates already available from Hallshill and Murton had been dated by the Isotope Measurements Laboratory at Harwell.

Waterbolk has stressed the importance of selecting a sample carefully in relation to the event it is meant to date (Waterbolk 1971, 1983), and he has put forward four categories for describing the degree of certainty with which samples are associated with the archaeological material they are intended to date, as well as for describing the age of the sample at the time of burial/discard. As I will be using these categories below, in section 5.2, they are briefly described here (following Waterbolk 1983, 58):

I. *Certainty of association*

(A) *Full certainty*: the sample came from the artefact itself, e.g. post of house, wagon wheel, grain.

(B) *High probability*: there is a direct functional relationship between the sample and the

archaeological finds, e.g. coffin dates finds in grave, charcoal dates urn, hearth dates occupation in house.

(C) *Probability*: the functional relationship is not demonstrable, but the quantity of organic material and the size of fragments argue in favour of it; e.g. charcoal concentration in a rubbish pit or occupation layer.

(D) *Reasonable possibility*: as above, but the fragments are small and scattered, e.g. 'dark earth' in an occupation layer, charcoal fragments in a grave.

While the samples submitted for accelerator dating could all be classified under category A, the samples submitted for conventional dating could not. No charcoal samples specifically collected for radio-carbon dating were available from the excavations, other than those from Hallshill and Murton which had already been dated, with the exception of one sample from Stanwick. Consequently, it was necessary to use the charcoal present in the flotation samples. This meant picking out large numbers of small fragments of charcoal and submitting these for dating. In these cases the samples had to be classified as category D, giving only a reasonable possibility that the sample was associated with the event to be dated.

In addition to the association between the sample and the event to be dated, the actual age of the sample at the time of burial or discard must also be taken into account (Waterbolk 1971, 1983). Samples are generally older than their contexts. While a post may date the construction of the house, the tree used for the timber must have been of some considerable age, and the re-use of old timber is another well known phenomenon. Below are listed the four categories of sample age as given by Waterbolk (1983, 58–59):

II. *Sample age in relation to burial/discard*

(A) *Small*: the difference in date is so small as to be negligible (less than 20 years), e.g. twigs, grain, leather, bone, outermost tree rings.

(B) *Medium*: the time difference can amount to several decades (over 20, less than 100 years), e.g. charcoal from short-lived wood species, outermost rings from long-lived wood species.

(C) *Large*: the time difference may amount to centuries (more than 100 years), e.g. charcoal from long-lived wood species, possibly subject to re-use.

(D) *Unknown*: the nature of the dated organic material is not precisely known, e.g. samples consisting of 'dark earth', 'ash', or 'soil'.

The samples submitted for accelerator dating can all be classified under category A, but most of the charcoal samples for conventional dating fall under category D. In the discussion of the results below this difference in the reliability of the samples will be considered.

In section 5.2 below, the results of the dating programme will be presented for each site separately and the results will be compared with the other types of dating evidence for the sites. The degree of certainty of association and the age of the sample at burial will be listed in the tables under columns I and II. The dates are given in radio-carbon years BP (i.e. the date is expressed in radio-carbon years BP, whereby the present is defined as AD 1950). A discussion of the results in relation to the objectives will be given in section 5.3, while section 5.4 will present the results calibrated into calendar years to allow comparison with historical events, in this case the occupation of the region by the Roman army.

5.2 Dating evidence for each site

5.2.1 Hallshill

Six conventional radio-carbon dates were already available from this site, giving early dates for some of the pits and a posthole, but slightly later dates for the 'hearth' or 'central area of burning' (Figure 4.4). The difference between the two groups of dates led the excavator to suggest that the central area of burning represented a later activity on the site, unrelated to the occupation of the house, as already mentioned in Chapter 4 (T. Gates, pers. comm.). Emmer and spelt wheat were present in both features. Three samples of spelt wheat were selected to date the earliest occurrences of spelt wheat and to test that the material did not derive from later occupation. One sample of emmer wheat was selected from the later feature, the 'hearth', in order to test whether it could have been derived from the earlier occupation. The results are given on page 60.

There is a very good agreement between the conventional and accelerator dates. The results suggest that no admixture between early and late contexts had taken place and that spelt wheat had been introduced into the region at the very beginning of the first millennium BC. There is a difference of *ca.* 100 years between the charcoal dates and the grain dates in context 23, but no difference was found between the two types of dates from contexts 8 and 25.

5.2.2 Murton

Three separate phases of occupation were recognized at Murton. The first one was an unenclosed settlement with timber-built round houses, undated, but possibly associated with a radio-carbon date of 2960±80 BP (HAR–6201). The second phase consists of an enclosed, defended settlement with timber-built houses of late Iron Age date (see HAR–6202 and HAR–6200 listed below). The third and final phase consisted of stone-built houses with a stone-built enclosure wall (Figure 4.5). This phase probably started in the early Roman period, and may have lasted till the late second or third century AD. The flotation samples all came from what were thought to be contexts from the timber-built, defended settlement. As the site had been occupied after the period from which the samples were available, it was important to establish whether the seeds could be intrusive, i.e. be derived from the later occupation, and three grain samples were selected to test this. A further three charcoal samples were selected in the hope of improving the date of the timber-built, enclosed settlement. The results are given on page 60.

The date on the charcoal sample from context 625 is far too old to relate to the settlement. It is important to note, however, that it came from the same flotation sample as the spelt and emmer chaff of samples OxA–1741 and 1742. Both HAR–6201 and GrN–15672 may represent material from the earlier, unenclosed settlement. The other dates agree with a late Iron Age date for the timber-built phase of occupation. There is no evidence for any intrusion of later material into earlier contexts. Context 623 represents the 'occupation earth' on the floor area of timber-built house T9, which was sealed by the paved floor of stone-built house S7 (Figure 4.5). The date for the emmer grain from context 623 suggests that this context represents the very end of the timber-built phase or the very beginning of the stone-built phase of occupation. The dates on the emmer and spelt wheat indicate that both crops were in use at the same time.

5.2.3 Dod Law

No radio-carbon samples were available, but two bronze fibulae dated to *ca.* AD 70–100 form a *terminus ante quem* for the botanical samples from the rubbish deposits in between the two ramparts (Figure 4.6). No archaeological date for the start of the occupation is available, but a starting date of *ca.* 300 BC is assumed (C. Smith, pers. comm.). Charcoal was selected from the flotation samples from context 45, the ground surface underneath the inner rampart, as well as from the flotation samples from contexts 48 and 30. Context 48 is one of the earliest rubbish layers in between the two ramparts, and context 30 is a rubbish layer accumulated against the outer rampart, overlying the rubbish deposits in between the two ramparts (Figure 4.6). Two samples of glume bases of spelt wheat were selected from contexts 40 and 48 (which, in fact, represent the same rubbish layer, but on either side of the baulk) to date the presence of spelt on this site and to test the

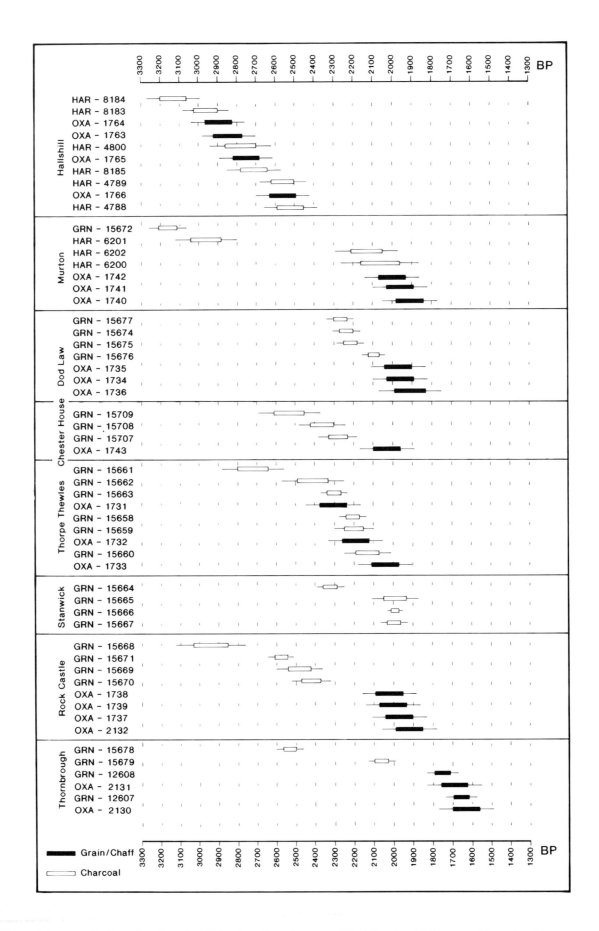

Figure 5.1   Radio-carbon dates for all sites in radio-carbon years BP, giving 1 and 2 sigma confidence levels.

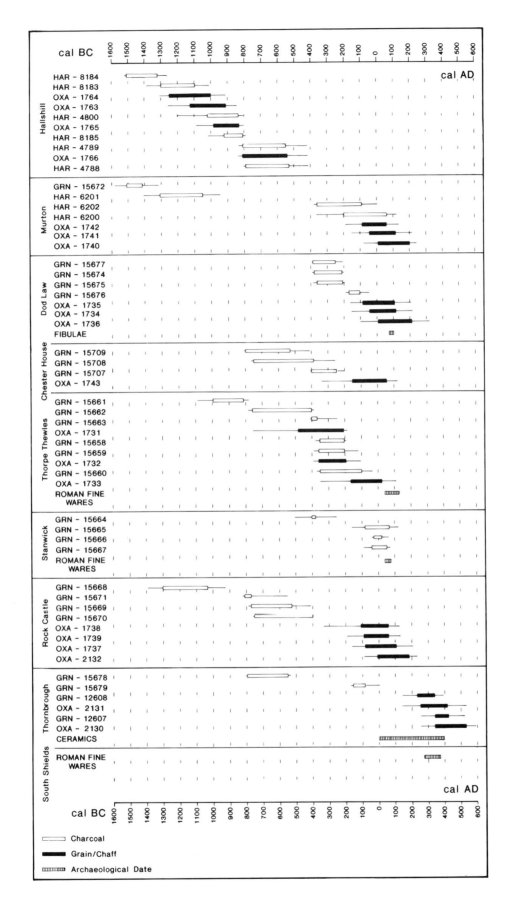

Figure 5.2   Calibrated radio-carbon dates for all sites in calendar years cal BC/AD, giving 1 and 2 sigma confidence levels, archaeological dating evidence, and thermoluminescence dates (1 sigma confidence level; first error term is a measure of precision; second error term is the overall error (Bailiff 1987, 71–72)).

material could not have been derived from later contexts. One sample of emmer glume bases was selected from context 30, stratigraphically the latest one from which plant remains were available, to test that they could not have been derived from earlier contexts. The results are given on page 60.

The dates from context 45 confirm a starting date for the site of *ca.* 300 BC. Most of the botanical samples, however, came from the rubbish deposits in between the two ramparts. The dates from these contexts suggest that the deposits were laid down within a relatively short period of time, somewhere between *ca.* 100 BC and AD 100, which agrees with the date for the fibulae. The charcoal date from context 30 is too old, and may represent the presence of residual material or the use of old wood. The accelerator dates of the emmer and spelt indicate that no admixture with older or younger material took place, and confirms that the two crops were in use at the same time.

5.2.4 Chester House

No radio-carbon dates or archaeological dates were available from this site. The stratigraphical sequence would appear to suggest that the circular houses could not be contemporary with the sub-rectangular enclosure (Figure 4.7). The sequence of an open, unenclosed settlement followed by a later, sub-rectangular enclosure, is preferred by the excavator (Holbrook 1988). Pottery from two lowland unenclosed sites in the region is dated to the early first millennium BC. The houses from Chester House may be of similar date. The enclosure probably belonged to a Romano-British settlement (N. Holbrook, pers. comm.).

One charcoal sample was selected from the flotation samples from each of the three houses to get some idea of the dating of the houses (Figure 4.7). One barley grain was selected from House 1 and emmer glume bases were selected from House 2 in order to date the plant assemblage from the site. One grain of bread/club wheat was also submitted for dating as it looked modern (the grain was quite large and the carbonization was different from that of the other grains). The sample was submitted to test this hypothesis. The results are given on page 60.

The barley grain from context 131 gave a very small beam in the accelerator and could, consequently, not be dated. The grain of bread/club wheat (context 142) gave a reading of 122.6% modern, which is the percentage of modern carbon-14 and indicates the presence of modern bomb carbon. This means that the grain post-dates AD 1945/1950, which confirms the hypothesis that the grain is intrusive and modern.

The remaining dates are difficult to interpret. There is a difference of 300–500 years between the charcoal and the emmer chaff samples. The two samples from context 117 (GrN–15708 and OxA–1743) differ by 300 years, with the emmer chaff sample being the younger of the two. This would suggest either the re-use of old wood, or the presence of old, residual charcoal on the site. Alternatively, the emmer chaff is intrusive and represents material from later occupation, possibly from the sub-rectangular enclosure. As there is no real evidence for the date of the sub-rectangular enclosure, and as the stratigraphical evidence does not rule out the possibility that the enclosure ditch preceded the open settlement (see section 4.4.2 above), this matter cannot be resolved.

In view of the rather unsatisfactory nature of both the stratigraphic and the dating evidence, it was necessary to regard the accelerator date on the emmer wheat from context 117 as the most reliable date for the total assemblage. In fact, context 117 produced the majority of the plant remains from this site. The assemblage is, consequently, assumed to be of late Iron Age date, although an earlier start to the material cannot be ruled out. It is not possible to determine with any certainty whether the assemblage belongs to an unenclosed settlement or to a rectilinear enclosure settlement, but both are examples of undefended settlements.

5.2.5 Thorpe Thewles

The settlement history on this site is complex. On the basis of the stratigraphy of the numerous features and the externally dated finds the occupation was divided into four phases (Heslop 1987). Phase I is only represented by two features which could not be dated, but they are stratigraphically the earliest features on the site (Figure 4.8). Phase II consisted of a sub-rectangular enclosure with a number of hut-circles inside. The starting date of this phase is unknown, but is thought to lie at *ca.* 300 BC, while the end date is *ca.* AD 0. The occupation of Phase III is thought to be of short duration, starting at *ca.* AD 0 and ending (on the basis of Roman fine wares) at *ca.* AD 80. This phase is characterized by the filling in of the enclosure ditch and the expansion of the settlement features outside the original enclosure. Phase IV consists of a rectangular enclosure ditch, but without associated hut circles within the excavated area. The presence of Roman fine wares dates this phase to *ca.* AD 80–130.

A total of twelve thermoluminescence (TL) dates were obtained on pottery sherds from phases II, III, and IV (Bailiff 1987). The TL dates (page 61) are given in years AD/BC and, being absolute dates, they require no secondary calibration. The associated error terms are given in years at the 68% level of confidence. The first is a measure of precision and is to be used when comparing TL dates from this site.

The second is the overall error and takes into account all known errors; it is to be used when comparing the TL dates with calibrated radio-carbon dates or calendar dates (Bailiff 1987,71–72).

The TL dates for Phase II suggest that the starting date of the occupation on the site was considerably earlier than previously thought, although this cannot be clearly demonstrated at the 95% level of confidence using the overall error (Bailiff 1987). The mean TL date for Phase III is also earlier than the archaeological dating evidence.

As the number of accelerator dates available within the present dating programme was limited, it was decided to concentrate these on samples from the early phases of occupation. Three samples of spelt wheat were selected, one from Phase I and two from Phase II, to date the earliest record of spelt on the site and, as no emmer wheat was found on the site, to date the period when spelt wheat had become the principal wheat crop. In addition to these samples of spelt wheat one sample of a grain of bread/club wheat was selected for dating. The grain looked modern and was assumed to be intrusive. The sample was submitted in order to test this hypothesis. A further six samples of charcoal (all from flotation samples) were selected for comparison with the other dating evidence. The results are given on page 60.

The radio-carbon dates from Phase I suggest that this phase is not much older than Phase II. The dates from Phase II suggest a starting date for the occupation at *ca.* 300 BC, with sample JS 7, from the bottom of the fill of the main enclosure ditch, dating the start of the deliberate infill of the enclosure ditch, the end of which marks the beginning of Phase III. This suggests that Phase III could have started at *ca.* 50 BC or even earlier. The date from sample LS 178 appears too old by *ca.* 100 years in comparison with the date on spelt grain from sample JS7. The results from sample JS 15, as well as samples LS 52 and LS 58 from Phase IV, demonstrate the presence of old charcoal in late features. Thus, while the radio-carbon dates have provided evidence regarding the starting date of the settlement, they did not throw much further light on the end of the occupation, due to the presence of residual material in the late contexts.

The three accelerator dates on spelt wheat from the earliest features on the site indicate that on this site spelt wheat had become the principal wheat crop by *ca.* 300 BC. The date from the grain of bread/club wheat confirmed its intrusive nature, although the grain was of medieval (13th century) rather than modern date.

A comparison between the radio-carbon dates and the TL dates can only be made after the radio-carbon dates have been calibrated. The calibration of all the dates is given below, in section 5.4, but the results for

the Thorpe Thewles dates are discussed here for convenience. A summary of the dating evidence for Thorpe Thewles is presented in Figure 5.3. The possibility of a much earlier starting date for Phase II, as suggested by the TL dates, is not confirmed by the radio-carbon dates, although the calibrated radio-carbon dates do allow a starting date of *ca.* 400 cal BC. The TL dates and radio-carbon dates both demonstrate the presence of much earlier material (charcoal and pottery) in the features of Phase III and especially in the rectangular enclosure ditch of Phase IV. The dates given by all these samples would appear to point to the presence of late Bronze Age occupation within the immediate vicinity of the settlement. The presence of fragments of saddle querns used in the wall slot of the main house of Phase II also points in this direction.

5.2.6 Stanwick

There is no dating evidence for the beginning of the occupation on this site, but it is generally assumed to start around the first century BC (or earlier) (C. Haselgrove, pers. comm.). The end is dated by a comparatively large collection of Roman imported fine wares, all dated to the period immediately prior to the Roman occupation of the region, i.e. *ca.* AD 40–70, although a continuation of the occupation after the supply of Roman pottery was interrupted cannot be ruled out.

A few potential samples for radio-carbon dating had been collected during the excavation, but none had yet been dated. It was decided to select two samples from stratigraphically early contexts, and two from late contexts, to provide evidence for the overall timespan of the occupation. No accelerator dates were obtained from this site. However, recent work on the stratigraphy, including the information obtained during the excavations in 1988 and 1989, indicates that the four contexts chosen for radio-carbon dates, do, in fact, all represent contexts which come late in the stratigraphical sequence (S. Willis, pers. comm.). This means that no information is available for the start of the occupation on this site. Further samples will be selected and submitted for dating as soon as the stratigraphical sequence of the samples has been finalized. The sample from context 1005 came from a piece of oak (*Quercus* sp.) charcoal, collected during the excavation. The remaining samples all consisted of small fragments of charcoal from the flotation samples. The results are given on page 61.

The date from context 2209 demonstrates the presence of older, residual charcoal on the site, or the re-use of old wood, or it may point to the starting date of the occupation. The remaining three dates are in agreement with the archaeological dating evidence.

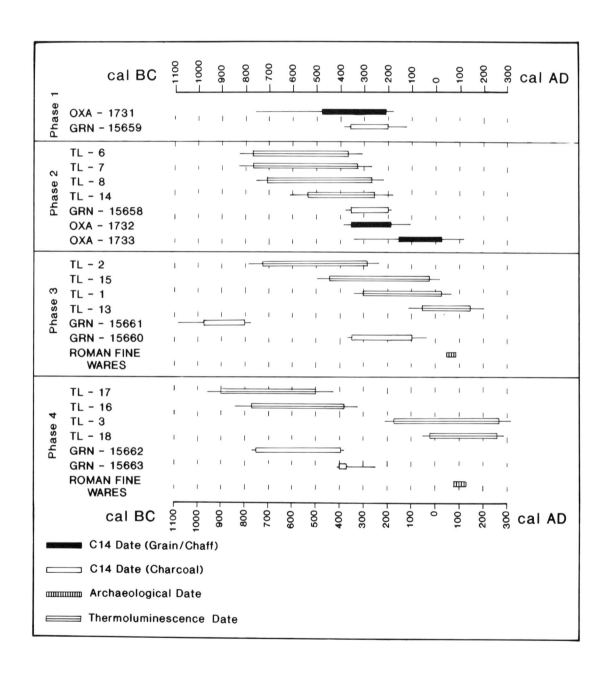

Figure 5.3   Dating evidence for Thorpe Thewles, calibrated radio-carbon dates (1 and 2 sigma confidence levels), and archaeological dating evidence.

5.2.7 Rock Castle

The occupation of this settlement is difficult to date. Sub-rectangular enclosures like these are usually dated to the late Iron Age or Romano-British period. The excavation, however, only produced a few pottery sherds of that date. Most of the pottery came from one pot, from context 12, the ring ditch of one of the phases of the round house. The pot is thought to date to the early Iron Age or even to the late second millennium BC (G. Ferrell, pers. comm.). The presence of flint tools in the topsoil points to the presence of Neolithic activity in the immediate vicinity of the settlement.

The crop marks on the aerial photograph show the presence of two, partially overlapping, enclosures (see Figure 4.12). The crop marks of the southern-most enclosure are the less substantial of the two, suggesting that they may represent a palisaded enclosure, as against the ditched enclosure of the excavated settlement. If so, and assuming that the fact that palisaded enclosures usually precede ditched enclosures also holds true here, the excavated remains of the round house, which is thought to be associated with the ditched enclosure, belongs to the later of the two settlements.

The actual period and length of the occupation is unclear. Samples were selected for conventional dating from contexts which, on stratigraphic grounds, were thought to represent early, middle and late occupancy on the site. No charcoal samples for radio-carbon dating were recovered during the excavation, so that all samples submitted consisted of small charcoal fragments from the flotation samples. Emmer wheat was not found in the assemblage. Two samples of spelt wheat were selected to date the period in which spelt wheat had become the principal wheat on this site. In one context, pit 50, a large number of rachis internodes of bread/club wheat were found. The pit was cut by what was thought to be a late feature, and the fill of the pit contained a sherd of possible medieval pottery. It was, therefore, possible that the bread/club wheat was intrusive. Two samples of bread/club wheat chaff were submitted for dating to test the hypothesis that the material was modern (or at least medieval) and intrusive. The results are given on page 61.

The results are extremely difficult to interpret. The dates on the charcoal all point to occupation in the early to middle part of the first millennium BC, while those on the grain suggest a date in the later part of the millennium. Unfortunately, there are no contexts with more than one sample, so direct comparisons between the dates are not possible. In deciding on how to interpret the results, the following observations were made. First of all, the dates on the grain are more reliable than those on the charcoal, as

the latter were based on small fragments of charcoal from flotation samples. Secondly, for the present study the date of the plant remains is more important than the date of the excavated features. Thirdly, the presence of another settlement, possibly earlier than the excavated one and partially underlying it, may explain the occurrence of large amounts of older charcoal (possibly representing the re-use of old timber), and early pottery on the site. Carbonized seeds and grains do not survive much mechanical damage and if they are not buried immediately, they are reduced to fine dust or at least become unidentifiable and are consequently not present or recognized during archaeobotanical analysis. Large pieces of charcoal, however, would disintegrate much more slowly and survive for much longer periods as small fragments, which are not recognizable during excavation, but which are concentrated during the flotation of bulk samples. On balance, therefore, the late Iron Age date of the grain samples is here taken as the most reliable date for the plant assemblage as a whole. The correctness of this decision can only be tested by dating many more grains and by excavating part of the other settlement, including those areas where the two settlements intersect one another. Both these steps are, unfortunately, outside the scope of the present study.

The two dates on the bread/club wheat demonstrate that the material is not intrusive, but belongs to the settlement under study. As the two dates are based on the same material, from the same context, they have been combined, giving a late Iron Age date for this first record of bread/club wheat in the region.

5.2.8 Thornbrough

The excavation produced a small number of datable artefacts (information from M. Millett, pers. comm.):

4 pottery sherds	late Iron Age or later
4 pottery sherds	1st to 2nd century AD
5 pottery sherds	2nd to 3rd century AD
1 pottery sherd	3rd to 4th century AD
2 fragments of a mortarium	4th century AD
1 melon bead	probably 2nd century AD

A total of four samples were submitted for conventional radio-carbon dating, but two of these (contexts 10 and 45) consisted of pure grain rather than charcoal, as the two contexts concerned produced large quantities of grain. The assemblage produced very small amounts of rye grain and chaff, a species not found in any of the other assemblages. As two of the contexts with rye (contexts 46 and 134) produced charcoal dates of Iron Age date, two samples of rye grain were submitted to date the earliest occurrence of rye in the region. The results are given on page 61.

Dating Evidence
Radio-carbon Dates

Hallshill

Context	Material	Stratigraphy	I	II	Lab.Code	Date BP
21	charcoal	-	C	C	HAR – 8184	3130±70
23	charcoal	early	C	C	HAR – 8183	2960±60
23	spelt chaff	early	A	A	OxA – 1763	2840±70
23	spelt grain	early	A	A	OxA – 1764	2895±70
25	charcoal	early	C	C	HAR – 8185	2710±70
25	spelt grain	early	A	A	OxA – 1765	2750±70
10	charcoal	-	C	C	HAR – 4800	2780±80
8	charcoal	late	B	C	HAR – 4788	2520±70
8	charcoal	late	B	C	HAR – 4789	2560±60
8	emmer grain	late	A	A	OxA – 1766	2560±70

Murton

Context	Material	Stratigraphy	I	II	Lab.Code	Date BP
-	charcoal	early	D	C	HAR – 6201	2960±80
617	charcoal	early	C	C	HAR – 6202	2130±80
633	charcoal	mid	C	C	HAR – 6200	2060±100
624	charcoal	mid	D	D	GrN – 15672	3160±50
623	emmer grain	mid	A	A	OxA – 1740	1910±70
625	charcoal	mid	D	D	GrN – 15673	4285±50
625	spelt chaff	mid	A	A	OxA – 1741	1960±70
625	emmer chaff	mid	A	A	OxA – 1742	2000±70

Dod Law

Context	Material	Stratigraphy	I	II	Lab.Code	Date BP
45(8)	charcoal	early	C	A	GrN – 15674	2235±35
45(9)	charcoal	early	C	A	GrN – 15675	2215±35
48(10)	charcoal	middle	C	D	GrN – 15676	2095±30
48(10)	spelt chaff	middle	A	A	OxA – 1734	1960±70
40(11)	spelt chaff	middle	A	A	OxA – 1735	1970±70
30(6)	charcoal	late	C	D	GrN – 15677	2265±35
30(6)	emmer chaff	late	A	A	OxA – 1736	1910±80

Chester House

Context	Material	Stratigraphy	I	II	Lab.Code	Date BP
127–H1	charcoal	'early'	D	D	GrN – 15707	2280±50
131–H1	barley	'early'	A	A	OxA –	no beam
117–H2	charcoal	'middle'	D	D	GrN – 15708	2360±60
117–H2	emmer chaff	'middle'	A	A	OxA – 1743	2030±70
140–H3	charcoal	?	D	D	GrN – 15709	2530±80
142	bread wheat	intrusive?	A	A	OxA – 1744	modern

Thorpe Thewles

Context	Material	Stratigraphy	I	II	Lab.Code	Date BP
LS 268	charcoal	Phase I	D	D	GrN – 15659	2200±50
LS 268	spelt chaff	Phase I	A	A	OxA – 1731	2305±70
LS 120	charcoal	Phase II	D	D	GrN – 15658	2205±35
LS 112	spelt chaff	Phase II	A	A	OxA – 1732	2190±70
JS 7	spelt grain	Phase II	A	A	OxA – 1733	2040±70
LS 178	charcoal	Phase III	D	D	GrN – 15660	2130±60
JS 15	charcoal	Phase III	D	D	GrN – 15661	2720±80
LS 52	charcoal	Phase IV	D	D	GrN – 15662	2410±80
LS 58	charcoal	Phase IV	D	D	GrN – 15663	2300±35
LS 239	bread wheat	intrusive?	A	A	OxA – 1745	720±70

Stanwick

Context	Material	Stratigraphy	I	II	Lab.Code	Date BP
2209	charcoal	early	D	D	GrN – 15664	2320±35
1095	charcoal	early	D	D	GrN – 15665	1990±60
1005	charcoal	late	C	C	GrN – 15666	1990±20
1013	charcoal	late	C	D	GrN – 15667	1995±35

Rock Castle

Context	Material	Stratigraphy	I	II	Lab.Code	Date BP
61	charcoal	early	D	D	GrN – 15668	2940±90
47	spelt grain	early	A	A	OxA – 1738	2020±70
74	spelt grain	early	A	A	OxA – 1739	2000±70
12	charcoal	middle	D	D	GrN – 15669	2480±60
60	charcoal	middle	D	D	GrN – 15670	2420±50
2	charcoal	late	D	D	GrN – 15671	2575±35
50	bread wheat(chaff)	late?	A	A	OxA – 1737	1970±70
50	bread wheat(chaff)	late?	A	A	OxA – 2132	1920±70
50					OxA – 1737 and 2132 combined	1955±50

Thornbrough

Context	Material	Stratigraphy	I	II	Lab.Code	Date BP
10	spelt+barley		A	A	GrN – 12607	1655±40
45	spelt+barley		A	A	GrN – 12608	1750±40
46	charcoal		D	D	GrN – 15679	2060±35
46	rye grain		A	A	OxA – 2130	1630±70
134	charcoal		D	D	GrN – 15678	2530±35
134	rye grain		A	A	OxA – 2131	1690±70

Thermoluminesence Dates

Thorpe Thewles

Phase II

TL 6	Main enclosure ditch	570 BC	± 200/260
TL 7	Main enclosure ditch	550 BC	± 220/280
TL 8	Main structure ditch	490 BC	+ 220/270
TL 14	Main enclosure ditch	400 BC	± 140/220
Mean date for phase II		485 BC	± 45/190

Phase III

TL 1	Masking layer	145 BC	± 165/210
TL 2	Masking layer	515 BC	± 220/275
TL 13	Circular structure K	40 AD	± 100/160
TL 15	Phase III partition	240 BC	± 210/260
Mean date for phase III		135 BC	± 110/190

Phase IV

TL 3	Late rect. enclosure	40 AD	± 220/270
TL 16	Late rect. enclosure	590 BC	± 195/260
TL 17	Late rect. enclosure	700 BC	± 200/270
TL 18	Late rect. enclosure	110 AD	± 145/175

Mean date cannot be calculated

The four dates on the grain demonstrate that the occupation on the site dates to the later Roman period. The presence of older charcoal (GrN–15678 and 15679) and the presence of some possible late Iron Age pottery suggests that there may have been some late Iron Age occupation within the immediate vicinity of the site. The two dates on the rye grain represent the first record of this species in the region. Note that the charcoal from the same two contexts as the rye grain is very considerably older than the grain.

5.2.9 South Shields

The deposits in the granary are dated by association with Roman pottery, coins etc., and no radio-carbon samples were submitted as these would not provide better dating evidence than was already available.

Deposit 12236 is dated to the last quarter of the third century (AD 275–300) through its association with a fire which destroyed much of the fort (P. Bidwell, pers. comm.). Deposit 12176 is dated to the mid fourth century (AD 325–375) by coins and pottery found within it (P. Bidwell, pers. comm.)

5.3 Discussion

5.3.1 Taphonomy of the materials dated

Before discussing the results of the dating programme in the light of the objectives set out in section 5.1 above, it is important to consider one aspect of the results in more detail, i.e. the difference in age between the dates based on charcoal and those based on grains and chaff. In many cases there is a difference of 100–200 years. In most of these cases this can be explained by the fact that there will have been a difference in sample age at the time of burial. While individual grains can never be more than one year old, and concentrations of grain in pits etc. are unlikely to represent more than one harvest, timber used for construction purposes must have originated from trees which had some considerable age, if thick posts or wide planks were used. In cases like this an age for the timber of between 100 and 200 years is not uncommon. However, this difference in sample age between grain and charcoal samples is not always recognized and has occasionally resulted in the questioning of the reliability of the results (Cunliffe 1986, Whittle 1989). Waterbolk (1983) and Millett (1983) both discuss methods for trying to estimate the age of the wood before it was cut to allow some 'correction' of the dates. As the exact origin of the charcoal samples used here is not known (the majority of the samples came from flotation samples), no such 'correction' is possible here. While

the difference in the dates based on charcoal or grain/chaff can in most cases be explained by assuming an age of *ca.* 100–200 years for the charcoal, there are a number of samples which cannot be explained in this way:

Anomalies in years

Murton:	GrN–15673	4285 ± 50 BP	2000
	GrN–15672	3160 ± 50 BP	1000
	HAR– 6201	2960 ± 80 BP	800
Chester House:	GrN–15709	2530 ± 80 BP	500
Thorpe Thewles:	GrN–15661	2720 ± 80 BP	600
	GrN–15662	2410 ± 80 BP	500
Rock Castle:	GrN–15668	2940 ± 90 BP	900
	GrN–15671	2575 ± 35 BP	550
	GrN–15669	2480 ± 60 BP	450
	GrN–15670	2420 ± 50 BP	400
Thornbrough:	GrN–15678	2530 ± 35 BP	800
	GrN–15679	2060 ± 35 BP	400

These dates point to the presence of charcoal of a much older date than that of the occupation of the site as indicated by the other radio-carbon dates and archaeological dating evidence. This suggests that there was earlier occupation within the immediate vicinity of the settlement. Alternatively, it is possible that charcoal from the settlement had become mixed with very old, e.g. late-glacial, charcoal. However, to increase the age of charcoal of 2200 BP to 2900 BP by admixture with charcoal of *ca.* 11,000 BP, an admixture of as much as 8 per cent of the late-glacial charcoal would be required (J. Lanting, pers. comm., Waterbolk 1983). While this is a possibility we cannot rule out, the large number of dates for which this pattern occurred, seems to point more in the direction of residual charcoal from earlier occupation, than to admixture with very old charcoal.

The fact that none of the grain/chaff samples gave old dates can probably be explained by the fact that grain and chaff fragments do not usually survive any mechanical movement and are therefore no longer present within the deposits if reworking had taken place. It has already been recognized that grain samples are often more reliable than charcoal samples because of their much smaller sample age. Their inability to survive transport may be added to their other advantages, as they are, consequently, more likely than charcoal to date the feature in which they are found. However, the presence of one medieval and one modern grain in the assemblages indicates that admixture with younger material does occur, probably through the actions of earthworms etc., but in both these cases the grain had already been identified as different.

5.3.2 Objectives

The first objective of the dating programme was to assess the likely contemporaneity of the plant assemblages from the nine sites under study. The results suggest the following timespan for the occupation:

Timespan of occupation after dating programme

	late BA	early IA	late IA	early RP	late RP
Hallshill	———				
Murton			———		
Dod Law			———		
Chester House		———			
Thorpe Thewles			———		
Stanwick			——		
Rock Castle			———		
Thornbrough					——
South Shields					——

(Abbreviations: BA = Bronze Age; IA = Iron Age; RP = Roman Period)

When we compare this information with that available before the dating programme (see section 5.1 above), it is clear that the timespan of occupation is now much more closely defined. Furthermore, the occupation at six sites (Murton, Dod Law, Chester House, Thorpe Thewles, Stanwick, and Rock Castle) is definitely contemporary, and dates to the late Iron Age to early Roman period. Two sites (Thornbrough and South Shields) date to the later Roman period, and one site (Hallshill) is dated to the late Bronze Age to early Iron Age.

The radio-carbon dates for all sites have been illustrated in Figure 5.1, giving the age ranges for both confidence levels. The difference in age between the charcoal and grain/chaff dates is clearly visible. The figure also shows the presence of the many older dates, which, as discussed above, have been interpreted as evidence for earlier occupation at these sites. This would suggest that there was a much greater continuity of settlement in the region than had previously been recognized.

The second objective was concerned with the aspect of residuality, i.e. could the spelt grains in early contexts have been derived from later occupation, or the emmer grains in late contexts be derived from earlier occupation? While the presence of older charcoal within the excavated features was a common phenomenon, this was not the case with the grain and chaff, which in all but two cases did belong to the deposit from which it was retrieved. The only two exceptions were the two samples of bread/club wheat grains from Chester House and Thorpe Thewles, which were shown by the accelerator dates to be modern and medieval respectively. In both cases, however, the grains had been suspected of

representing intrusions, and they were submitted for dating to test this hypothesis. The results confirmed the hypothesis.

The third objective of the dating programme was to date the earliest records of spelt wheat, bread/club wheat, and rye in the region, and to obtain dates for the period when spelt wheat had become the principal wheat crop. The results were as follows: the earliest record for spelt wheat is 2895 ± 70 BP (OxA–1764) from Hallshill; the earliest record of bread/club wheat is 1970 ± 70 BP (OxA–1732) from Rock Castle; and the earliest record for rye is 1690 ± 70 BP (OxA–2131) from Thornbrough. Spelt wheat had become the principal wheat crop by 2190 ± 70 BP (OxA–1732) at Thorpe Thewles. For a discussion of these results see Chapter 6.

The fourth objective concerned the comparison of the dates with historical events, in this case the Roman occupation of the area. This aspect will be discussed in the next section.

5.4 Calibration of the radio-carbon dates

All dates mentioned above are given in 'conventional radio-carbon ages', which means that the following assumptions and conventions were observed (Gillespie 1984, Van der Plicht and Mook 1987):

(1) a half-life for carbon-14 of 5568 years (the so-called 'Libby half-life')

(2) the natural specific carbon-14 activity has always been equal to a value defined by the 0.95 NBC oxalic acid modern standard

(3) the carbon-14 activity is corrected for isotope fractionation to $\delta^{13}C = -25^{\circ}/_{\circ\circ}$

(4) the age is expressed in years BP, whereby the present is defined as AD 1950

As the true half-life of carbon-14 is 5730 ± 40 years rather than 5568 years, and as the carbon-14 content in the atmosphere has not been constant through time, it is necessary to correct the 'conventional radio-carbon age' for these two factors before the dates are compared with historical events dated by calendar years, or with artefacts which are dated by calendar years (such as Roman pottery). A number of different correction or calibration curves have been constructed over the last twenty years, by dating the carbon-14 content of tree rings, which are independently dated by dendro-chronology. This has indicated that the size of the correction needed varies considerably, depending on the age of the sample.

Unfortunately, the calibration of radio-carbon dates into calendar dates does not usually produce a narrowly defined calendar date. There are not only long-term oscillations in the calibration curve, but medium term variations, the so-called 'wiggles', have also been identified (Van der Plicht and Mook 1987). These 'wiggles' in the curve can cause radio-carbon dates to correspond with more than one calendar date, while horizontal parts of the curve cause the clustering of dates (Van der Plicht and Mook 1987).

At present the high-precision calibration curves published by Stuiver and Pearson (1986), Pearson and Stuiver (1986) (together covering the period 2500 BC to AD 1950), and by Pearson *et al.* (1986) (extending these series to 5210 BC) are the curves recommended for use by the 12th Radio-carbon Conference held at Trondheim (Stuiver and Kra 1986). All the dates discussed here have been calibrated using a computer program developed by Van der Plicht and Mook (1987), which uses these curves. The program produces a calendar age probability distribution. One graph shows the normalized cumulative probability distribution, from which one can determine the probability of any calibrated age range or peak. A second graph plots the same probability distribution, but indicates on this the location of the 68.3% and 95.4% confidence levels (i.e. one and two standard deviations). The calibrated age ranges corresponding to these levels are also printed out (for details see Van der Plicht and Mook 1987). These calibrated age ranges for both confidence levels are given in Table 5.1 for each of the dates discussed here. The probability distribution graphs are only presented here for those dates where the calculation of the probability of a certain age range was important (see below). To allow visual comparison of all the calibrated dates, the maximum calibrated age ranges for both confidence levels are given in Figure 5.2. The age ranges for the Roman pottery from Thorpe Thewles, Stanwick, Thornbrough and South Shields, and the date for the fibulae from Dod Law are also indicated on Figure 5.2.

As we can see from comparing Figures 5.1 and 5.2 (i.e. the results of the dates in radio-carbon years and in calendar years respectively), the relative position of the dates has not changed, but the overall timespan covered by the dates has been extended by the calibration. This is largely due to the presence of a horizontal stretch in the calibration curve, between 800–400 cal BC.

One of the reasons for calibrating the radio-carbon dates is to assess the relationship between the date of the plant assemblages and the arrival of the Roman army. It is difficult to determine the exact time period from which the presence of the Roman army started to influence the native economy. Here the date of AD 90 has been chosen. This date represents the moment in time when most of the forts along the main Roman roads through the region were built, but before the withdrawal of the army from Scotland took place (see also Chapter 2). Excluding the two Roman

period sites (Thornbrough and South Shields), there are two samples from Murton, three from Dod Law, and two from Rock Castle whose dates, at two standard deviations, stretch beyond the cal AD 90 date. For these seven dates the normalized cumulative probability distribution plots are given, in Figures 5.4 to 5.9. From these we can calculate the probability of the date representing the period before AD 90. The results are as follows:

Probability of the date lying before cal AD 90

Murton:	OxA–1741	1960±70 BP	80%
	OxA–1740	1910±70 BP	51%
Dod Law:	OxA–1735	1970±70 BP	83%
	OxA–1734	1960±70 BP	80%
	OxA–1736	1910±80 BP	51%
Rock Castle:	OxA–1737	1970±70 BP	83%
	OxA–2132	1920±70 BP	57%
	OxA–1737)	1955±50 BP	85%
	OxA–2132)		

These results suggest that at both Murton and Dod Law there is a real possibility that a small part of the assemblage belongs to the very early Roman period, rather than the late Iron Age. In the case of Murton, the sample for OxA–1740 came from context 623, which represents the 'occupation earth' on the floor area of timber-built house T9, sealed by the paved floor of stone-built house S7. The date of the grain indicates that the change from timber-built to stone-built houses at Murton took place at the very end of the first century or the very beginning of the second century cal AD, which is in agreement with the date postulated by the excavator on the basis of other evidence (Jobey 1987). The sample for OxA–1736 from Dod Law came from context 30, a rubbish deposit accumulated against the outer rampart, which is stratigraphically later than the other rubbish deposits (between the ramparts). The latter were dated by a *terminus-ante-quem* of AD 100 by two bronze fibulae (C. Smith 1990). The very early second century cal AD date for context 30 is in agreement with this. One of the two dates on bread/club wheat from Rock Castle could also belong to the early Roman period. However, as the two dates on the bread/club wheat originate from the same sample of rachis internodes (context 50) and represent the same event, these two dates have been combined to improve the precision of the calibration. The combined date (1955 ± 50 BP) suggests that a very late Iron Age date is slightly more likely (Figure 5.9).

A small part of the plant assemblage from Thorpe Thewles is also dated to the very early Roman period. The material from Phase IV at this site is dated by Roman fine wares to *ca.* AD 80–130.

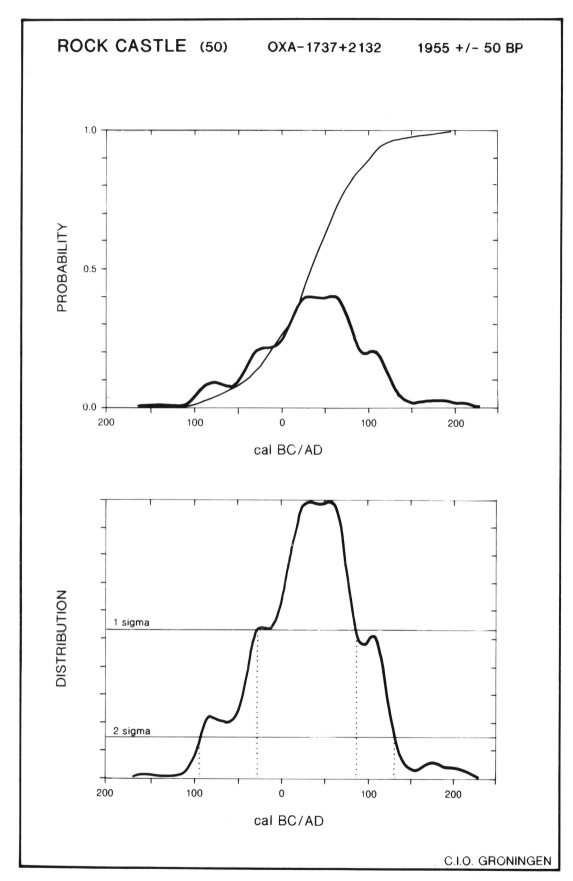

Figure 5.6   Calibration for OxA-2132. The top graph shows the calibrated calendar age distribution (thick line) and the normalized cumulative probability distribution (thin line). The bottom graph shows the calibrated calendar age distribution (thick line) and the levels corresponding to 1 and 2 sigma confidence levels (thin lines). (see also Van der Plicht and Mook 1987).

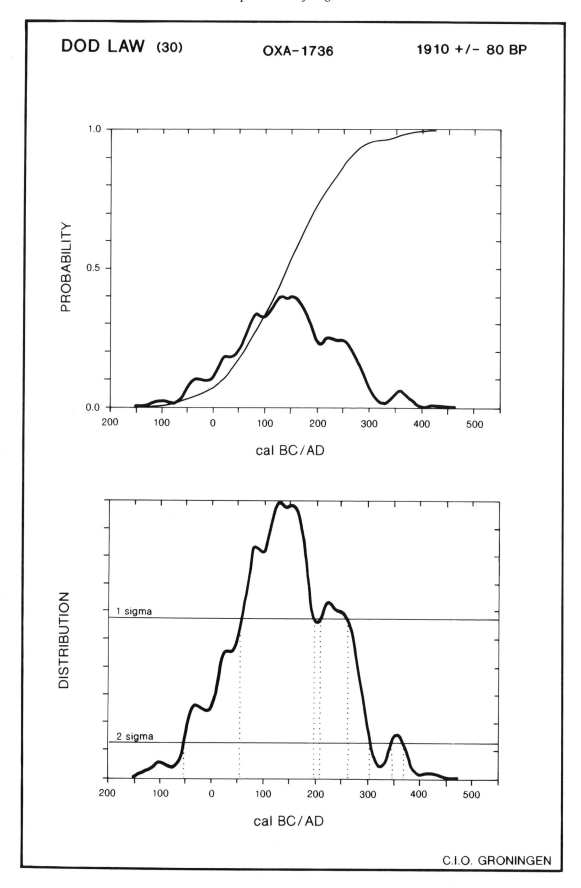

Figure 5.7 Calibration for OxA-1736. The top graph shows the calibrated calendar age distribution (thick line) and the normalized cumulative probability distribution (thin line). The bottom graph shows the calibrated calendar age distribution (thick line) and the levels corresponding to 1 and 2 sigma confidence levels (thin lines). (see also Van der Plicht and Mook 1987).

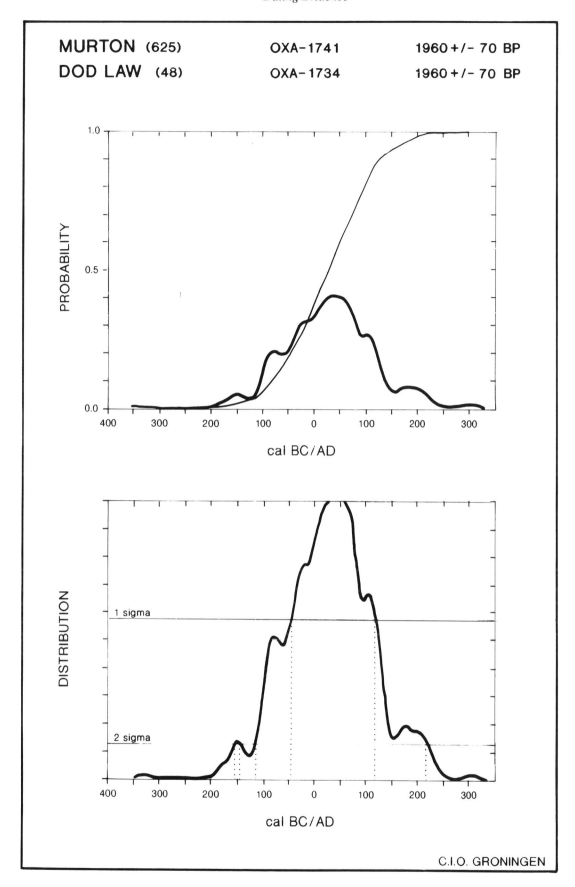

MURTON (625) OXA-1741 1960 +/- 70 BP
DOD LAW (48) OXA-1734 1960 +/- 70 BP

C.I.O. GRONINGEN

Figure 5.8 Calibration for OxA-1740. The top graph shows the calibrated calendar age distribution (thick line) and the normalized cumulative probability distribution (thin line). The bottom graph shows the calibrated calendar age distribution (thick line) and the levels corresponding to 1 and 2 sigma confidence levels (thin lines). (see also Van der Plicht and Mook 1987).

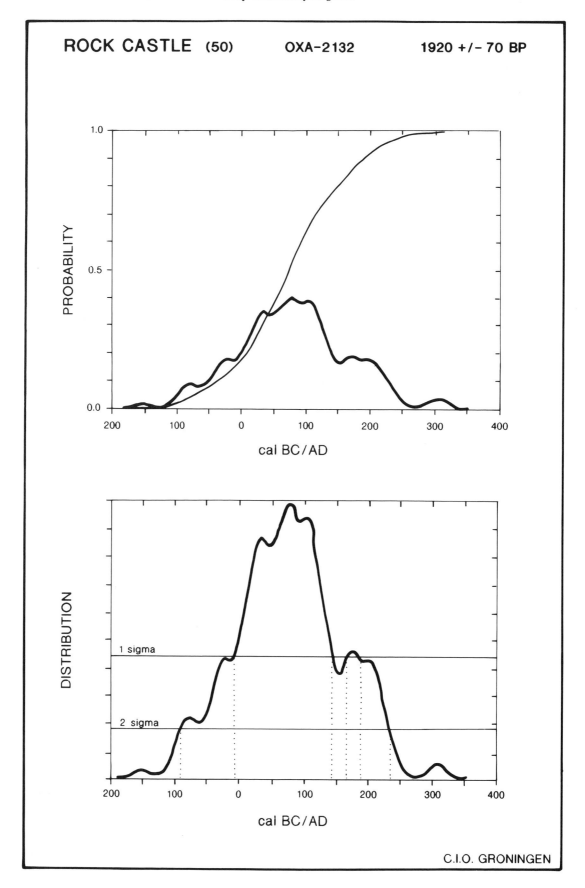

Figure 5.9   Calibration for OxA-1737 and OxA-2132 combined. The top graph shows the calibrated calendar age distribution (thick line) and the normalized cumulative probability distribution (thin line). The bottom graph shows the calibrated calendar age distribution (thick line) and the levels corresponding to 1 and 2 sigma confidence levels (thin lines), (see also Van der Plicht and Mook 1987).

Thus, while on four sites (Murton, Dod Law, Thorpe Thewles, and Rock Castle) the occupation does include the very early Roman period, the emphasis of the occupation lies within the late Iron Age. As there is no recognizable difference in the composition of the assemblages from the earlier and later contexts on each of these sites, the assemblages are taken to represent native arable farming practices, in existence before any Roman influence on the economic base of the region was felt.

5.5 Conclusion

The grain and chaff samples formed ideal material for radio-carbon dating. Firstly, they allowed the dating of the actual objects studied here, i.e. the spelt, rye, etc. Secondly, due to their very small sample age, i.e. just one year, they can date the contexts more closely than charcoal samples. And thirdly, as grains and chaff fragments do not survive reworking and transport, they do not lie around for any length of time, and, consequently, in contrast to charcoal, they are very unlikely to be older than the feature in which they are found. While grain dates tend not to give dates which are too old, they do, occasionally, represent younger material, as demonstrated by the dates on the bread/club wheat from Chester House and Thorpe Thewles. These grains probably represent material which has fallen down cracks in the soil, or which has been brought down by earthworms etc.

The results of the dating programme confirm that charcoal from flotation samples is not ideal material for dating, as its origin and sample age are unknown. This material was only used here because no other charcoal was available. In combination with the dates on the grain and chaff, however, the dates did provide the required information regarding the timespan of the settlement occupation.

The difference of 100–200 years between the dates based on charcoal and grain/chaff forms an important aspect of the results. It probably represents the difference in sample age of the two types of material. This is, obviously, a factor to be taken into account when calibrating radio-carbon dates based on charcoal in order to compare them with historical dates. A discrepancy of *ca.* 100 years is, in many cases, to be expected, a phenomenon which, to the writer's knowledge, is rarely acknowledged in the archaeological literature (but see Millett 1983 and Waterbolk 1983). The same problem, i.e. that of the sample age, also needs to be taken into account when calibrating dates based on peat. As one or two cm of peat from a peat core can represent as much as 200 years (Turner 1981a), calibration of these dates can give a false sense of accuracy (see Chapter 2 above).

The approximate calibrated age ranges for the seed assemblages from the nine sites under study are given below, based on the radio-carbon dates of the grain and chaff, and on the dates provided by the Roman fine wares at Thorpe Thewles, Stanwick, and South Shields.

Approximate calibrated age ranges for the seed assemblages

Hallshill	1200 cal BC – 800 cal BC (or 500 cal BC)
Murton	200 cal BC – 200 cal AD
Dod Law	200 cal BC – 200 cal AD
Chester House	200 cal BC – 100 cal AD
Thorpe Thewles	400 cal BC – 150 cal AD
Stanwick	100 cal BC – 100 cal AD
Rock Castle	200 cal BC – 150 cal AD
Thornbrough	200 cal AD – 500 cal AD
South Shields	275 AD – 375 AD

6.   General Results

In this chapter some general aspects of the results of the archaeobotanical analysis are presented. Section 6.1 discusses the presence and relative importance of the various crop plants, and compares these results with the evidence from southern Britain. Sections 6.2 and 6.3 discuss the arable weeds and 'other' plant remains respectively, while in section 6.4 the overall composition of the carbonized seed assemblages is considered. The results are summarized in three tables: Table 6.1 gives the total number of seeds for each species and each site; Table 6.2 gives the relative proportions for each species and each site within each category of data; and Table 6.3 gives the number of seeds for each species and each site per one litre of sieved sediment.

6.1   Crop plants

A total of six different crop plants were found in the samples: *Triticum dicoccum*, emmer wheat, *Triticum spelta*, spelt wheat, *Triticum aestivo-compactum*, bread/club wheat, *Hordeum vulgare*, six-row barley, *Secale cereale*, rye, and *Linum cf. usitatissimum*, flax, although this last identification is tentative only. The presence or absence of these species in the nine sites is indicated below:

	Emmer Wheat	Spelt Wheat	Bread Wheat	Six-row Barley	Rye	Flax
Hallshill	*	*	-	*	-	*
Murton	*	*	-	*	-	-
Dod Law	*	*	-	*	-	-
Chester House	*	*	-	*	-	-
Thorpe Thewles	-	*	-	*	-	-
Stanwick	-	*	-	*	-	-
Rock Castle	-	*	*	*	-	-
Thornbrough	-	*	-	*	*	*
South Shields	-	*	*	*	-	-

It is difficult to assess the relative importance of the various crops, but numerically emmer wheat is more important than spelt wheat at Hallshill, Murton, Dod Law and Chester House, in that there are more glume bases of emmer than of spelt on these sites. Emmer wheat is absent on the remaining three prehistoric sites (Thorpe Thewles, Stanwick, and Rock Castle). Bread/club wheat is present at Rock Castle and South Shields. Barley is present on all sites, and, with the exception of Hallshill and South Shields, occurs in similar proportions to wheat. At Hallshill barley is present as a crop, but in smaller quantities than wheat, while at South Shields the assemblage consists almost entirely of spelt and

bread/club wheat. The few barley grains within this assemblage are interpreted as contaminants. Rye was present in small quantities at Thornbrough, and flax was present at Hallshill and Thornbrough.

The three most striking aspects of these results are, first of all, the presence of spelt wheat at Hallshill, as early as 2895 ± 70 BP (OxA–1764). Secondly, the difference in the importance of spelt wheat on contemporary sites (spelt wheat appears to play only a minor role at Murton, Dod Law and Chester House, but forms the dominant wheat crop at Thorpe Thewles, Stanwick and Rock Castle). And thirdly, the occurrence of bread/club wheat at Rock Castle, rye at Thornbrough, and the large amount of bread/club wheat at South Shields.

Apart from an enigmatic find of spelt wheat at the Neolithic site of Hembury, Dorset (Field *et al.* 1964), no finds of spelt wheat are known in Britain until the first millennium BC. Only a few of these records have been directly dated by radio-carbon analysis: small amounts of spelt were present on the late Bronze Age site at Black Patch, Sussex, the occupation of which is dated to 3020 ± 70 to 2780 ± 80 BP (Drewett 1982); a trace of spelt was found in a grain deposit in an early Iron Age pit at Godmanchester, Cambridge, with a radio-carbon date of 2880 ± 80 BP (M. Jones 1981); spelt wheat was found at Potterne, Wiltshire, in a midden associated with an archaeomagnetic date of 750 BC (Carruthers 1986); waterlogged spelt wheat was found at Runnymede, Surrey, associated with radio-carbon dates ranging from *ca.* 800 – 700 BC (Greig 1990); spelt wheat from a deposit sealed by the inner rampart of Dinorben fort, north Wales, was dated by radio-carbon analysis to between 2495 ± 65 and 2390 ± 45 BP (M. Jones 1981); and finally during the excavation of a circular house at Peel Castle, Isle of Man, a large quantity of spelt wheat was found, dated to 2250 ± 50 BP (Tomlinson forthcoming). The dates on the spelt wheat from Hallshill in Northumberland (2895 ± 70 BP (OxA–1764), 2840 ± 70 BP (OxA–1763), and 2750 ± 70 BP (OxA–1765)) indicate that spelt wheat was introduced into the north of Britain as early as into the southern part of the country.

Emmer wheat was the principal wheat crop in Britain during the Neolithic and Bronze Age, but after the introduction of spelt wheat in the first half of the first millennium BC, spelt wheat rapidly replaced emmer as the most important wheat species. In the most recent synthesis of the development of crop husbandry in Britain (M. Jones 1981) it was suggested that the replacement of emmer by spelt wheat did not take place in the Highland Zone, and that in the north and west of the country emmer wheat continued as the principal wheat crop. The results from Thorpe Thewles, Stanwick, and Rock Castle indicate that spelt wheat had, in fact, replaced emmer wheat in the Tees lowlands by at least 300 BC. The absence of seed assemblages dated to the middle part of the first millennium prevents a more detailed assessment of its history at the present moment.

The dominance of emmer wheat over spelt wheat in the assemblages from Murton, Dod Law, and Chester House, all in Northumberland, does, however, demonstrate that this replacement of emmer by spelt wheat was not uniform across the region. While spelt wheat is present in the assemblages of Murton, Dod Law, and Chester House, it is secondary in numerical importance to emmer wheat at all three sites. The radio-carbon dates from these sites indicate (Chapter 5) that the difference between these two groups of sites cannot be 'explained' by chronological differences. We will come back to this in Chapters 10 and 11.

The history of bread/club wheat in Britain is still poorly understood. Free-threshing species such as bread/club wheat tend to be under-represented in the charred seed record, and this hampers the interpretation of the results. There are occasional records of one or two grains of this species from the Neolithic period onwards, but the species is not thought to have become a crop in its own right until the late Roman or Anglo-Saxon period (Helbaek 1952, Hillman 1981a, M. Jones 1981). As a crop it has many advantages: it is winter-hardy, high yielding, and free-threshing. This last feature does, however, leave the crop much more prone to attack from birds and fungi (M. Jones 1981). It is also a poor competitor with weeds and needs a greater soil fertility than other wheat species, which has led to the suggestion that it only became an important crop when 'it became possible and desirable to invest the greater amount of fertilizer and man-hours, in the form of cultivation and weeding, that would be necessary to obtain the high yield potential of bread wheat' (M. Jones 1981, 107). Recently, two late Iron Age sites have yielded assemblages in which bread wheat occurred in fairly substantial amounts, i.e. at Barton Court Farm, Oxfordshire, and Bierton, Berkshire (Jones 1981, 1984a, 1986). This suggests that this development might have started locally earlier than was previously thought.

In the assemblages from north-east England, only that from South Shields contained large amounts of bread/club wheat. In this Roman granary, it occurs in similar proportions to spelt wheat. However, this site is a Roman fort, and the grain might, therefore, have been brought in from afar, rather than have been produced by the inhabitants of the site. This point

will be discussed in Chapter 11.

At Rock Castle one context contained a large number of rachis internodes of bread/club wheat. They were initially interpreted as medieval contamination, because the context was cut by a late feature and contained a sherd of possible medieval date. The material was, however, dated to the late Iron Age by two radio-carbon samples on the rachis internodes (1970 ± 70 BP (OxA–1737) and 1920 ± 70 BP (OxA–2132)).

On several other sites occasional grains or rachis internodes of bread/club wheat were found, but they always looked very recent, being much larger than prehistoric or Roman fragments and generally only partially charred. They have always been interpreted as modern contaminants and are not listed in the tables. To test whether they were indeed modern, two grains were selected for accelerator dating, one from Chester House and one from Thorpe Thewles. The Chester House grain was modern (122.6 % modern, OxA–1744) and that from Thorpe Thewles medieval (720 ± 70 BP, OxA–1745), which confirmed the hypothesis. The implications are that the rare occurrences of bread/club wheat in the literature could represent medieval or later contaminants. It would be desirable to submit some of them for dating, to test the authenticity of these records.

Six-row barley, *Hordeum vulgare*, was found on all sites. At Hallshill, Murton, Dod Law, and Chester House a trace of naked barley was present, although the identifications at Murton and Chester House are very tentative. The vast majority of the barley grains belonged to the hulled variety. The hulled variety is thought to have replaced the naked variety during the early part of the first millennium BC (Helbaek 1952, Hillman 1981a), a development also found in The Netherlands and Denmark. In The Netherlands naked barley is absent from the records after *ca.* 500 BC (Van Zeist 1968), while in Denmark the increase of hulled barley at the cost of the naked variety took place during the first half of the first millennium AD (Helbaek 1952).

In naked barley the grains are loose in the spikelets, while in the hulled variety the palea and lemma do not just firmly enclose the grain, but, in fact, are fused with it. They cannot be removed by threshing. Consequently, the change from a naked to a hulled variety does not appear to be a logical one, as the removal of the hulls, not essential but certainly preferable prior to human consumption, requires extra processing time.

As yet, no explanation of this development has been put forward, although Helbaek mentions the possibility of a climatic factor (Helbaek 1952). It could well be that the increasing wetness of the climate in the early part of the first millennium BC (see Chapter 2) played a role. The spikelet structure of naked barley is more open than that of hulled barley, which means that in wet weather the ears hold more water. This could encourage fungal diseases and sprouting in the ear. In hulled barley the grains are protected from wetness and disease by the hulls.

If the change in the climate was the only reason for this development, however, one would expect it to occur at more or less the same time across Europe. While the dating evidence from The Netherlands is compatible with that from Britain, in Denmark the replacement took place almost a millennium later. This would suggest that the climatic factor, while almost certainly important, was not the only reason for the change. An additional reason may be a change in the utilization of barley, from a crop principally for human consumption to a crop principally for fodder. The removal of the hulls is unnecessary if the grain is used for animal fodder. Gill & Vear mention that 'barley was formerly used largely as a cereal for human food, and many of the forms cultivated were naked barleys, ..., in regions such as Britain where barley is now used either for stock feeding or for malting, only hulled forms, ..., are used' (Gill and Vear 1980, 90). The suggestion that the replacement of naked barley by hulled barley could, at least partially, have been caused by a change in the use of barley, has implications for our understanding of the development of animal husbandry (the supply of grain in addition to leaf fodder or grazing), but this falls outside the scope of the present study.

To summarize, the change-over from naked to hulled varieties of barley during the first millennium BC was probably a result of two factors, one climatic (increase in rainfall), and one agricultural (change from human to animal food), although the two are likely to have been interdependent. It is not suggested, however, that barley was exclusively used for fodder. The frequent presence of barley in carbonized seed assemblages suggests that it remained an important human food as well.

Rye, *Secale cereale*, was only present at Thornbrough, and only as a minor component. This species, like other free-threshing cereals, is generally under-represented in the archaeobotanical record. It is usually assumed to be a late introduction, with records starting to appear in the late Iron Age and Roman period, though generally as a minor component of the seed assemblages (Hillman 1981a, Jones 1981, Green 1981). It did not become an important crop until the Saxon and medieval period, although even then its importance was far from universal (Green 1981). Recently, both pollen and individual carbonized grains of rye have been recorded from deposits dated to the second millennium BC (Chambers 1989). Whether these occurrences refer to the use of rye as a crop, or as an arable weed, remains unclear. The rye from

Thornbrough is dated by two radio-carbon dates (1690 ± 70 BP (OxA–2131) and 1630 ± 70 BP (OxA–2130)).

Flax, *Linum usitatissimum*, was present as a tentative identification on two sites, Hallshill and Thornbrough. This occurrence during the Bronze Age and Romano-British period, but not during the intermediate period, is similar to that in other parts of the country, where its absence in the Iron Age has also been noted (Jones 1981). It has been suggested that this absence during the Iron Age might be associated with the fact that flax is not very tolerant of frost and heavy rain (Jones 1981, quoting Kirby 1965). However, if the plant was used for its fibres rather than the oil in its seeds, the chances of recovery in carbonized seed assemblages are small (Green 1981).

6.2 Arable weeds

In Table 6.4 the species found in the present analysis are listed and an indication is given as to the degree to which they are, at present, associated with arable fields. The pattern is very similar to the study carried out by M. Jones for assemblages from central-southern Britain (M. Jones 1984b: Table 9.1), i.e. the great majority of species have arable fields as one of their principal habitats. The one group for which this is not the case is the group of plants listed in the data tables under the category 'other'. This group consists of plants from three separate habitats, i.e. woodland edge and hedgerows (presumably collected fruits and nuts), moorland vegetation (presumably plants collected for bedding and/or thatching), and water plants (probably from ditch fills). This group represents only one per cent of the total number of identifications (see section 6.4 below). They will be discussed in section 6.3 below.

There are a few species which are listed under the category of 'weeds' in the data tables, but which are not found in arable fields today. Most of these are plants of damp ground, such as *Ranunculus flammula*, *Stellaria palustris*, *Veronica cf. scutellata*, *Galium palustre*, *Isolepis setacea*, *Juncus* spp., *Eleocharis* sp., and *Carex* spp. Of these, *Stellaria palustris*, *Veronica cf. scutellata*, *Galium palustre*, and *Isolepis setacea* occur only very rarely, but the others, especially *Carex* spp., are relatively common in the seed assemblages studied. The possibility of these species having originated from sources other than arable fields cannot be ruled out, but is unlikely for the following reasons. First, most of these species also occur in the Iron Age assemblages from central-southern England (Jones 1984b, 1988a, 1988b), and secondly, some of them (e.g. *Eleocharis* and several species of *Carex*), were found in the granary deposits at South Shields associated with stored cereal crops, which points to their occurrence as crop weeds. M. Jones (1984b, 1988a, 1988b) also noted that these two taxa have been found in grain stores, as well as in stomach contents of Iron Age corpses from Denmark. The association of these species with grain stores and their regular occurrence in carbonized seed assemblages from different parts of the country suggest strongly that, in the past, they occurred as weeds in the arable fields. Their absence in fields today is probably the result of the much improved drainage of these fields (M. Jones 1984b, 1988a, 1988b).

The only other important species not found in arable fields today, but here listed under the category of arable weeds, is *Sieglingia decumbens*, heath grass. This species is a densely tufted, perennial grass, approximately 10–40 cm tall. Clapham *et al.* (1962) describe its habitat as acid grassland, though it is locally also found on damp base-rich soils. Hubbard (1984) describes it as a grass of no agricultural value, occurring on sandy and peaty soils in most parts of the British Isles, and frequently on moorland, heaths, or wet places.

Hillman (1981a) found large numbers of seeds of *Sieglingia* associated with spelt chaff in the carbonized seed assemblages from the late Iron Age site Cefn Graeanog in Wales. He interpreted the presence of *Sieglingia* as an arable weed and explained it by the change in ploughing practice since the prehistoric period. The practice of mouldboard ploughing, introduced sometime during the medieval period, eliminates most perennial and biennial weeds, while ard ploughing does not. The absence of *Sieglingia* from arable fields today may, therefore, be the result of mouldboard ploughing, rather than of the inability of the species to grow in such a habitat (Hillman 1981a, 1982). Behre (1981b) has also stressed the impact of changes in cultivation methods on the arable weed flora, and, equally, associated the disappearance of perennial weeds with the introduction of the mouldboard plough (see also Chapters 9 and 10).

The seeds of *Sieglingia* in the assemblages studied here are, as at Cefn Graeanog, strongly associated with spelt chaff, i.e. spelt glume bases (see Chapter 10). Furthermore, seeds of this species have been found in the granary deposits of South Shields (section 4.9 and Table 4.18), associated with spelt and bread/club wheat grains. The present evidence points to the occurrence of *Sieglingia* as an arable weed in the past, forced out of this habitat by the introduction of different tillage methods. Consequently, *Sieglingia* has been listed under the category of 'weeds' in the data tables.

The weed assemblages from the different sites will be studied in more detail in Chapter 10, but it is worth pointing out here that two new arable weeds,

Agrostemma githago and *Centaurea cf. cyanus*, appear with the introduction of the free-threshing cereals, *Triticum aestivo-compactum*, bread/club wheat, and *Secale cereale*, rye, in the Roman period assemblages from Thornbrough and South Shields. The association of these two weed species with bread/club wheat and rye has been noted by several authors (Godwin 1975, Jones 1984b, Pals and Van Geel 1976), even though the exact reasons for this association are still poorly understood.

6.3 'Other' plant remains

As already described above, the plant species grouped under this category are those which today are not associated with arable fields. They occur only very rarely in the carbonized seed assemblages and in total consist of only 1.2 per cent of the total number of identifications (see section 6.4 below). They can be divided into four distinct groups.

First of all, there are the fruits and nuts which may have been collected from the wild to supplement the diet. This group includes *Corylus avellana*, hazelnut, *Rubus* spp., blackberry/raspberry, *Prunus spinosa*, sloe, *Crataegus monogyna*, hawthorn, *Sambucus nigra*, elderberry, and *Rosa* sp., rosehips, and forms 0.3 per cent of the total assemblage. These fruits could all have been collected from the woodland edge or from hedgerows. The inedible parts, i.e. the nut shells, stones or pips, may have been thrown into the domestic fires, where they became carbonized.

The second group consists of plants which were probably also deliberately collected, for bedding and/or thatching material, such as *Calluna vulgaris*, heather, *Erica cf. cinerea*, bell heather, and *Pteridium aquilinium*, bracken. The few seeds of *Vaccinium myrtillus*, bilberry, and *Empetrum nigrum*, crowberry, both moorland species, were probably brought in with the heather and bracken. They form 0.9 per cent of the total assemblage. It is possible that the *Juncus* species, rushes, also came in in this way, though they may also have grown in wet areas of the arable fields.

The third group is that of the waterplants, consisting of *Lycopus europaeus*, gipsy wort, *Caltha palustris*, marsh marigold, *Potamogeton* spp., pondweed, *Viola* Subgenus *Viola*, violet, and *Menyanthes trifoliata*, bogbean. They are very rare in the assemblages (0.01 per cent). All of them are plants commonly found in ditches, marshes and ponds, and their carbonized remains may represent the cleaning-out of the settlement ditches, although they may have grown at the edges of arable fields if these were located near very damp ground.

The final group consists of a few tree buds which may have fallen off the collected firewood, and one fruit of *Thelycrania sanguinea*, dogwood, which was recovered from the granary at South Shields. The taphonomic origin of this fruit is unknown.

6.4 Composition of the assemblages

The composition of carbonized seed assemblages is remarkably uniform through space and time. Knörzer (1971a), having analysed a large number of samples from sites in the Rhine area, noted that most samples contained cereal grains, chaff and weed seeds. M. Jones (1984b, 1988b, 44) characterized Knörzer's findings as follows:

(1) carbonized assemblages always include finds of cereal grain.

(2) samples with large quantities of seeds of wild plants also include large quantities of chaff.

(3) most of the wild plants present in the assemblages, grow today as weeds in cereal crops.

Recent work has entirely corroborated this. M. Jones (1984b, 1988a, 1988b), working on Iron Age assemblages from central-southern England, concluded that carbonized seed assemblages, in virtually all cases, represent the harvested grain crops and their associated impurities. Other types of plant remains are rarely present and occur generally only in very small quantities. In areas where firewood is scarce, however, such as in parts of the Near East, an additional source for carbonized seeds must be reckoned with. In these areas animal dung is frequently used as fuel, and seeds included in the dung survive in carbonized state (Bottema 1984, Miller and Smart 1984). This phenomenon is, however, unlikely to occur commonly in Britain.

The proportions of grains, chaff, weeds, and 'other' plant remains within the total of identifications, for each of the sites studied here, is given below (see also Figures 6.1 and 6.2):

	Grains %	Chaff %	Weeds %	Other %	Indet. %
Hallshill	15.1	73.6	6.9	3.2	1.2
Murton	23.1	20.8	49.9	1.1	5.1
Dod Law	33.9	23.1	28.8	12.8	1.4
Chester House	22.8	36.2	32.1	2.7	6.2
Thorpe Thewles	10.1	26.7	61.5	0.1	1.6
Stanwick	19.1	16.1	61.0	1.8	2.1
Rock Castle	8.3	22.3	65.8	1.3	2.3
Thornbrough	60.4	32.7	6.4	0.3	0.2
South Shields	92.8	2.4	4.7	0	0.1
TOTAL	50.0	18.0	29.8	1.2	1.0

These results confirm previous findings that carbonized seed assemblages in most cases largely consist of harvested cereal crops and their waste products. We will return to this aspect in Chapter 9.

6.5 Summary

The general aspects of the results discussed in this chapter indicate that the assemblages have much in common with other carbonized seed assemblages from Britain. The results can be summarized as follows:

(1) *Triticum spelta*, spelt wheat was introduced into the north east of England at the very beginning of the first millennium BC, which is as early as in southern Britain. The radio-carbon dates on spelt from Hallshill are 2895, 2840, and 2750 ± 70 BP.

(2) *Triticum spelta* had replaced *Triticum dicoccum*, emmer wheat, as the principal wheat crop by 2190 ± 70 BP, at least in part of the region (i.e. the Tees lowlands). In northern Northumberland, however, emmer maintained its role as an important wheat crop.

(3) The free-threshing wheat, *Triticum aestivo-compactum*, bread/club wheat, was probably introduced towards the very end of the Iron Age. The single find of this species, from Rock Castle, is dated by two radio-carbon dates to 1970 and 1920 ± 70 BP.

(4) *Secale cereale*, rye, was introduced in the region sometime during the Roman period, which is not much later than in southern Britain. The two dates on rye from Thornbrough are 1690 and 1630 ± 70 BP.

(5) These two new, free-threshing cereals were, as elsewhere, accompanied by two new weeds, *Agrostemma githago* and *Centaurea cf. cyanus*.

(6) Flax, *Linum usitatissimum*, was present as a trace in the Bronze Age and the Romano-British assemblages, but was not found in the intermediate period.

(7) Like most other carbonized seed assemblages the samples studied here consisted of the remains of the harvested grain crops and their associated impurities (chaff and weed seeds). In total the cereal grains took up 50 per cent, the chaff 18 per cent, and the arable weeds 29.8 per cent of the total number of identifications. The plant remains which could not have derived from the arable fields consisted of only 1.2 per cent of the total. One per cent of the seeds could not be identified. Obviously, these proportions vary considerably from site to site (see section 6.4 above and Figures 6.1 and 6.2).

(8) The vast majority of the wild plants present in the samples are species which are still found in arable fields today. A small number, however, are not (e.g. *Ranunculus flammula, Stellaria palustris, Veronica cf. scutellata, Galium palustre, Isolepis setacea, Juncus* spp., *Eleocharis* sp., *Carex* spp.). These are plants

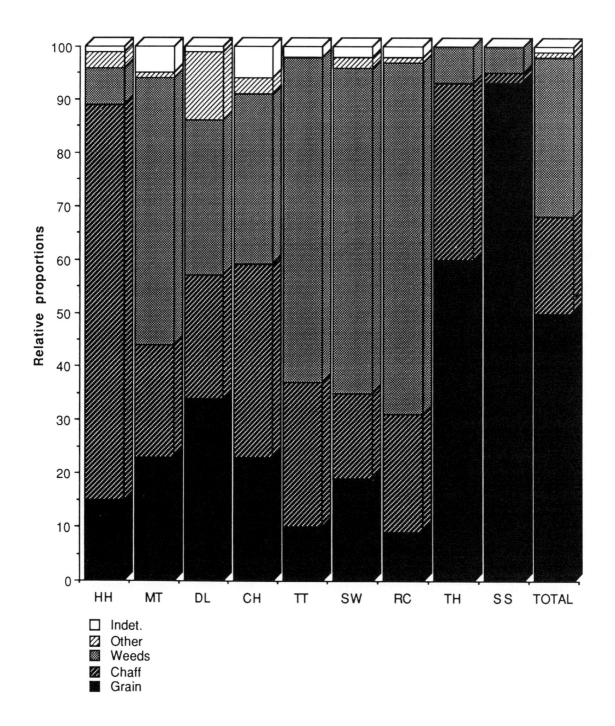

Figure 6.1   Composition of the carbonized seed assemblages for each site: relative proportions of major components (HH = Hallshill, MT = Murton, DL = Dod Law, CH = Chester House, TT = Thorpe Thewles, SW = Stanwick, RC = Rock Castle, TH = Thornbrough, SS = South Shields).

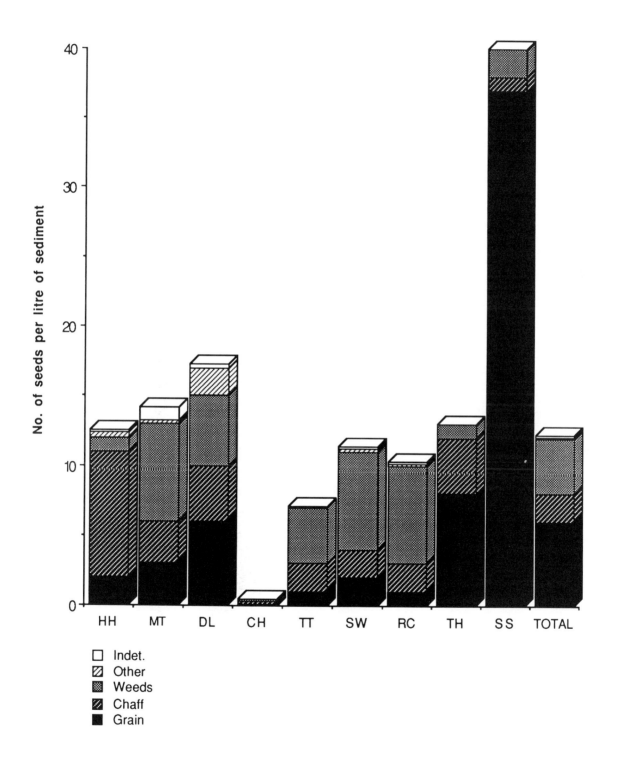

Figure 6.2   Composition of the carbonized seed assemblages for each site: number of seeds per litre of sieved sediment for the major components (see Figure 6.1 for key to site codes).

commonly found on damp ground. Most of these were also found in the Iron Age assemblages from central-southern England and their presence appears to point to the fact that the arable fields in the past were much less well drained than today. *Sieglingia decumbens* is another species not found in arable fields today. It occurred commonly in the samples, especially in association with spelt glume bases. It has been suggested that its absence from arable fields today is the result of the introduction of different cultivation techniques.

(9) In addition to the production of cereal crops in arable fields, the assemblages provide evidence for the exploitation of two other ecological zones: moorland, for the collection of bedding and/or thatching material, and the woodland edge or hedgerows, for the collection of wild fruits.

7. The Taphonomic Role of Crop Processing

7.1 Introduction

It has been recognized for some time that the plant assemblages from archaeobotanical samples cannot be compared with one another directly, without first screening each sample for taphonomic processes such as mode of preservation and crop processing stage. The first factor does not concern us here, as only assemblages preserved by one mode of preservation (i.e. charring) are studied. We have already mentioned, in Chapter 6 above, that carbonized seed assemblages consist largely of remains of harvested grain crops and their associated impurities. The second factor is important, however, as the processing of the crop after the harvest and before storage or consumption alters the original composition of the assemblage.

During the 1970s great emphasis was placed on the type of archaeological context (storage jar, oven, midden etc.) from which the archaeobotanical sample derived, as it was recognized that the composition of the sample varied considerably between different contexts (Dennell 1972, 1974, 1976, Renfrew 1973). The 1980s are characterized by an emphasis on the effect of crop processing on the sample composition. Both Hillman (1981a, 1984) and G. Jones (1983a, 1983b, 1984, 1987) have demonstrated how ethnographic models based on observations of present-day traditional agriculture allow a more accurate analysis of the internal composition of the samples. This method is independent of the archaeological context, which, as few archaeobotanical samples derive from primary (*in situ*) contexts, is an additional advantage (G. Jones 1987).

Both Hillman (1981a, 1984) and G. Jones (1984) have argued that the processing of the crops can only be achieved in a limited number of ways, given a traditional technology, as the physical properties of the plants determine the type and sequence of processes required: 'though the details may vary and, in particular, the implements used, the processing stages remain essentially the same and so, more importantly, do the effects on composition' (G. Jones 1984, 46). This argument does rely on the uniformitarian principle that the technology available in the prehistoric period was similar to the 'traditional' or non-mechanized one observed in recent ethnographic studies, and that human attitudes regarding the purpose of crop processing have not changed through time. While ethnographic and ethno-historical accounts of crop processing show little variation across space and time (Hillman 1981a, G. Jones 1984), suggesting that the assumptions

underlying the model are essentially correct, it is still important to be aware of the existence of these underlying assumptions.

In very simplified form the sequence of processing for free-threshing cereals (such as barley, bread/club wheat and rye) is as follows (see also Table 7.1 (A)):

Harvesting	to remove the crop from the field
Threshing	to release the grain from the straw and chaff
Winnowing	to remove the light chaff and straw fragments, and the light weed seeds
Coarse sieving	to remove the weed heads, large weed seeds, unthreshed ears, and straw nodes
Fine sieving	to remove the small weed seeds from the grain

In contrast to free-threshing cereals, the grains of the glume wheats (such as emmer and spelt) are enclosed by their glumes in a vice-like grip and, consequently, require additional stages of processing in order to release the grains. To dehusk the glume wheat grains the following additional stages are introduced, usually after coarse sieving (see also Table 7.1 (B)):

Parching	to render the glumes brittle
Pounding	to release the grains from the glumes
2nd Winnowing	to remove the light chaff fragments and light weed seeds
2nd Coarse sieving	to remove remaining weed heads, large weeds, straw nodes etc.
Fine sieving	to remove the glume bases and small weed seeds

Thus, in order to obtain the final product (i.e. clean grain without any contaminants) the harvested crop goes through a series of processing stages, each creating products and by-products, which can be identified by their varying proportions of grains, chaff, straw, and weeds. Many of these products and by-products are short-lived or are mixed with other by-products. Others, however, are relatively long-lived and may, consequently, come into contact with fire. These are the ones likely to be preserved archaeologically (see also Table 7.2 (A)):

Winnowing by-product
Coarse sieve by-product
Fine sieve by-product
Fine sieve product

To summarize, the sequence of crop processing stages acts as a filter on the original composition of the harvested plant assemblage. Each stage alters the composition and creates a product and a by-product, some of which may survive in the archaeological record. It is important to recognize which product or by-product the archaeobotanical samples represent, in order to compare like with like when analysing the data for crop husbandry practices etc. (G. Jones 1984, 1987).

In this chapter all samples will be examined in order to assess which crop processing stage they represent. Two methods will be applied. The first method uses the relative quantities of the major constituents (i.e. grain, chaff, weeds) and is based on the qualitative information available in Hillman's flow-diagrams (Hillman 1981a: Figures 5, 6, and 7). The second method uses the relative proportions of different weed seed categories and is based on Jones' statistical analysis of the behaviour of different weed seeds through the crop processing sequence (G. Jones 1984, 1987). The results of the analysis will determine which samples will be selected for the analyses in Chapter 10.

7.2 Method 1: Ratios of major sample constituents

7.2.1 Description of method 1

This method uses the ratios of the major sample constituents (cereal grains, certain chaff elements, and weed seeds). Three ratios are calculated:

Ratio 1: The number of glume bases to glume wheat grains

The samples contain two glume wheat species (emmer and spelt). In samples where both species occur their glume bases and grains have been combined before calculating the ratio. The ear of emmer and spelt is made up of spikelets, each of which contain two grains and two glumes, i.e. a ratio of 2 : 2 = 1 (in reality some spikelets can contain three grains, but others contain only one, so that on average the ratio is 2 : 2). If the ratio is considerably higher than 1 (more glume bases than grains) the sample is likely to represent a fine-sieving residue (the by-product of Hillman's stage 12). A ratio of much less than 1 (more grains than glume bases) points to a cleaned product (the product of Hillman's stage 12, 13 or 14). A ratio of *ca.* 1 either represents a sample consisting of complete ears or, more likely, a spikelet store in wet climates (and occasionally in dry ones as well) the glume wheats are often stored as semi-cleaned spikelets, rather than as dehusked grain, and the dehusking takes place piecemeal throughout the year, rather than immediately after the

harvest; see also Table 7.1 (B)).

Ratio 2: The number of rachis internodes to grains

This ratio is calculated for the free-threshing cereals only. The free-threshing cereals present in the samples are six-row barley, bread/club wheat and rye. In six-row barley there are three spikelets, each containing one grain, at each node, i.e. a ratio of 1 : 3 = 0.3. A ratio of *ca.* 0.3 represents a sample with complete ears. A ratio of much more than 0.3 (more rachis internodes than can be expected from a complete ear) points to the presence of early processing (winnowing/coarse sieving) residues. A ratio of much less than 0.3 suggests the presence of barley grains from the later crop processing stages. These ratios are calculated for bread/club wheat and rye as well, but in the whole plant they are 1 : 2 (or 1 : 3) for rye (i.e. two or three grains per internode), and 1 : 2–6 for bread/club wheat (i.e. two to six grains per internode).

Ratio 3: The number of weeds to the number of grains

If there are many more cereal grains than weed seeds, the sample probably represents a cleaned product. If there are many more weed seeds than grains, this points to the presence of cleaning residues.

7.2.2 Application of method 1

These three ratios were calculated for all samples (but excluding those with fewer than 50 identifications, see Chapter 3 above). If fewer than 10 items were available the ratio was not calculated. When samples contained both wheat and barley and indeterminate grains the following procedure was used to calculate the actual numbers used in the ratios: the number of indeterminate grains was divided into two according to the proportion of wheat and barley grains in the sample, and these were then added to the wheat and barley grains respectively. Emmer and spelt grains and glume bases were combined. The weed seeds refer to the sum of weed seeds listed in the tables under weeds, but excluding the rhizomes of Gramineae.

It is important to point out here that the various cereal components are subject to differential preservation under certain charring regimes, which can seriously affect the values of the ratios calculated. Experimental work has shown that rachis internodes of free-threshing cereals are one of the first components to burn away completely, and that glume bases of glume wheats, under certain conditions, also burn away before the grains are destroyed (Boardman and Jones 1990). This means that, while the absence of rachis internodes or glume bases cannot necessarily be taken to reflect the original assemblage, the presence and, especially, the predominance of these components can, as their predominance cannot be the result of differential survival. Cereal grains always survive charring as well as, or better than, these chaff components (Boardman and Jones 1990).

7.2.3 Results of method 1

The values for each sample and each ratio are listed in Table 7.3.

Ratio 1: Prehistoric sites

The vast majority of samples have a value for ratio 1 of well above 1, i.e. representing fine-sieve residues. There are 13 exceptions, but six of these are based on very low numbers of fragments (fewer than 25) making them rather unreliable. The values for ratio 1 of the remaining seven samples are not really low enough (i.e. there are too many glume bases) to represent dehusked grain. The samples may represent spikelets with the glume bases under-represented due to differential preservation:

dehusked grain (?):	Hallshill, one sample (8)
spikelets (?):	Thorpe Thewles, 5 samples (JS2, JS3, LS422, JS20, LS52)
	Stanwick, 1 sample (1023)

Ratio 1: Roman sites

The Roman samples have very different values for ratio 1. The majority of the samples from South Shields have a value of well below 1, pointing to a dehusked crop. But those from Thornbrough were more varied:

dehusked crop:	one sample (43)
spikelets:	ten samples (10, 40, 120, and possibly spikelets 5, 39, 41, 46, 49, 54b, 127)
fine sievings:	five samples (45, 54a, 58, 125, 134)

Ratio 2: Prehistoric sites

Of the 95 samples for which this ratio could be calculated, *ca.* half (i.e. 51 samples) had a ratio of well below 0.3 (i.e. below 0.2), that is containing barley from one of the later processing stages (fine-sieve by-product or product). The remaining 44 samples had ratios either close to 0.3 (0.2 – 0.4) or greater than 0.4, i.e. representing either an unprocessed crop or early processing residue (winnowing or coarse-sieve by-product). The values close to 0.3 (suggestive of an unprocessed barley crop) may, in fact, also point to the presence of early crop processing residues with an under-representation of the rachis internodes due to differential preservation. Some of these ratios were based on very low numbers of fragments (i.e. below 25) and are consequently not very reliable. The 22 remaining samples contained evidence for early processing residues (in the case of sample 50 from

Rock Castle this is true for both barley and bread/club wheat):

Murton:	3 samples (623, 624, 625)
Dod Law:	6 samples (40(10), 40(11), 38(7), 25(5), 30(6), 51(13)
Chester House:	1 sample (117)
Thorpe Thewles:	9 samples (PF1, LS120, JS13, LS465, JS6, JS12, JS14, LS69, ML14)
Stanwick:	1 sample (2045)
Rock Castle:	2 samples (74, 50)

Ratio 2: Roman sites

The values of ratio 2 for all samples suggest that the barley remains at Thornbrough and the bread/club wheat remains at South Shields represent the later crop processing stages; but two samples from Thornbrough (10 and 45) contain early processing residues from rye.

Ratio 3: Prehistoric sites

The problem with calculating this ratio is that many samples contain both wheat and barley and it is not possible to determine with which crop the weeds came in. Most samples contained more weed seeds than grains, suggesting they contained a cleaning residue. There are only four exceptions, in the form of samples with more grains than weeds, pointing to a cleaned or semi-cleaned crop:

Hallshill:	3 samples (8, 25L, 25U)
Dod Law:	1 sample (25(3))

Ratio 3: Roman sites

None of the samples contained many weed seeds, which suggests that they all represent cleaned crops. Only nine samples may contain small quantities of cleaning residues:

Thornbrough:	4 samples (54a, 44, 58, 120)
South Shields: (12236)	1 sample (24)
South Shields: (12176)	4 samples (14, 21, 28, 30)

7.2.4 Summary of method 1

The vast majority of the prehistoric samples consisted of fine-sieving residues of the glume wheats and residues of the later processing stages of barley (probably also fine-sieving by-products), although a number of samples (notably from Murton, Dod Law and Thorpe Thewles) contained early processing residues of barley, and, in the case of sample 50 from Rock Castle, early processing residues of bread/club wheat. Only six samples contain some evidence for glume wheat spikelets, and one for a fully processed crop. The Roman samples are quite different, in that most of the samples from South Shields represent a fully processed crop (not very surprising considering the fact that the samples came from a granary), while those from Thornbrough also consist of cleaned crops, though in the case of spelt wheat the grain is present as spikelets, rather than dehusked grain. A few samples from this site did contain fine-sieving residues of spelt wheat, but the very low number of weed seeds suggests that the grain had already been fine-sieved prior to dehusking. This might also have been the case with the samples from Hallshill.

7.3 Method 2: Weed seed categories

7.3.1 Description of method 2

This method uses the results of an ethnographic study of non-mechanized crop processing carried out by Jones at Amorgos, Greece (G. Jones 1983a, 1983b, 1984, 1987). The crops present were free-threshing cereals, e.g. *Triticum aestivum, Triticum durum, Hordeum vulgare* and *Avena sativa*, as well as pulses. Samples were collected from the by-products of winnowing, coarse-sieving, and fine-sieving, and from the product of fine-sieving. Using discriminant analysis Jones demonstrated that the samples from the various crop processing groups could be separated successfully using the relative proportions of crop seeds, chaff elements and weed seeds. The archaeological samples cannot be compared directly with the ethnographic ones, as they do not consist exclusively of free-threshing cereals, but contain the glume wheats *Triticum dicoccum* and *Triticum spelta*, the chaff fragments of which behave quite differently during processing. However, the crop processing groups could also be clearly distinguished by using the weed seeds alone, and, more importantly still, by using weed seed categories (G. Jones 1983a, 1983b, 1984, 1987). By redefining the weed seeds into behavioural categories, direct comparisons can be made between the ethnographic and the archaeological samples. This is of great archaeobotanical relevance, as the weed species found in archaeological samples will rarely, if ever, consist of exactly the same range of species as those found in the ethnographic study.

The new weed seed categories or weed types created by Jones, are based on characteristics relevant to crop processing (G. Jones 1984, 54):

(1) *size of seed* – this is most relevant to fine-sieving since small seeds tend to pass through the sieve, while large seeds are retained.

(2) *tendency of seeds to remain in heads*, spikes, or clusters despite threshing (sometimes because the seeds are slightly immature) or to retain large projections – this is most relevant to coarse sieving since the seeds in heads etc. tend to be retained by the sieve while free seeds pass

through.

(3) *aerodynamic qualities of the seeds*, including density, shape, and presence or absence of features such as wings or hairs – this is most relevant to winnowing since light seeds and seeds with a pappus tend to blow away during winnowing.

The weed seeds were redefined into categories, such as big/headed/heavy (BHH), small/free/light (SFL), etc., so as to take account of these three characteristics simultaneously (see Table 7.2 (B)). Jones found that the different crop processing groups could be successfully separated by discriminant analysis, using these new weed seed categories and the ratio of weed seeds to crop seeds (i.e. ratio 3 of Method 1) as the new variables. Out of the 216 ethnographic samples 84.3 per cent were correctly reclassified (G. Jones 1984). This is illustrated in Figure 7.1.

7.3.2 Application of method 2

In order to compare the archaeological samples with those from the four crop processing groups, the weed seeds were redefined according to the same characteristics (see Table 7.4). It was not always easy to determine the right type for each species. In some cases, when the identification was insufficiently precise (e.g. Leguminosae indet.), the species was omitted from the analysis. The capsules of *Raphanus raphanistrum* were categorized as BHH (big/headed/heavy), though it was recognized that when the capsules are mature/ripe they tend to break up and may be better categorized as BFH (big/free/heavy). Both were tried in the preliminary analysis, but no changes took place in the classification of the samples, so that in the final analysis the capsules were defined as BHH.

The seeds of the category 'small grasses' were also difficult to classify. These seeds are mostly free-threshing and are classified as 'small' and 'free'. As far as the third characteristic, the aerodynamic qualities of the seeds, is concerned, they could be classified as either heavy or light. They were initially classified as 'heavy', but it was recognized that, as they are very small, they could in some cases be blown away during winnowing, which means that they could also be classified as 'light'. Unfortunately, they were not present in the ethnographic study, so that we do not know how they behaved there, but it is likely that they occurred in both winnowing *and* fine-sieving residues. Consequently, it was decided to run the analysis twice, once with the small grasses as SFH (small/free/heavy), and once as SFL(small/free/light).

The procedure consists of running a discriminant analysis using the samples from the ethnographic

Crop Processing Groups – Ethnographic Data

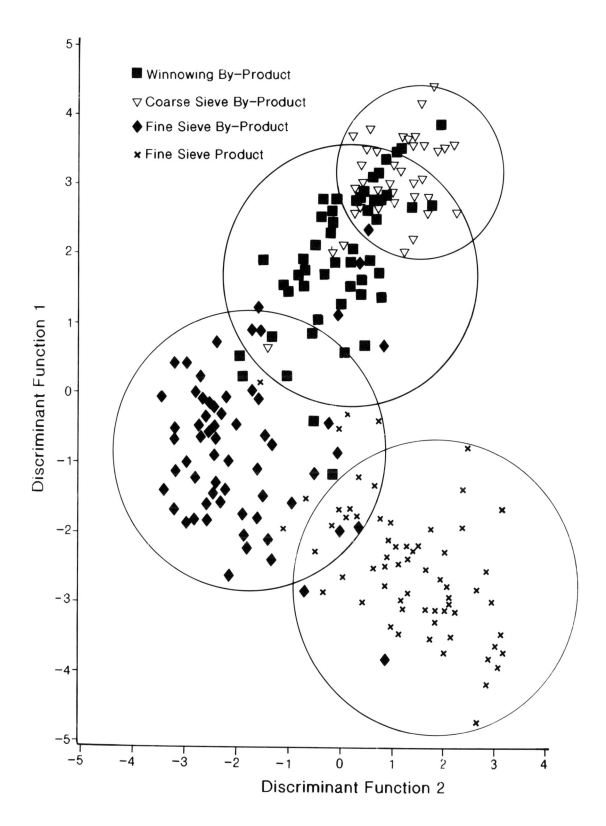

Figure 7.1   Discrimination of crop processing groups for ethnographic data from Amorgos, Greece (G. Jones 1983a, 1984, 1987), using the percentages of weed seed categories. The large circles enclose 90 per cent of the samples of each group.

85

processing groups as the control groups to extract the discriminant functions (the data were provided by Glynis Jones). These discriminant functions are then used to classify the archaeological samples. The discriminant analysis classifies each archaeological sample to the processing group it most closely resembles. The advantage of the method is that, as Jones indicated, 'it is not necessary to assume at the outset that all archaeobotanical samples are derived from one of the crop processing stages considered in this [the ethnographic] study. The analysis does recognize unusual samples, contaminated samples, and samples of ambiguous status' (G. Jones 1987, 321).

7.3.3 Results of method 2

The discriminant analysis produces both a graphic and a tabular output. The classification of each sample is given in Table 7.5. The graphical output (scatter plots) is, for reasons of space, only presented for the samples of two sites (Rock Castle and South Shields, deposit 12236) to illustrate the results (Figures 7.2 and 7.3).

Table 7.5 lists the classification of each sample for both analyses (i.e. with the small grasses grouped as SFH and SFL). Column A gives the classification for each sample, column B gives the probability of the classification, and column C gives, in brackets, the next most probable classification. Most samples from the prehistoric sites are classified as belonging to group 3 (fine-sieve by-product); most samples from South Shields are classified as belonging to group 4 (fine-sieve product), while the samples from Thornbrough consist of a mixture of groups 3 and 4, with some samples allocated to group 1 (winnowing by-product). If we look at the difference which results from moving the seeds of the small grasses from SFH to SFL, the following observations can be made:

(1) The probabilities of the classification in the first analysis are mostly very high. The classifications of 155 out of the 158 prehistoric samples and 40 out of the 60 Roman samples had a probability of 0.9 or above.

(2) When the classification of a sample remains the same from analysis 1 to analysis 2, the probability of the second classification remains high, though is usually lower than the first one (the probability is, of course, identical in those cases where the sample did not contain any seeds of small grasses). In 20 Roman samples, however, the probability increases slightly.

(3) When the classification of a sample changes from analysis 1 to analysis 2, the probability of the second classification is always lower, and usually considerably lower (i.e. from *ca.* 0.9 to

ca. 0.5). There is only one exception to this, and that is with sample 54a from Thornbrough. Here the probability increases slightly from 0.8874 to 0.8926.

(4) The changes in the classification that do occur (only 17 out of a total of 218) are easy to interpret. A change from group 3 to group 4 or *vice versa* is not surprising, as these are two closely related processing groups. The change from group 3 to group 1 is a direct result of moving the small grasses from weed type SFH to SFL.

(5) When the probability of the classification is low, this points either to the presence of a mixture of more than one processing group, or to the absence of sufficient numbers of weed seeds. The analysis relies heavily on the weed seeds and samples with few weed seeds are likely to produce spurious results. Most of the low probabilities occur with the Roman samples, which have few weed seeds.

(6) In the scatter plots, the archaeological samples classified as fine-sieve by-products are usually plotted as extreme cases (they lie on the outside of the group range; see Figure 7.2). Several of the archaeological samples used by Jones to test the method also lie slightly outside the group range (G. Jones 1987: Figure 1). This can probably be explained by the fact that in the archaeological samples the weed seed content (i.e. both numbers and categories) is not identical to that in the ethnographic samples (G. Jones, pers. comm.).

7.3.4 Summary of method 2

There is little difference between the results of the two analyses, suggesting that the data are fairly robust. On balance, however, the first analysis would appear to give better results (i.e. the probabilities of the classification are higher), and is therefore used in the comparison with those of Method 1 below. The results of Method 2 (analysis 1) suggest that all the prehistoric samples consist of fine sieving residues, that most of the samples from South Shields consist of fully processed crops, but that those from Thornbrough consist of both fine-sieve by-products and products and one sample of winnowing by-product.

7.4 Conclusion

The aim of this chapter was to identify the taphonomic processes which might have transformed the original plant assemblages, in order to filter out any variation due to these processes prior to using the archaeological samples for further analysis in Chapter 10. The main taphonomic process

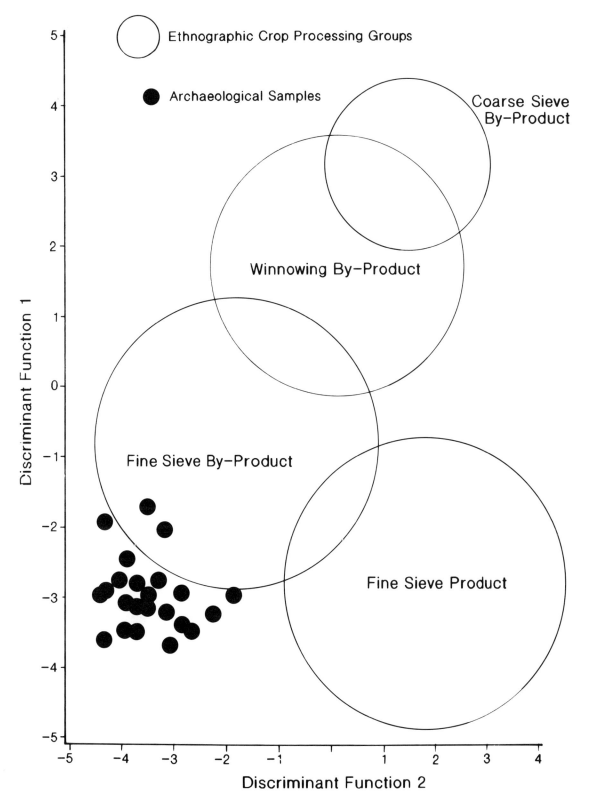

Figure 7.2   Discriminant analysis using the four crop processing groups as the control groups, to which archaeological samples are compared. The large, open circles represent the ethnographic processing groups (see Figure 7.1). The small, solid circles represent the archaeological samples from Rock Castle, classified as fine sieve by-products.

Crop Processing Groups – South Shields Deposit 12236

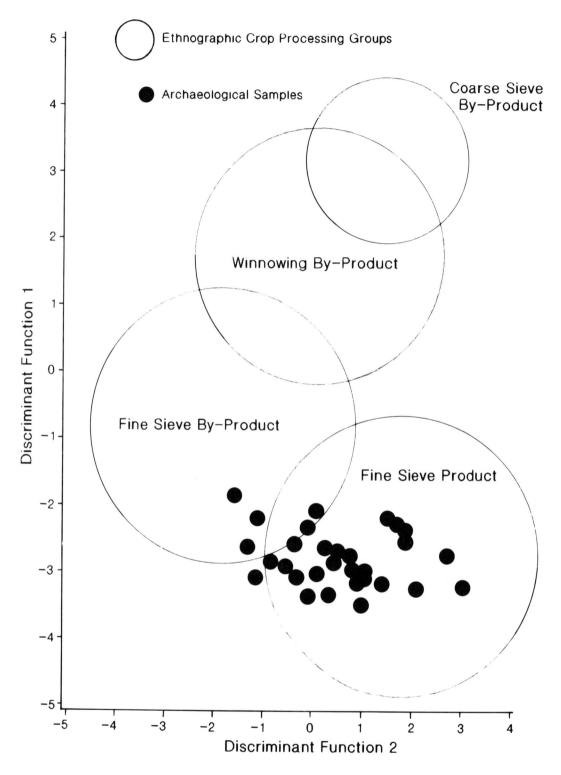

Figure 7.3   Discriminant analysis using the four crop processing groups as the control groups to which archaeological samples are compared. The large, open circles represent the ethnographic processing groups (see Figure 7.1). The small, solid circles represent the archaeological samples from South Shields, deposit 12236, classified as fine sieve product (27 samples) and fine sieve by-product (3 samples).

influencing the composition of carbonized seed assemblages is that of crop processing. Consequently, each sample has been examined using two different methods, to assess which crop processing stage it represented.

There is a large amount of agreement between the two methods. Both identify most prehistoric samples as consisting of fine-sieving residues and most Roman samples as fine-sieve products, with the exception of some samples from Thornbrough. Some divergence between the two methods does, however, exist, in that those samples which were identified by Method 1 as containing some early processing residues of barley, were not recognized as such by Method 2. In fact, the probability of the classification of these samples as fine-sieve by-product was high (0.9 or above). Even in the second discriminant analysis, in which the seeds of the 'small grasses' were categorized as SFL (small/free/light), the weed type relevant to winnowing, only one out of the 22 samples identified by Method 1 as containing early processing residues of barley, was classified as winnowing by-product (LS465 of Thorpe Thewles). The classification of the remaining 21 samples remained the same, i.e. group 3 (fine-sieve by-product).

The following suggestions can be put forward to explain this:

(1) The weeds from the fine-sieve by-product of the glume wheats dominate the samples and 'swamp' those from the early processing by-products of barley.

(2) The weeds belonging to the early processing by-products have not been preserved. Many of them, for example, are light, with feathery attachments, and as a result they stay high in the fire and often get burnt away (Hillman 1984).

To conclude, the results produced by both methods indicate that all samples from the prehistoric sites can be used in the analyses of the weed species in Chapter 10, as they all contain the remains of the same crop processing group. The fact that some of the samples also contain some early processing waste of barley is unlikely to influence the outcome of the analyses, as the barley rachis internodes do not appear to be associated with many weed seeds.

The results of the two methods also indicate that it will not be possible to make direct comparisons between the samples from the Roman and the prehistoric sites, as they consist of different remains, i.e. remains from different crop processing groups. Any comparison between the two groups of samples (Chapter 10) will, therefore, have to be of a more indirect nature.

8. Economic Classification of the Sites

8.1 Introduction

As we have seen in Chapter 1, it has been suggested that the economic base of the later prehistoric settlements in the north of England was founded largely on animal husbandry or pastoralism, rather than arable farming, especially in upland areas of the region. An important objective of this study of the charred plant remains from these sites must, therefore, be to try and establish whether the arable crops present in the samples were grown by the inhabitants themselves, or whether they were imported from somewhere else. That is, can we determine whether the sites were producers or consumers of cereal crops?

This question is of wider relevance, as during the later prehistoric and Roman period a number of developments took place which resulted in economic specialization. During this period new levels of social organization developed, giving rise to social élites, the creation of large, non-agricultural settlements, and the growth of long-distance trade (Bradley 1984, Champion *et al.* 1984). This encouraged the production of food surpluses, both for export and to sustain those not actively involved in food production. In the North East the settlement at Stanwick has been identified as the possible residence of members of such an élite (Haselgrove 1982, 1990), and the arrival of large numbers of Roman soldiers in the area is also relevant here, as the Roman forts are likely to have been consumer sites, as far as arable produce is concerned.

In this chapter the carbonized seed assemblages will be examined for evidence concerning the role of the arable crops at each settlement. There are two models available in the archaeobotanical literature which address this issue. One is based on ethnographic data collected by both Hillman and G. Jones (Hillman 1981a, 1984, G. Jones 1983a, 1984), and emphasizes the importance of the crop processing sequence. The other is based on archaeological data (M. Jones 1985), and stresses the concept of complementarity. Here both models will be applied, in order to assess whether the nine sites can be described as either producer or consumer sites.

8.2 Model A: The ethnographic approach

8.2.1 Description of model A

Model A is based on ethnographic studies of crop processing using traditional methods, as described in Chapter 7 above (Hillman 1981a, 1984, G. Jones 1983a, 1984). These studies suggest that the difference between producer and consumer sites is likely to lie in the presence or absence of by-products from the early stages of crop processing (winnowing and coarse sieving). On producer sites the entire crop processing sequence takes place. In contrast, consumer sites will obtain their grain in fully processed form (free-threshing cereals) or in the form of semi-cleaned spikelets (glume wheats). On these sites the residues of the early stages of processing are absent.

If the crop present is a free-threshing cereal, such as barley, bread/club wheat or rye, then the assemblage from a consumer site will probably consist of prime grain with small quantities of weed seeds, mostly those with a diameter similar to that of the grains, i.e. those which could not be removed by sieving (e.g. *Bromus* sp., *Avena* sp., *Agrostemma githago*). On these sites the grain would not normally be charred unless a grain store had caught fire, and samples would normally consist of just the occasional grain and large weed seed (Hillman 1981a). It is, however, also possible that these sites imported grain in semi-cleaned state (i.e. before fine-sieving had taken place), in which case there would be more weed seeds (both large and small ones) in the samples. On producer sites the assemblage would, in addition to the grain, also consist of rachis internodes, straw nodes, and a range of different weed types, i.e. the residues typical of early processing stages (but see below).

The situation with glume wheats (emmer and spelt) is different in that the grain is often only processed in bulk up to (or including) the stage of coarse sieving, after which the grain is stored in the form of semi-cleaned spikelets. Consumer sites will often, though not necesarily, receive glume wheat grains in spikelet form and be expected to dehusk the grain themselves, on a piecemeal basis throughout the year. This means that the glume bases of emmer and spelt can be present on both types of settlement. Producer sites can only be recognized by the presence of straw nodes and weed seed types characteristic of the winnowing and coarse sieve by-products.

These residues of the early processing stages are, however, rarely found in carbonized seed assemblages (Green 1981, Hillman 1981a, G. Jones 1987, M. Jones 1985). This may be explained by the fact that the early stages of processing take place away from sources of fire (Hillman 1981a, 1984, M. Jones 1985). What is more, the rachis internodes of free-threshing cereals, and light weed seeds, when exposed to fire, do not survive as well as the components characteristic of the later stages, such as grains and glume bases (Boardman and Jones 1990, Hillman 1981a, 1984). This means that the indicators for producer sites are commonly under-represented in the archaeobotanical record, making a distinction between producer and consumer sites difficult (Hillman 1981a, 1984, G. Jones 1987).

8.2.2 Application of model A

The identification of remains from the early processing stages of both types of crop relies on the same two methods used in Chapter 7, i.e. the calculation of ratios of the major sample constituents and the analysis of different weed seed categories present in the samples. The outcome of both these methods has already been listed in Tables 7.3 and 7.5. Here the data are not re-analysed, but the results are restated, though in terms of the assemblages of sites rather than individual samples.

8.2.3 Results of model A

It is clear from the values of ratio 2 (i.e. the ratio of barley rachis internodes to barley grains) in Table 7.3 that all prehistoric sites contain evidence for early processing waste of barley (and bread/club wheat in the case of Rock Castle), although the evidence for Hallshill is rather meagre due to the low number of barley remains present on this site. At Murton, Dod Law and Chester House the majority of the samples contain early processing waste of barley, while at Thorpe Thewles, Stanwick and Rock Castle only a small number of samples contain this evidence. All these sites also contain straw nodes (= culm nodes), but in such small quantities that they cannot be taken as evidence either way. The remains of the glume wheats (ratio 1, the number of glume bases to glume wheat grains) offer no information on the issue, while the weed seed types all point to the presence of fine sieve residues, which can be present on both types of site.

None of the samples from the two Roman settlements contains evidence for the presence of early processing waste of barley or bread/club wheat, but two samples from Thornbrough do contain early processing waste of rye. The evidence of the glume wheats points to the presence of semi-cleaned spikelets and some fine-sieve residues at Thornbrough and fully dehusked grain at South Shields. There were some straw nodes at Thornbrough, but none at South Shields. The weed seed types and the values for the weed to grain ratio (ratio 3) suggest that the samples at Thornbrough consisted of fully cleaned barley and semi-cleaned spikelets of spelt wheat, while those from South

Shields consisted of fully processed spelt and bread/club wheat and a small amount of fine-sieve residues.

8.2.4 Summary of model A

When using Model A the evidence suggests that all prehistoric sites were producer sites, although the evidence was only there for the free-threshing cereals, barley, and bread/club wheat in the case of Rock Castle. The evidence from Thornbrough was difficult to interpret, but it appears that some production of arable crops (i.e. rye) did take place, while South Shields was identified as a consumer site.

8.3 Model B: The complementarity approach

8.3.1 Description of model B

The second model is that developed by M. Jones, based on his analysis of a number of late Iron Age and Romano-British settlements in the Upper Thames Valley, central-southern England (M. Jones 1985). He has put forward the thesis that producer sites are likely to be characterized by the presence of large quantities of grain. He argued that while it was not surprising to find assemblages dominated by weeds and chaff (the waste products of the harvest), it was unusual to find assemblages dominated by grain: 'The grain is the part of the harvest least likely to be treated as waste material, and for that reason its deposition as charred debris is least expected. The most likely place for this unlikely event to occur is at its place of production ... With further processing and transportation, the perceived unit value of the crop accrues, while its quantity at any single point lessens. The chance of the prime product itself being discarded into a fire consequently drops. A non-producer site receiving the harvest product through exchange is likely to allow only the waste material from any final processing to be discarded into the settlement fires.' (M. Jones 1985, 120).

On the basis of this supposition and with the help of a considerable amount of background information regarding his sites (i.e. geographical location, insect and other environmental evidence, the nature of the archaeological features, etc.) Jones was able to distinguish two types of plant assemblages, one characterizing consumer sites, and one characterizing producer sites, see Figure 8.1.

His producer assemblage is characterized by large quantities of plant remains per litre of sediment (indicated by the larger diameter of the circles on the scatter plot), and by relatively high proportions of cereal grains in the samples. Most samples contain

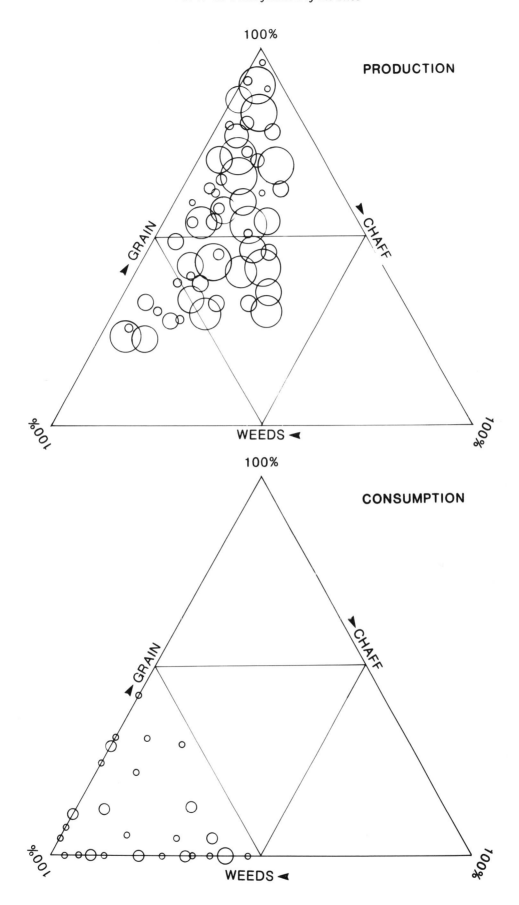

Figure 8.1   Triangular scatter plots showing the relative proportions of cereal grains, chaff and weed seeds for generalized producer and consumer assemblages (after M. Jones 1985). Each circle represents one sample. For key to size of circle, see Figure 8.3.

over 30 per cent cereal grains and a few contain as much as 80 to 100 per cent cereal grains. No sample has more than 50 per cent chaff fragments or 70 per cent weed seeds. In contrast, his consumer assemblage is characterized by low quantities of seeds per litre of sediment (mainly small circles on the scatter plot), and by a low proportion of cereal grains. Samples in this assemblage may be composed of up to 50 per cent chaff or 100 per cent weed seeds, and cereal grains make up no more than 50 per cent of any individual sample. Several samples contain no grain at all. The assemblage is dominated by weed seeds.

8.3.2 Application of model B

To compare the seed assemblages from the sites under study with these two generalized triangular scatter diagrams, similar plots were drawn up for each of the nine sites. The three categories of plant remains plotted on the diagrams were counted in the following manner: Grain: each grain (wheat, barley, rye, unidentified cereal grain) counts as one; fragments are combined to make up whole grains. Chaff: each glume base, rachis internode and straw node counts as one, a spikelet fork counts as two, unless the rachis internode is still attached, in which case it counts as three; awn fragments are not included, nor are glume fragments other than the basal parts. Weeds: each weed seed counts as one; all seeds under the category 'Weeds' in the tables are included with the exception of the rhizomes of Gramineae. The diameter of the circles on the scatter plots represents a measure of the density of seeds (the sum of the cereal grains, chaff fragments and weed seeds) in the deposit, i.e. the number of seeds per litre of sediment.

8.3.3 Results of model B

The triangular scatter plots drawn up for each site are given in Figures 8.2, 8.3 and 8.4. When we compare these with those of the model the following observations can be made:

(1) None of the plots from the prehistoric sites resembles that of the producer assemblage. Very few samples from these sites contain more than 50 per cent grain.

(2) The plots from Hallshill, Murton, Dod Law and Chester House do not resemble the consumer assemblage either, as none of them is dominated by weed seeds.

(3) The number of samples from Hallshill, Murton, Dod Law and Chester House is low, and additional samples may alter the present plots.

(4) The plots from Thorpe Thewles, Stanwick, and Rock Castle resemble the consumer assemblage, with most of the samples plotted in the left hand

corner of the diagram, meaning that they are dominated by weed seeds. There are, however, two differences. Firstly, the archaeological samples are generally much richer in plant remains (i.e. the size of the circles is larger) than those in the model, and secondly, the archaeological assemblages contain fewer samples with no grains or no chaff.

(5) The plots from the Roman assemblages resemble the producer assemblage in that they are characterized by large quantities of cereal grains, i.e. many samples are plotted in the top corner of the diagram, but, with the exception of South Shields deposit 12176, they differ in that the samples are concentrated into a more constricted part of the diagram.

8.3.4 Summary of model B

The application of this model to the data did not produce any close matches between the archaeological assemblages and those of the model. The assemblages from Thorpe Thewles, Stanwick and Rock Castle show some similarities to that of consumer sites, those from Thornbrough and South Shields show some similarities to that of producer sites, but the assemblages from Hallshill, Murton, Dod Law and Chester House could not be classified within the present model.

8.4 Discussion

The application of the two models has resulted in contradictory information regarding the economic classification of the sites. Model A classified all prehistoric sites as producer sites, but Model B classified them either as possible consumer sites (e.g. Thorpe Thewles, Stanwick, and Rock Castle), or did not classify them at all. Furthermore, Model A classified the Roman sites as consumer sites (although recognizing some production of rye at Thornbrough), while Model B classified both sites as possible producer sites. This has raised a number of points regarding both models which are discussed below. Those relating to Model A are considered first.

The distinction between producer and consumer settlements in Model A relies heavily on the presence or absence of the waste products of early crop processing. The reduced likelihood of exposure to fire and the differential preservation of certain cereal components has already been mentioned a number of times in relation to the difficulty of identifying these waste products. As the rachis internodes of free-threshing cereals are rarely preserved, one might argue that the presence of even small quantities of these elements is significant and may be taken as evidence for production. One could go one step

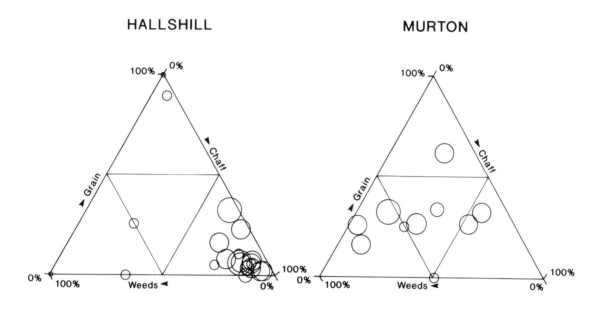

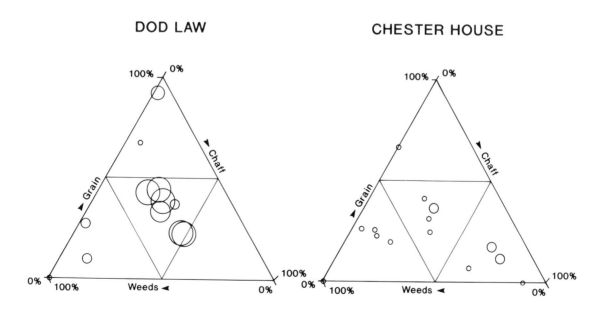

Figure 8.2　　Triangular scatter plots showing the relative proportions of cereal grains, chaff and weed seeds for four of the prehistoric assemblages. Each circle represents one sample. For key to size of circle, see Figure 8.3.

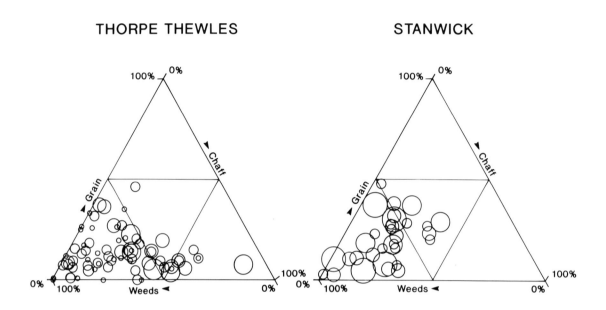

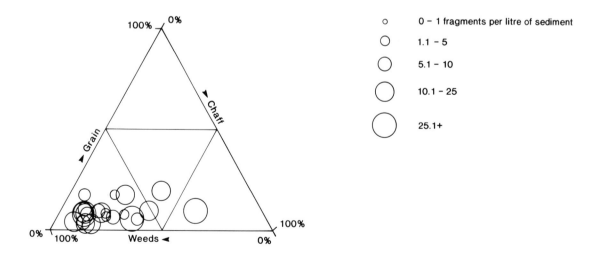

Figure 8.3   Triangular scatter plots showing the relative proportions of cereal grains, chaff and weed seeds for three of the prehistoric assemblages. Each circle represents one sample.

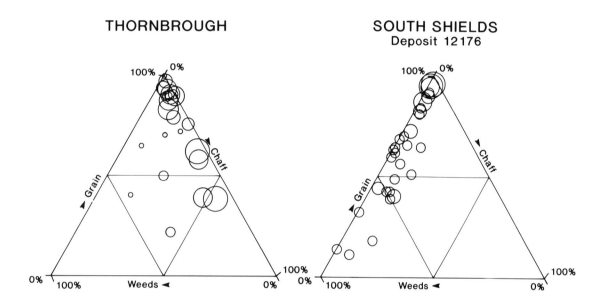

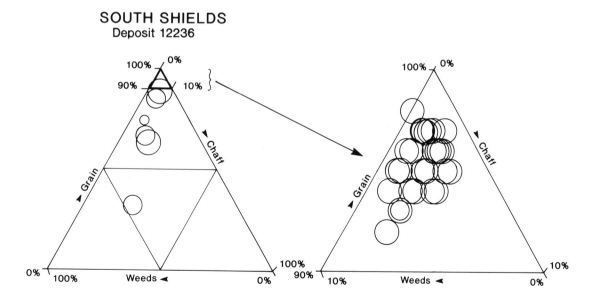

Figure 8.4   Triangular scatter plots showing the relative proportions of cereal grains, chaff and weed seeds for the Roman assemblages. Each circle represents one sample. For key to size of circle, see Figure 8.3.

further and argue that even a value of *ca.* 0.3 for ratio 2 (the ratio of barley rachis internodes to barley grains) was evidence for early processing waste, but relying on negative evidence like this is unwise. Here, the presence of just a few samples with values for ratio 2 of 0.4 or above has been taken as evidence that the assemblage represents a producer assemblage.

The situation regarding the glume wheats is more complicated. There are no chaff fragments of these crop species characteristic of early processing residues, other than the straw nodes. The presence of straw nodes is difficult to interpret, however, as they cannot be identified to species, and straw could, theoretically, have been imported for bedding or fodder (this is, of course, also the case with rachis internodes of free-threshing cereals). This means that as far as the glume wheats are concerned, it may not be possible to identify producer sites. Certain types of consumer sites may, however, be recognized. While Hillman (1981a, 1984) has suggested that the glume wheats were usually traded in spikelet form (resulting in the presence of glume bases on consumer sites), there is now evidence from a number of Roman sites for the presence of fully processed crops. Dehusked spelt grain was found in the granary deposit of the Roman fort at South Shields, Tyne and Wear (Table 4.18); in the granary of the Roman villa at Droitwich, Warwickshire (V. Straker, pers. comm.); in Roman London (Straker 1984); in Roman Colchester, Essex (Murphy 1984 and forthcoming); and in the Roman fort at Rocester, Staffordshire (Moffett 1989). Hillman (1984) did, in fact, point out that the existence of manorial farms during the Roman period may have meant that the processing of the glume wheats at such farms was carried out indoors, right up to the stage of grain dehusking and cleaning, and resulted in the trade of fully processed grain. The present evidence supports this.

The presence of fully dehusked grain on Roman period consumer sites may also partly explain the evidence from insect studies for the serious damage to grain by grain weevils and beetles during the Roman period (Buckland 1978, Kenward 1979). The absence of any evidence for serious crop loss through insect infestation prior to the Roman period may possibly be linked to the fact that up to that period grain was normally stored in spikelet form, rather than in dehusked state, with the glumes protecting the grains from any insect damage. The increase in the scale of grain transport and storage during the Roman period will also have played a role, by allowing insect infestation to spread more easily.

It may be possible to identify by-products from the early processing stages of both types of crops by looking at the different weed types present in the samples, i.e. by applying Method 2 described in Chapter 7. But when the samples contain both wheat and barley, as is the case in most of the archaeological samples studied here, the weeds are likely to consist of a mixture from both crops, which may hinder their interpretation. For example, as we have seen in Chapter 7, this method did not identify the presence of early processing waste of barley, probably because the samples contained large amounts of weed seeds from the fine sieving residues, which swamped the other weed types, and, possibly more importantly, because the light weed seeds associated with winnowing tend to burn away and be under-represented.

There is disagreement between the models regarding the type of site on which large quantities of grain can be expected. Hillman has suggested that the burning down of grain stores, which results in deposits dominated by cereal grains, are accidental, and can occur on both types of sites (Hillman 1981a). M. Jones, while agreeing with the accidental nature of the event, maintains it is more likely to occur on producer sites (M. Jones 1985). However, the assemblages dominated by cereals at Ashville, Oxfordshire, (the site on which the producer assemblage of the model is largely based) all originate from pits, probably storage pits, and the same is true at Danebury, Hampshire (M. Jones 1984b, 1985). Thus, these assemblages originate from storage contexts rather than general refuse contexts, and they are similar to storage contexts on consumer sites such as the granary at the Roman fort at South Shields. Assemblages dominated by cereal grain are probably best interpreted as representing storage contexts, rather than regarded in terms of production or consumption.

Most of the assemblages consist of a mixture of glume wheats and barley. While these may be treated separately when the ratios of chaff and grains are calculated (ratios 1 and 2), they are treated together both when we look at the different weed types (see above), and when we calculate the relative proportions of grain, chaff and weed seeds for the triangular scatter plots. This does mean, however, that the plots on the triangular diagrams may be located in different parts of the diagram depending on the relative proportions of wheat and barley grains at any one site. As very few of the barley chaff fragments are normally preserved (e.g. no barley glumes are preserved in charred assemblages), variations in the relative proportions of wheat and barley can alter the scatter plot quite considerably (G. Jones 1987, 1989). For example, the charred remains of unprocessed six-row barley would contain a ratio of three grains to one chaff fragment (i.e. one rachis internode), while for glume wheats this ratio would be *ca.* two grains to three chaff fragments (i.e. two glume bases and one rachis internode), resulting in different points on the scatter plot. Thus, the difference between the scatter plots from Murton, Dod Law and Chester House, and those from Thorpe Thewles, Stanwick and Rock Castle may be at least partly related to variations in the proportions of barley and wheat remains at those sites, rather than to differences in the role of arable husbandry.

The division of settlement sites into two types only, producer and consumer sites, is rather simplistic. Even a redefinition into four categories is arbitrary, though possibly slightly more realistic:

> *site producing for subsistence (small farmstead)*
> *site producing for a surplus (village or Roman villa)*
> *small consumer site (pastoral settlement)*
> *large consumer site (town or Roman fort)*

Previous attempts to explain the differences between the scatter plots of Model B and those of archaeological sites (Thorpe Thewles and Stonea) by inferring that the producer assemblage of Model B actually reflects production for a surplus, while the archaeological assemblages characterized production for subsistence, remain unsatisfactory (Van der Veen 1987a and 1991).

8.5 Conclusion

It follows from the discussion above that Model B could not, at present, be successfully applied to the data. While the model has the advantage of providing a visual characterization of the assemblages, which greatly facilitates comparisons between sites, and has pointed towards differences in the assemblages of the sites studied here, it has not offered any clear interpretation of the assemblages. The model relies too heavily on the premise that large quantities of grain are associated with producer sites, it cannot take sufficient account of differences in the relative proportions of barley and glume wheats in the assemblages, and does not offer well defined ways of interpreting plots which do not match those of the model.

The main disadvantage of Model A is that it relies on the presence of cereal components and weed types which are known to be under-represented in the archaeobotanical record due to their differential preservation during the charring event. However, this problem can, to a certain degree, be accounted for as the direction of the loss is known.

To conclude, the evidence from Model A is taken to be more reliable than that of Model B, and all seven prehistoric sites under study have been identified as producer sites. This has important implications for the research question set out in Chapter 1, regarding the role of arable farming in northern England during the Iron Age. All six late Iron Age sites studied here clearly did produce their own cereal crops. The similarity of the assemblage from Stanwick to those of the other prehistoric sites in this respect, and especially the similarity to those of Thorpe Thewles and Rock Castle, was slightly surprising as the archaeological remains suggest that the nature of the settlement at Stanwick was, at least for a short period, very different from those of the others (see Chapters 2 and 4). We will come back to this aspect in Chapter 11.

The assemblage from South Shields was identified as a consumer site, and this was in line with our expectation, as the assemblage came from a granary in a Roman fort. The evidence from Thornbrough is difficult to interpret. There is evidence for the production of rye, but the role of the spelt and barley crops could not be established with certainty, though the evidence does suggest that they may have been brought in from elsewhere.

9. Weed Ecology

9.1 Introduction

Weed ecology can be described as the study of the 'household affairs' of weeds, that is, the study of their relationship with one another and with their environment (Willis 1973). The environment incorporates a number of different aspects, such as climatic factors (temperature, light etc.), edaphic factors (moisture, pH, nitrogen), biotic factors (other living organisms), and, most importantly in the case of weed species, anthropogenic factors (soil disturbance, weeding, manuring etc.) (Willis 1973). By studying the weed ecology of the species present in the archaeobotanical assemblages we can obtain information about these environmental factors.

Within the study of ecology two different analytical traditions can be identified: those of synecology and autecology. Synecology is defined as the study of the relationships between plant communities and their environment, and is usually referred to as plant sociology or phytosociology. Autecology is defined as the study of the ecology of individual species. To some extent these two approaches have grown up as national traditions with the synecological approach more prominent in Germany, The Netherlands and France, while in Britain the autecological approach has had a greater following (M. Jones 1988a, Whittaker 1973, Willis 1973). Both approaches have been applied to archaeobotanical data and their suitability in interpreting this type of data will be discussed below.

9.2 Analytical traditions

9.2.1 Phytosociology

The synecological or phytosociological approach, emphasizing the study of plant communities, was developed by Braun-Blanquet (1964), and the Zürich-Montpellier School or Braun-Blanquet Approach (Westhoff and Van der Maarel 1973) has become one of its best known representatives. This approach has found a widespread following in Europe (Ellenberg 1950, 1979, Oberdorfer 1962, Tüxen 1950, Westhoff and Den Held 1975). A detailed account of the methodology of the approach is given by Westhoff and Van der Maarel (1973). The approach is based on the co-occurrence of species in the field and consists of a classification of vegetation communities into a hierarchical system. The three most essential concepts underlying the approach are, first of all, that the classification and interpretation of communities should be based on their full floristic composition. Secondly, that the classification is

based on the presence of diagnostic species whose relatively narrow ecological amplitude characterizes communities and their environments. And thirdly, that communities are organized into a formal, hierarchical classification.

The plant communities or syntaxa, are characterized by three different types of diagnostic species: first of all, character species, which are centred on or relatively restricted to a particular syntaxon compared with all others. Secondly, differential species, which are species which distinguish two closely related syntaxa by their presence in most samples of the one and their absence in most samples of the other. And finally, constant companions, species which are not restricted to a given syntaxon, but help to characterize it and indicate its relationship to higher syntaxa. Vegetation communities are described by recording the presence and abundance of species in a series of uniform stands (relevés or Aufnahmen). The size of these stands varies with the type of vegetation studied, from 5–10 square metres in pastures, 25–100 square metres in arable fields, to 100–500 square metres in temperate, deciduous forests. Tables are drawn up to explore similarities and dissimilarities, and plant associations are constructed. A hierarchical classification is adopted in which Associations (names ending in -etum) are grouped into Alliances (ending in -ion), Alliances into Orders (ending in -etalia), and Orders into Classes (ending in -etea). The Classes are grouped into Formations, which correspond to distinct types of vegetation, often characterized by particular life forms (woodland, water plants, grassland etc.).

The weeds of arable fields and waste places are grouped under the Formation of the 'herbaceous plants of disturbed ground'. The two Classes of this Formation most commonly encountered in the archaeobotanical literature are the Secali(n)etea and the Chenopodietea (but see section 9.4 below). The Secalietea consist of weed species which occur most commonly in fields of winter-sown cereal crops. They are annual weeds which germinate in the autumn, survive the winter as small plants, and develop to the stages of flowering and seed setting during the following spring and summer, alongside the cereal crop. The Chenopodietea consist of weed species which occur in fields of summer-sown cereal crops or in fields of 'row' crops (e.g. root crops or pulses, in general 'garden' crops). This Class also includes weeds of ruderal and waste places. These weeds need relatively high temperatures for germination and therefore generally germinate in the spring or early summer. They often have a fairly short life cycle, with the timespan between germination and seed setting being less than six months.

A number of different factors are thought to underlie this divergence of weed communities (Ellenberg 1950):

(1) garden crops are generally better manured (many of the typical garden weeds are distinctly nitrogen-loving)
(2) the leaf-growth of garden crops produces a different type of shade from that of cereals
(3) garden-crops are hoed and weeded more often than cereal crops
(4) perhaps most importantly of all, the rhythm of their growth and development and, consequently, of the agricultural operations associated with them, is different.

Ellenberg (1950) has illustrated this last point by describing the cultivation regime for winter rye, summer oat and row crops such as beet and potatoes. For winter rye the field is ploughed and prepared in the autumn. Winter annuals, i.e. weeds which can germinate at low temperatures, develop at the same time as the cereal plants and overwinter with them (Ellenberg 1950; but see M. Jones 1984b for a critical discussion of the variety of dormancy-breaking mechanisms displayed by weed species). In spring these winter annuals have an immediate advantage over the summer annuals, and occasionally have already set seed when the summer annuals are still developing. Early summer annuals take up the space left over by the winter annuals and cereal plants. Rye plants develop early and cast a lot of shade, so that many late germinating summer annuals do not get the chance to develop. The winter annuals remain the dominant weed type throughout. If the fields are hoed or weeded during the spring, however, this encourages the development of summer annuals, but the winter annuals continue to predominate. Rye fields are, in fact, rarely harrowed, hoed or weeded in spring as the plants develop so early and the root system is quite shallow (Ellenberg 1950).

For summer oat the field is ploughed and prepared in early spring, and this kills many of the winter annuals which had developed over the autumn and winter. Consequently, winter annuals do not have a head start, but develop at more or less the same pace as summer annuals. By harvest time summer and winter annuals occur in similar numbers. However, at the time of Ellenberg's study it had become common practice in many regions to harrow, hoe and weed fields of summer cereals during April and May, and this damages the winter annuals more than the summer annuals. Thus, the weed flora of summer

cereals tend to be slightly richer in summer annuals than those of winter cereals (Ellenberg 1950).

For row-crops such as beet and potatoes, the fields are prepared in mid or late April. By that time the soil is already warm enough for most warmth-loving summer annuals to germinate. They can develop without being immediately checked by the shade of the crop plants. They quickly become dominant, as they develop faster, grow taller and have shadier leaves, thus out-competing the winter annuals. Hoeing affects both types of weeds, but the summer annuals can recover more quickly, so that by harvesting time they predominate (Ellenberg 1950).

Thus, the cultivation regime plays an important role in the development of the arable weed flora. The weeds of winter and summer cereals differ only in the proportion of Chenopodietea species present, and this is influenced by whether or not there is harrowing and weeding during spring. The weed flora of a field of winter cereals which was harrowed and weeded during the spring, will be very similar to that of a field of summer cereals which was not harrowed and weeded during the spring.

Ellenberg (1950) recognized that the relative proportions of summer and winter annuals could vary not only with the type of crop and cultivation, but, to a lesser extent, also with the type of soil and climatic conditions. On certain warm calcareous soils ('Muschelkalk-böden') the distinction between garden and cereal weed communities is minimal. The clearest separation can be found on sandy soils in areas of cool summers and mild winters, especially where winter rye and garden crops are cultivated side by side (Ellenberg 1950).

Thus, the two communities are not fundamentally different from one another. They vary only in the different amounts of summer annuals and nitrogen-loving species present, and this is influenced by soil conditions, climate, cultivation regime, and type of crop (see also section 9.4 below). Ellenberg has identified many transitional communities and suggests that we should see the two classes as separate aspects of the same community (Ellenberg 1950). The regional variability in the degree to which these two communities can be successfully distinguished is reflected in the manner in which they have been grouped in the phytosociological classification. Some authors recognize the two weed communities at the highest syntaxonomic level of Class (Chenopodietea and Secalietea: Ellenberg 1979, Oberdorfer 1962, Westhof and Den Held 1975), while others see them as Orders within a single Class (Chenopodietalia albi and Centauretalia cyani within the Class Stellarietea mediae: Tüxen 1950; Polygono-Chenopodietalia and Centauretalia cyani within the Class Stellarietea: Silverside 1977; see also section 9.4 below).

9.2.2 Autecology

Autecology is the study of the behaviour of individual plant species in relation to the environment. Information regarding the ecological preference of individual species can be obtained either by observations in the field, or by laboratory experiments. Most information regarding the influence of environmental factors on plant species comes from field observations. Controlled laboratory experiments can provide detailed insights into the physiological preferences of plants, but these preferences cannot always be directly translated into ecological behaviour (Ellenberg 1979, Holzner 1978). The occurrence of plant species is determined by both the presence or absence of the preferred physiological conditions *and* the presence or absence of competition from other species. Thus, a species may be most abundant in habitats which are suboptimal to their physiological requirements, as that is where they are not out-competed by rival species (Ellenberg 1950, 1979, Holzner 1978). A plant species rarely grows within the whole range of its physiological, ecological amplitude (Ellenberg 1979). The interaction of environmental factors and the effects of competition can only be studied by field observations. In this sense the approach is closely connected to the study of communities.

The regular association of plant species with certain environmental conditions is used as the basis for assigning indicator values. The most detailed autecological study is that by Ellenberg (1979). In this study of the vascular plants of Central Europe Ellenberg has summarized the autecological behaviour of about 2000 species, based on *ca.* 40 years of research by a number of different people, into so-called 'indicator values'. The behaviour of these plants with regard to three climatic factors (i.e. light, temperature, and continentality) and three edaphic factors (i.e. moisture, pH, nitrogen) is recorded on a scale of 1 to 9, while indifferent behaviour or unclear results are marked by the sign 'x'. In the case of moisture the scale was increased by three steps to 12 to allow for water plants (see Table 9.1).

In addition to studying the ecology of individual species, the autecological approach has also been used to study groups of species. In his 1950 study Ellenberg allocated plant species to 'ecological groups', using their indicator values for the edaphic factors (moisture, pH, nitrogen) and for temperature as criteria for allocation to a particular group.

In a recent study, autecological information was used to create similar 'ecological groups' for the Dutch flora (Runhaar *et al.* 1987). This division is based on biotic and abiotic characteristics such as vegetation structure, stage of succession, salinity, substratum, moisture regime, nutrient availability, acidity, and dynamics of the ecosystem (e.g. sand drift, trampling etc.). Each characteristic has been subdivided into several classes and combinations of these classes are used to define different habitat types (see Table 9.2). Examples are 'grassland on dry acid soil of low nutrient availability' or 'woodland on wet soil of high nutrient availability'. The assignment of plant species to ecological groups has been based on ecological literature (including Ellenberg) and has been tested by using *ca.* 20,000 relevés of Dutch vegetation. In contrast to Ellenberg's study, the plant species in this classification system have been assigned to as many ecological groups as is necessary to explain two thirds of the occurrence of the plant species in The Netherlands (Runhaar *et al.* 1987).

9.3 Application to archaeobotany

9.3.1 The phytosociological approach

The application of the phytosociological approach to the interpretation of archaeobotanical data sets is common on the continent, but rare in Britain. While most authors are aware of the fact that archaeobotanical samples cannot be equated with phytosociological 'Aufnahmen' or 'relevés', it is generally assumed that a particular Alliance or Order occurs in the archaeobotanical data set, if a reasonable number of species of that particular syntaxon is present in the samples. The method of application varies considerably from publication to publication, with some studies recognizing the fact that species can occur in more than one syntaxon (e.g. Behre 1986b, Greig 1988b, Jacomet 1980, Lundstrom-Baudais 1984, Pals 1984, Van Zeist 1974, 1988, Van Zeist and Palfenier-Vegter 1979 and 1981, Van Zeist *et al.* 1986 and 1987), while others tabulate the species under the syntaxon to which it shows the highest fidelity or to which they think it is most likely to belong (e.g. Behre 1981a, 1986a, Jacomet *et al.* 1988, Jaquat 1986, Knörzer 1973b, 1984a, 1984b, 1987, Lynch and Paap 1982, Paap 1984, Wasylikowa 1981, Willerding 1983a, 1983b, 1984). In some cases the species are tabulated under broader categories such as vegetation units, or a combination of vegetation units and phytosociological syntaxa (Knörzer 1976, Küster 1985a), or categories decribed as 'ecological groups' (Bakels 1981, Jacomet 1987b, Lange 1988), although these are, strictly speaking, neither ecological groups *sensu* Ellenberg 1950 (i.e. they are not autecological groups), nor syntaxa *sensu* Braun-Blanquet 1932, but are based on a mixture of ecological and sociological criteria (Arnolds and Van der Maarel 1987).

Once the presence of certain syntaxa at a particular site has been established, conclusions are drawn regarding the environment and crop husbandry practices of the past. For example, Van Zeist (1974) was able to determine for a series of settlements in the coastal region of The Netherlands whether they were situated in a brackish or fresh-water environment. Both Behre (1975) and Van Zeist (1988, and Van Zeist *et al.* 1987) have used the relative proportions of weeds from the Chenopodietea and Secalietea to analyse the occurrence of autumn and spring-sown crops in the coastal areas of northern Germany and The Netherlands. The areas in which the arable fields of settlements in this region are thought to have been located are prone to flooding, especially during high tides in winter. The dominance of Chenopodietea species was taken to confirm the likelihood that the crops were cultivated as summer crops (Behre 1975), while the presence of some Secalietea species at Leeuwarden was interpreted as evidence for the import of crops, e.g. from the sandy soils to the south east of the settlement (Van Zeist 1988, Van Zeist *et al.* 1987). Willerding (1981) has pointed out that the predominance of Chenopodietea species (i.e. nitrogen-loving species) in the early prehistoric assemblages suggests that the fields were not lacking in nitrogen, which has implications for the commonly held view that soil exhaustion was a recurrent problem in the past and formed the reason behind shifting cultivation.

Consistent dissimilarities between the species composition in archaeobotanical samples and those in present-day weed communities led Knörzer (1971a) to suggest that a new weed Association could be created, i.e. the Bromo-Lapsanetum praehistoricum, typical for arable fields of the early Neolithic Bandceramic and Rössen cultures, but nowadays extinct. In imitation of Knörzer, Pals (1984) has postulated the existence of an Althaeo-Atriplicetum praehistoricum Association in the northern part of Holland during the Neolithic. Lange (1988) took a slightly different approach. In an application of Correspondance Analysis to his archaeobotanical data set he found that while most species from the same 'ecological group' clustered together in the ordination, some species were lying far away from their 'ecological groups'. He suggested that these outliers should be interpreted as species which in the past had a different ecological distribution to that of today.

Charred seed assemblages often contain species which today are not classified as weeds of arable habitats, but, instead, occur in grassland or wetland habitats. This pattern has already been discussed in Chapter 6 above. In the literature we can find two opposing views on how to interpret the presence of these species. On the one hand their presence is taken to mean that the archaeological samples contain

refuse of a mixture of different activities (e.g. Bakels and Van der Ham 1981, Lange 1988, Knörzer 1973a, 1984a, Küster 1985a, Van Zeist 1981, Van Zeist and Palfenier-Vegter 1979). While some authors simply accept the presence of mixed materials (Bakels and Van der Ham 1981, Küster 1985a, Van Zeist 1981, Van Zeist and Palfenier-Vegter 1979), others try to explain how such mixtures could have occurred. Knörzer (1973a, 1984a) has suggested that the many grassland plants in his assemblages might have originated from grass used as tinder or collected for bedding. He rules out the possibility of hay being represented, as most of the grass seeds were fully ripe. Lange (1988) suggests that the large number of perennial plants in his samples may originate from the burning of the ruderal vegetation around the settlement after the abandonment of the farm buildings, or, alternatively, may derive from hay or animal dung.

However, the opposite view is that charred seed assemblages are made up primarily of species from one single plant community, that of arable fields, and that the presence of grassland and wetland species points to the fact that past arable weed communities were very different from those of today (Hillman 1981a, M. Jones 1984b, 1988a, 1988b, Knörzer 1971a, Küster 1988, Pals 1984, Van der Veen 1987a). The presence of *Eleocharis palustris* in Iron Age assemblages from central-southern England and of *Sieglingia decumbens* in late Iron Age assemblages from Wales and north-east England has been interpreted as evidence for the fact that the fields were poor in drainage and fertility (M. Jones 1984b, 1988b, Van der Veen 1987a, see also Hinton 1991). The presence of grassland plants, such as *Plantago lanceolata*, *Prunella vulgaris*, *Stellaria gramineae*, *Knautia arvensis*, and *Sieglingia decumbens*, has been interpreted as pointing to different levels of soil disturbance compared to today (Hillman 1981a, Knörzer 1971a, Küster 1988, Pals 1987). There are, of course, exceptional circumstances in which plants from other habitats do enter the charred seed assemblage, such as through the use of animal dung as fuel in areas where firewood is scarce (Bottema 1984, Miller and Smart 1984), but in western Europe this phenomenon is unlikely to have occurred on any scale.

As has already been discussed in Chapter 6 above, the writer supports the latter view and regards the majority of wild plants present in the samples as having derived from arable fields. Though central to the interpretation of charred plant assemblages, this issue is rarely discussed in detail in the literature. The fact that many authors regard charred seed assemblages as the product of several activities rather than of the cereal harvest alone, may, in fact, have more to do with their deeply rooted grounding in phytosociology and their consequent difficulty in regarding a present-day grassland species as anything else but a grassland species, than with their conscious rejection of the above statement.

9.3.2 The autecological approach

Until recently, the autecological approach was used less often in the archaeobotanical literature than the phytosociological approach. The first two publications in which Ellenberg's indicator values were used are those by Wasylikowa (1978) and Willerding (1978). Both studies use the autecological information alongside a phytosociological classification. In fact, Wasylikowa (1978) uses Ellenberg's indicator values to gain information about the ecological conditions represented by the phytosociological syntaxa.

Willerding (1978) coined the term 'eco-diagram' and was the first to use these diagrams: bar-graphs showing the number of species in an archaeobotanical assemblage for each indicator value of each of Ellenberg's three edaphic and climatic factors. He has demonstrated how these eco-diagrams can facilitate an inter-site comparison of environmental conditions (Willerding 1978, 1980). These eco-diagrams have subsequently been used by Behre (1986b), Jacomet (1987b), Van der Veen (1987a), Wasylikowa (1981), and Van Zeist (1981, *et al*. 1986). The way eco-diagrams have been drawn up does, however, vary considerably from author to author. Willerding (1978, 1980), Van Zeist (1981, *et al*. 1986), and Van der Veen (1987a) use all herbaceous species present in the assemblage. Behre (1986b) only uses the species tabulated under the syntaxa Chenopodietalia albi and Secalietea. Jacomet (1987b) also only uses the segetal and ruderal species. Küster (1985a) has drawn up tables with indicator values for arable weeds, ruderals and grassland plants separately. Consequently, though the eco-diagram was meant to facilitate inter-site comparison, in practice this is not always the case.

Ecological groups (sensu Ellenberg 1950), rather than Ellenberg's indicator values, have, as far as the writer is aware, been used in only two cases (G. Jones 1983a and Wasylikowa 1981).

In most cases the autecological approach is used alongside the phytosociological one. In only three cases, to the writer's knowledge, has the autecological approach been explicitly preferred to the phytosociological one: Van Zeist 1981 (although this author has continued to use phytosociology in subsequent publications), Küster 1985a, and M. Jones 1984b and 1988a.

9.4 Problems of applying modern ecological data to past weed communities

9.4.1 Introduction

There is a growing awareness of the problems associated with the application of modern ecological information to archaeobotanical data. Many authors mention the fact that changes through time have occurred and that the vegetation, and especially the weed flora, will not have been static. More than any other plant community, an anthropogenic vegetation such as that of arable fields has been modified by the activities of man (digging, ploughing, manuring, weeding, sowing, etc.) (e.g. Helbaek 1977, M. Jones 1988a, 1988b, Knörzer 1971a, 1987, Van Zeist and Neef 1983, Van Zeist *et al*. 1986, Willerding 1988). While some authors have consequently decided not to classify past vegetation types below the highest syntaxonomic units (Behre 1986b, Van Zeist and Palfenier-Vegter 1979), others have started to place more emphasis on autecological information, e.g. the use of Ellenberg's indicator values and ecological groups (e.g. G. Jones 1983a, Küster 1985a, Van Zeist 1981, Willerding 1978 and 1980). In Britain the application of the Braun-Blanquet approach to certain seed assemblages (Greig 1988a, 1988b, M. Jones 1984b, 1988a, Lambrick and Robinson 1988, Robinson 1989) has, ironically, been introduced at a time when on the continent this approach is receiving a critical reassessment (Küster 1989, Willerding 1988).

There are a number of reasons why the application of modern ecological models to archaeobotanical data is problematic. Here these reasons have been ordered under four different factors, although several of these factors are interrelated.

9.4.2 Climatic factors

The existence of regional variability in the diagnostic value of character species was recognized by the Braun-Blanquet Approach from the beginning (Westhoff and Van der Maarel 1973). While the ecological amplitude of a species usually shows great uniformity within regions of uniform climate and geology, most plant species, in fact, occur in larger, climatically and geologically more heterogeneous areas, and vary accordingly in their ecology (Westhoff and Van der Maarel 1973).

Holzner (1978) has identified a number of climatic gradients along which plant communities, especially weed communities, vary in floristic composition.

One group of segetal species, Holzner's type 'A' weeds (species of the Order Secalietalia, following the classification by Westhof and Den Held 1975), are thermophilous weeds of southern origin, which occur as weeds in winter crops. As one moves south and east the number of species characteristic of this group increases, but towards the northern and western limit of its range the species diversity gradually decreases (Holzner 1978), mainly as the result of declining summer temperatures. There are, however, no marked boundaries to this distribution, and the transition is a smooth one. Also, towards the northern and western limit of their range these species show a typical preference for calcareous soil (Caucalion Alliance), while in their optimal climatic region they are indifferent to this soil factor (Holzner 1978, Westhoff and Van der Maarel 1973).

Another group of segetal species, Holzner's type 'B' species (species of the Order Aperetalia), show a similar gradient, not from warm to cool, but from an oceanic to a continental climate. They have an oceanic distribution centre and are acidophytes. In their optimal range they can compete with other species on rich or even neutral soils, but in continental areas they lose their competitive edge and become restricted to acid and poor soils (Holzner 1978). Again, this gradient is a gradual one.

Holzner (1978) also distinguishes two groups of species in the weeds of ruderal and arable habitats (Order of Polygono-Chenopodietalia). Type 'a' weeds are thermophilous species requiring high summer temperatures (Alliances Eragrostidion and Panico-Setarion). In their optimal climatic areas they occur in arable as well as ruderal habitats, but the cooler the climate, the more they are restricted to ruderal sites only. In these habitats the micro-climate is warmer, nutrients and lime are more widely available, and there is less competition. Type 'b' species have a distribution centre in subatlantic or even atlantic areas. In their optimal climatic regions they occur as arable and ruderal weeds, but in more continental climates they cannot compete well with other species in these habitats and they are forced into very shady sites, gardens or even forests (Holzner 1978).

Silverside (1977) identified a climatic factor behind the poor separation between weed communities of the Chenopodietea and Secalietea in Britain, which he only recognizes at the lower syntaxonomic unit of Order (Polygono-Chenopodietalia and Centauretalia cyani). He has suggested that the maritime influence on the climate and consequently the smaller contrast between the seasons, has extended the period during which weed species may germinate, and this has resulted in a blurring of the division between the two syntaxa in Britain. Like Holzner, Silverside (1977) recognized that because many Centauretalia species are

thermophilous the Order is poorly represented in northern Europe, and shows a southern and eastern distribution in Britain.

These local differences in the behaviour of weed species, caused by climatic variations across Europe, mean that one cannot rely on one single phytosociological classification, and this has given rise to a proliferation of associations described from a purely local point of view (Holzner 1978, Westhoff and Van der Maarel 1973). This, of course, defeats the original aim of the phytosociological approach of providing an objective, uniform framework of plant communities. The climatic variation across Europe also means that one needs to use local autecological studies, to avoid the introduction of climatic variables.

9.4.3 Edaphic factors

Ellenberg (1950) has already pointed out that certain soil conditions favour certain classes of weeds. The relative proportions of summer annuals (e.g. Chenopodietea) and winter annuals (e.g. Secalietea) vary with local soil conditions. Bannink *et al.* (1974) has stressed that it is difficult for winter annuals to germinate in wet soils (such soils are too cold), so that on these soils weeds of the Secalietea are rare. Thus, the scarcity of Secalietea species in the coastal areas of The Netherlands and Germany (Behre 1975, Van Zeist 1988, Van Zeist *et al.* 1987) may be related as much to the predominance of wet, cold soils in the region, as to the practice of spring-sowing (Pals 1987). The richness of the soil, i.e. the availability of nitrogen, appears to be another major factor, if not the dominant factor, in determining the occurrence of the cereal and garden crop communities. On rich soils a strong divergence between these two communities does not normally occur (the winter annuals (Secalietea) are generally under-represented here, as they cannot compete with the summer annuals (Chenopodietea)) (Bannink *et al.* 1974).

9.4.4 Biotic factors

We have already discussed above how the crop can influence the weed flora, both by the type and amount of shade cast by its leaves, and by the rhythm of its development and the consequent agricultural operations (Ellenberg 1950). Modern studies of weed communities are based on weeds growing among crops that were not grown in the past. The phytosociological classification of weeds into the Chenopodietea and Secalietea is largely based on observations in fields of winter rye, summer oat, and beet grown in Central Europe during the first half of this century. Certain winter annuals (Secalietea) are regarded as particularly characteristic of winter rye.

The continuous cultivation of rye in parts of Central Europe has created a selection process in favour of these species (Ellenberg 1950). Winter rye is, however, not really characteristic of winter sown cereals. It develops more quickly than the other cereals, casts more shade than the others earlier on in the growing season, and, at least at the time when these observations were made (the beginning of this century) it was a crop not normally harrowed and hoed in spring. As a result, winter rye only tolerates species which do not require high temperatures for germination, and which can take advantage of the short period in early spring when there is not yet much shade cast by the rye plants (Ellenberg 1950). Of the winter cereals rye is by far the greatest enemy of the summer annuals or Chenopodietea (Ellenberg 1950).

Rye and oat did not, in fact, become important crops in western Europe until the first millennium AD, and even then they were of regional importance only (Green 1981, Knörzer 1984a, Willerding 1979, Van Zeist 1968). The application of a classification based so strongly on species which were not cultivated during the prehistoric period must by definition be problematic, and in many areas misleading. The contrast between the weed communities of crops like emmer, spelt and barley may have been much less than (or at least very different from) that between winter rye and summer oat. In fact, Ellenberg (1950) already pointed out that rye and oat both represented extremes, with winter wheat and summer wheat taking more intermediate positions.

Furthermore, while Ellenberg (1950) recognized that the difference between the two weed communities was the product of a number of different factors (see section 9.2.1 above), in the archaeobotanical literature there has been a tendency to equate Chenopodietea simply with spring sowing and Secalietea with autumn sowing, ignoring the other variables underlying the divergence.

9.4.5 Anthropogenic factors

The influence of human actions is regarded as one of the most important ecological factors in the formation of weed communities, and, consequently, their composition has been subject to strong alterations through time, a process still going on today (Holzner 1978). The number of arable weeds present in the archaeobotanical record has been found to increase rapidly through time (Knörzer 1987), and there is an enormous regional and temporal variation in the occurrence of certain weed species (Küster 1985b, Helbaek 1977), which is due to anthropogenic factors (introduction of 'exotic' weeds with grain imports) as well as climatic factors. The range of agricultural operations to which the fields are

subjected (digging, ploughing, sowing, weeding, manuring, etc.) all influence the composition of the weed flora. Intensive manuring is likely to favour nitrogen-loving species (Chenopodietea) (Bannink *et al.* 1974, Pals 1987, Warington 1924), intensive harrowing, hoeing and weeding also favours summer annuals (Chenopodietea) (Ellenberg 1950, Wasylikowa 1981), while the absence of soil disturbance or crop rotation with fallow tends to encourage the growth of grasses and other perennial weeds (Behre 1981b, Knörzer 1971a, Küster 1985a, Wasylikowa 1981). The introduction of the mouldboard plough during the medieval period is thought to have resulted in the disappearance of many perennial weed species from the arable fields (Behre 1981b, Hillman 1981a), through the destruction of their roots or rhizomes, or through inversion and burial. Furthermore, mouldboard ploughing has encouraged certain other perennial weeds, i.e. those that can regrow from small pieces of their fragmented rhizomes, e.g. *Agropyron repens*, *Ranunculus repens*, *Convolvulus arvensis*, and *Potentilla reptans* (Hillman 1981a). Consequently, twentieth century weed communities can be very different in their composition from those in the past.

9.5 Choice of approach

9.5.1 Critique of the phytosociological approach

In attempting to provide an objective, uniform framework for present-day plant communities, the phytosociological approach largely ignores the two most fundamental principles underlying the discipline of archaeobotany, i.e. the influences of man and of time, making it an analytical technique ill-suited to a historical discipline like archaeobotany. The application of the phytosociological approach to archaeobotanical data sets implies a uniformitarian view of past weed communities, which, in the light of the overwhelming evidence for changes in the composition of weed communities through time and space, is highly inappropriate. While a phytosociological application to archaeobotanical data could be used to highlight differences between past and present communities (e.g. Knörzer 1971a, Lange 1988, Pals 1984), it is, in fact, rarely used in this manner. More often than not the approach is applied in a search for similarities, rather than dissimilarities.

There are two further reasons why the phytosociological approach is unsuitable for archaeobotanical data. First of all, a methodological point. It is important to realize that the individual archaeobotanical samples or even archaeobotanical assemblages cannot be directly compared or equated with phytosociological 'relevés' or 'Aufnahmen' (M. Jones 1988a). The content of a sample or an assemblage may derive from more than one field and, therefore, does not represent one vegetation unit. Nor does the sample or assemblage contain all the species that were once present in the field. Some species would not have been in seed at the time of harvest, others will have been removed during crop processing. Thus, neither the presence, nor the abundance of the species in the original community can be assesssed in the same way as can be done with modern communities. Rare exceptions to this rule may occur when actual vegetation units are recovered, such as the fossil turves found at the coastal settlements of northern Germany (Körber-Grohne 1967, Willerding 1988). Consequently, the presence (or absence) of certain syntaxa cannot be determined using the standard phytosociological methodology. The creation of past plant 'associations' in the sense of Braun-Blanquet, as put forward by Knörzer (1971a: Bromo-Lapsanetum praehistoricum) and Pals (1984: Althaeo-Atriplicetum praehistoricum) should, therefore, be avoided, as it is not based on sound methodological principles. The recognition of such communities is, of course, important, but they should be identified as consistent patterning in the archaeobotanical record, rather than as phytosociological communities growing in ancient arable fields.

Finally, phytosociology studies the occurrence of plant communities, while in archaeobotany we are primarily concerned with the relationship between plant species and the activities of man, rather than with plant communities as such, their relationships with one another, or the relationship between one species and another (G. Jones 1983a). For all these reasons it is felt that the phytosociological approach is not appropriate for archaeobotanical data sets.

9.5.2 Advantages of the autecological approach

The autecological approach is concerned with the behaviour of plant species with regard to a range of environmental factors. As far as an application to archaeobotany is concerned, this approach has a number of advantages. First of all, it is exactly these environmental factors that we, in archaeobotany, are interested in. Secondly, by using indicator values we can use all species present in the archaeobotanical assemblage (provided that they are identified to species level), rather than just those species which are character species or differential species of a particular syntaxon (which can reduce the number of species to be used to a mere handful). There are no specific methodological problems associated with the application of the autecological approach, as long as

we recognize that we are dealing with patterning in the archaeobotanical record rather than with actual weed communities growing in the field. However, the general problems associated with applying modern ecological analogues to the past do, of course, remain, and these are addressed below.

As far as climatic variation in species behaviour is concerned, this problem can be overcome by using local studies. Unfortunately, to date, few detailed autecological studies are available. Ellenberg's work on the indicator values of the vascular plants of Central Europe remains the classic work (Ellenberg 1979). No such studies exist specifically for the British flora, but the '*Atlas of the Wild Flowers of Britain and Northern Europe*' (Fitter 1978) does contain autecological information regarding the factors of wetness, acidity, fertility and shade. Under these four headings the plant species have been classified using a 5-point scale (see Table 9.3). The atlas, however, leaves out the grasses, sedges and rushes. Detailed ecological studies of certain plant species of the British flora are also available in the *Journal of Ecology* under the section of the *Biological Flora of the British Isles*. Publications relevant to the present study are those by Cavers and Harper 1964 (*Rumex crispus* and *R. obtusifolius*), Grieg-Smith 1948 (*Urtica urens*), Kay 1971 (*Anthemis cotula* and *A. arvensis*), Harper 1957 (*Ranunculus* subgenus *Ranunculus*), Hutchinson and Seymour 1982 (*Poa annua*), McNaughton and Harper 1964 (*Papaver* spp.), New 1961 (*Spergula arvensis*), Pfitzenmeyer 1962 (*Arrhenatherum elatius*), Sagar and Harper 1964 (*Plantago* spp.), Simmonds 1945 (*Polygonum lapathifolium* and *P. persicaria*), Sobey 1981 (*Stellaria media*), Walters 1949 (*Eleocharis palustris*), Welch 1966 (*Juncus squarrosus*), and Williams 1963 (*Chenopodium album*).

Regarding the edaphic factors, the evidence suggests that there has been not so much a change in the preference for certain edaphic factors (moisture, pH, nitrogen), but a change in the part of the physiological amplitude that is occupied due to competition by other species (Ellenberg 1979, Willerding 1983b, 1988). Ellenberg (1979) stresses that the indicator values (especially low values) should not be regarded as depicting actual soil requirements of a species, but as reflecting the soil conditions which can be tolerated. For example, *Bromus erectus* does not prefer very dry conditions, but tolerates them well compared to other species. Another example is *Luzula luzuloides* which is only found on soils with a pH range of 3.5 to 5.5, even though in laboratory experiments it was found to grow successfully on soils with a pH range of 3.5 to 8.5, with an optimal productivity at about pH 6.5. In other words, *Luzula* behaves as an acidity indicator, but is not acidophilous, it does not 'like' very acid soils (Ellenberg 1979).

The influence of the crop plants on the composition of the weed flora (biotic factors) is a more difficult factor to assess. No information is available for ancient crop plants such as emmer and spelt wheat and experimental work is urgently required. The influence exerted by the crops themselves is not, however, restricted to the amount of shade cast by their leaves (Ellenberg 1950). Most of the influence exerted by the crop plants is connected with the type of cultivation regime required by the crop (spring or autumn sowing, manuring, weeding etc.) (Ellenberg 1950), which are aspects we are specifically interested in studying. The influence of the cultivation regime is here dealt with under the heading of anthropogenic factors. The biotic factor does, of course, include the influence of man as well as plants and animals, but, as this study deals specifically with arable weed communities, I have chosen to elevate the influence of human actions to a separate environmental factor.

The fact that human activities in the form of agricultural operations such as digging, ploughing, sowing, weeding, manuring, etc. have an influence on the composition of the weed flora is beneficial to archaeobotany, as this means that we can study human activities and past crop husbandry practices by analysing the weed assemblages in the archaeobotanical samples. Autecological studies such as those on the influence of manuring on the weed flora of arable land (Warington 1924), the influence of tillage on the weed flora (Pollard and Cussans 1976), or the effect of seed-bed preparation on the weed flora (R. Jones 1966) provide crucial information towards a recognition and understanding of these human factors in the past.

It is acknowledged that the possibility of temporal change in the ecological behaviour of plant species does exist, but the autecological approach largely uses elements of the plant's behaviour which are genetically determined and which, consequently, are less likely to change, or, at least, which change much less rapidly than elements such as co-occurrence in the field (M. Jones, pers. comm.). Furthermore, it is felt that this problem can largely be overcome by looking at the behaviour of several species simultaneously, i.e. by analysing the behaviour of all species in the archaeobotanical assemblage together, rather than by looking at individual species, as it is unlikely that all species will have changed in the same direction (G. Jones, forthcoming). It is, in fact, commonly accepted that the use of single species as indicators is unsatisfactory, as the amplitude of a species for any particular factor is usually far too wide (Ellenberg 1950). Only when a number of species all with similar requirements are found together, can we make certain inferences regarding the environmental (including anthropogenic) conditions (Ellenberg 1950, G. Jones 1983a, Westhoff and Van der Maarel 1973).

Thus, as long as we recognize that we study patterning in the archaeobotanical record rather than actual ancient weed communities, analyse groups of species rather than individual species, use local autecological studies, and treat the indicator values of species as levels of behaviour and tolerance rather than actual physiological requirements, the autecological approach can be successfully applied to archaeobotanical studies. In the writer's opinion this approach is to be preferred above the phytosociological approach, and in the next chapter the archaeobotanical data set will be analysed using the autecological approach.

10. Multivariate Analysis of the Weed Assemblages

In this chapter the composition of the assemblages will be analysed in more detail, using the autecological approach. It was already established in Chapter 5 that there were no differences in the chronology of the prehistoric assemblages, with the exception of Hallshill, which is earlier in date than the other sites. In Chapter 7 it was established that there were no taphonomic differences between the prehistoric assemblages, all representing the same crop processing group, i.e. fine sieve by-products, but it was demonstrated that the Roman assemblages were different (i.e. representing largely fine sieve products). These will, therefore, be analysed separately.

First of all, the prehistoric assemblages will be analysed using a number of multivariate techniques, while differences detected in the assemblages will be interpreted using autecological information in the following section. The final section will consider the Roman assemblages, and compare these with the results of the prehistoric ones.

10.1 The prehistoric assemblages

10.1.1 Multivariate analysis

Gauch describes multivariate analysis as 'the branch of mathematics that deals with the examination of numerous variables simultaneously' (Gauch 1984,1). Classical statistical analyses deal with the relationships between one or two variables, and are concerned with the testing of hypotheses, most commonly in the form of calculating the probability of a null hypothesis being true (Gauch 1984, quoting Williams and Gillard 1971). Archaeobotanical data consist of a two-way matrix of samples (= cases) and species (= variables), and archaeobotanical studies are typically concerned with the analysis of the relationships between all variables simultaneously. In contrast to classical statistics, multivariate analyses do not begin with any hypotheses, but are, instead, exploratory techniques, eliciting from the data some internal structure from which, at a later stage, hypotheses may be generated (Gauch 1984).

Multivariate analyses are applicable in situations in which the data set consists of a two-way matrix and has a size of at least 10 x 10 or 15 x 15 (Gauch 1984). The data set analysed here consists of a matrix of samples by species which, even after the exclusion of certain samples and species (see Chapter 3) still consists of 156 samples by 40 (or 44) species, which means 6240 (or 6864) entries. As a raw data set this information is too bulky and too complex to be

assimilated directly by the human mind. Multivariate analyses summarize the data, either by representing the samples and species relationships in a low-dimensional space (ordination), or by classifying the samples and species into clusters (classification) (Gauch 1984).

Thus, the purpose of multivariate analysis is to summarize the data and reveal its structure (Gauch 1984). Its performance is usually assessed by the extent to which it provides interpretable results, and the extent to which it is consistent in providing similar results despite differing techniques.

Here three multivariate techniques have been applied to the data (Principal Components Analysis, Cluster Analysis, and Discriminant Analysis), in order to analyse the structure of the data set. The analyses have been carried out after standardization and transformation of the data as described in Chapter 3, and using SPSSx version 3.1 (Norusis 1985).

10.1.2 Principal components analysis (PCA)

Principal Components Analysis is an ordination technique, that is, it arranges the species and samples in a low-dimensional space such that similar entities are close together and dissimilar entities are far apart (Gauch 1984). Linear combinations of the observed variables are formed and the first principal component (or axis) represents the combination that accounts for the largest amount of variance in the samples. The second principal axis accounts for the next largest amount of variance, etc. The variance explained by each component is expressed by its 'eigenvalue', which in turn can be expressed as a percentage of the total variance. It is difficult to determine how many axes need to be considered in the analysis. Frequently, the first three eigenvalues may account for 40 – 90 per cent of the total variance. In some cases, however, the first two axes account for as little as five per cent of the total variance, but are quite informative, while in other cases 90 per cent of the variance may be accounted for, but the results may be meaningless or distorted (Gauch 1984). In general, the assessment of the PCA results are based on interpretability. The actual percentage of variance accounted for has not been found to be a reliable indicator of the quality of the results (Gauch 1984). The output of a principal components analysis consists of factor loadings for each variable on each axis. Variables with high loadings on any one axis contribute strongly to that axis. Variables with low loadings contribute little to the variation. Variables with high opposite loadings (i.e. positive and negative) are located at opposite ends of the axis.

Four PCA analyses were carried out:

Analysis	Variables	Transformation
PCA–1	grain/chaff/weeds (weeds>10%)	square root
PCA–2	grain/chaff/weeds (weeds> 5%)	square root
PCA–3	grain/chaff/weeds (weeds>10%)	octave scale
PCA–4	grain/chaff/weeds (weeds> 5%)	octave scale

The results of the analyses are summarized in Table 10.1, listing the eigenvalues and percentages of variance for the first three axes, and the five highest positive and negative loadings defining the orientation of the first axis.

There is no appreciable difference between the outcome of the four analyses. In each case the first axis separates samples with emmer wheat (both grains and glume bases) plus *Chenopodium album*, *Polygonum lapathifolium/ persicaria* and either small grasses, *Atriplex* spp., or *Spergula arvensis* from samples with spelt wheat (glume bases) and *Sieglingia decumbens*, *Montia fontana*, *Bromus mollis/secalinus*, and *Arrhenatherum elatius*. As there is so little difference between the four analyses, the complete information for the first three axes is only given for analysis PCA–1. To display the information provided by the factor loadings, bar graphs have been constructed (Figures 10.1, 10.2, and 10.3). The results indicate that the main variation in the samples on the first axis concerns a division between samples with emmer wheat and spelt wheat (Figure 10.1). These two crops appear to be associated with different weed species, i.e. *Chenopodium album*, *Polygonum lapathifolium/ persicaria*, and small grasses with emmer wheat, and *Sieglingia decumbens*, *Montia fontana*, and *Bromus mollis/secalinus* with spelt wheat. The second axis (Figure 10.2) suggests that the weeds *Vicia/Lathyrus*, *Galium aparine*, *Tripleurospermum inodorum*, and *Stellaria media* are strongly associated, but tend not to occur together with emmer wheat or *Polygonum lapathifolium/persicaria*. The third axis (Figure 10.3) suggests that barley (both grains and rachis internodes) is associated with *Carex pilulifera*, *Potentilla cf. erecta* and *Rumex acetosella*, and tends not to occur together with wheat (either emmer or spelt).

10.1.3 Cluster analysis (CA)

In order to test whether the data contain any other underlying differences a classification technique, Cluster Analysis, was used. Cluster analysis is an agglomerative hierarchical technique, which groups samples into bigger and bigger clusters until all

samples are members of a single cluster. Samples which cluster early on in the analysis, i.e. below fusion coefficient 15, are very similar, those which cluster late on in the analysis, i.e. above fusion coefficient 15, are quite dissimilar. The method chosen for combining the clusters was Ward's method, which means that at each step the two clusters that merge are those which result in the smallest increase in the overall sum of squared within-cluster distances (Norusis 1985).

Initially four cluster analyses were carried out:

Analysis	Variables	Transformation
CA–1	grain/chaff/weeds (weeds>10%)	square root
CA–2	grain/chaff/weeds (weeds> 5%)	square root
CA–3	grain/chaff/weeds (weeds>10%)	octave scale
CA–4	grain/chaff/weeds (weeds> 5%)	octave scale

As the results of the four analyses were very similar (see Table 10.2), only the dendrogram of Analysis CA–1 is illustrated here (Figure 10.4). Figure 10.4 shows that samples containing emmer form a separate cluster from those not containing emmer wheat. The two clusters are only joined right at the top of the dendrogram, at fusion coefficient 25, although they were formed at much lower fusion coefficients. One sample from a site which did not contain emmer wheat (Thorpe Thewles PF1) is grouped with the 'emmer samples'. It contains neither emmer nor spelt wheat, but does contain a large number of barley rachis internodes, a feature it has in common with some of the samples which contain emmer wheat. When the octave scale transformation is used there is also one sample that is grouped 'wrongly', i.e. is not grouped within its original group (Murton 630).

Thus, these cluster analyses confirm the pattern observed by the principal components analyses, i.e. that the main variation in the samples lies in the presence or absence of emmer wheat. To test whether it is solely the presence/absence of emmer wheat, or whether the difference is also reflected in the associated weed species, as the principal components analyses suggested, the cluster analyses were repeated, but this time only using the weed species as the variables, that is, excluding the cereal grain and chaff from the analysis. For ease of reference the sites containing emmer wheat are from now on referred to as Group A, and those without emmer as Group B. The following cluster analyses were run:

Analysis	Variables	Transformation
CA–5	weeds (>10%)	square root
CA–6	weeds (> 5%)	square root
CA–7	weeds (>10%)	octave scale
CA–8	weeds (> 5%)	octave scale

Principal Components Analysis - Prehistoric Assemblages

PCA - 1

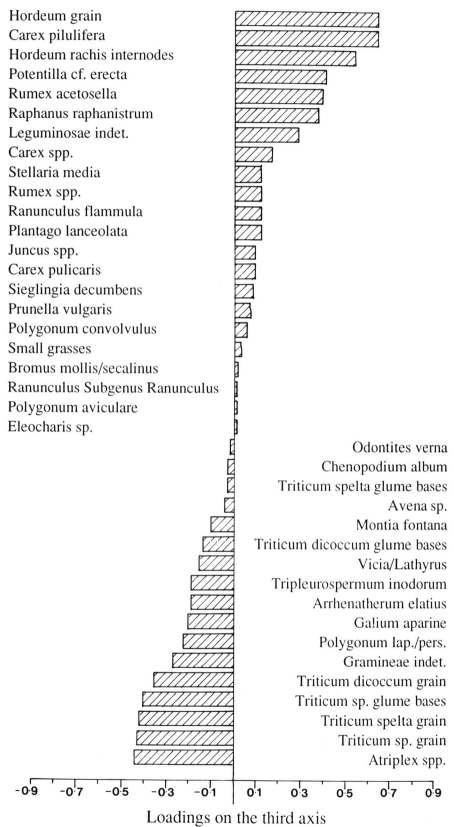

Hordeum grain
Carex pilulifera
Hordeum rachis internodes
Potentilla cf. erecta
Rumex acetosella
Raphanus raphanistrum
Leguminosae indet.
Carex spp.
Stellaria media
Rumex spp.
Ranunculus flammula
Plantago lanceolata
Juncus spp.
Carex pulicaris
Sieglingia decumbens
Prunella vulgaris
Polygonum convolvulus
Small grasses
Bromus mollis/secalinus
Ranunculus Subgenus Ranunculus
Polygonum aviculare
Eleocharis sp.

Odontites verna
Chenopodium album
Triticum spelta glume bases
Avena sp.
Montia fontana
Triticum dicoccum glume bases
Vicia/Lathyrus
Tripleurospermum inodorum
Arrhenatherum elatius
Galium aparine
Polygonum lap./pers.
Gramineae indet.
Triticum dicoccum grain
Triticum sp. glume bases
Triticum spelta grain
Triticum sp. grain
Atriplex spp.

-0·9   -0·7   -0·5   -0·3   -0·1   0·1   0·3   0·5   0·7   0·9

Loadings on the third axis

Figure 10.3   Principal components analysis of the prehistoric assemblages using both cereals and weed species. Factor loadings on the third axis.

The results are summarized in Table 10.3. Again, the results are very similar, so that only the results of analysis CA–5 are illustrated (Figure 10.4). The samples from Group A and Group B form separate clusters and only join together at the top of the dendrogram at fusion coefficient 25. Only two samples from Group A (Murton 630 and Chester House 117) are grouped with Group B. In analysis CA–7 one sample from Group B is also classified 'wrongly' (Rock Castle 47). The results indicate that there are major differences between the samples of Group A and Group B, differences which do not solely consist of the presence or absence of emmer wheat, but are also expressed in the weed assemblages associated with the crop plants.

10.1.4 Discriminant analysis (DA)

As the data could be divided into two separate groups, it was decided to run a discriminant analysis, to explore in more detail which variables are causing most of the variation between the groups. Discriminant Analysis (also known as Canonical Variates Analysis) is a type of ordination technique which can be used when the samples are known to belong to *a priori* groups (Digby and Kempton 1987). It seeks linear combinations of the variables that have the greatest between-group variation relative to their within-group variation. For an analysis between two groups the value of Wilk's Lambda is the proportion of the total variance in the discriminant scores not explained by differences between the groups. A Lambda of 1 occurs when the mean of the discriminant scores is the same in both groups and there is no between-group variability. Thus, the closer Wilk's Lambda is to 0, the better the discrimination between the two groups. The percentage of correctly reclassified samples is another measure of the success of the discrimination. The value of the discriminant scores (= pooled within group correlations) for each variable provides an indication of the importance of that variable in the separation of the two groups.

Initially four analyses were carried out:

Analysis	Variables	Transformation
DA–1	grain/chaff/weeds (weeds>10%)	square root
DA–2	grain/chaff/weeds (weeds> 5%)	square root
DA–3	grain/chaff/weeds (weeds>10%)	octave scale
DA–4	grain/chaff/weeds (weeds> 5%)	octave scale

The results of the four analyses were, again, very similar to one another (Table 10.4); only the results of analysis DA–1 are presented in detail (Figures 10.5 and 10.6). The value of Wilk's Lambda lies close to 0 (0.04), which means that most of the

variation in the data is explained by the differences between the two groups. The percentage of correctly reclassified samples is 100% (Figure 10.5). Again, the difference between the two groups is identified as being between emmer and spelt (Figure 10.6).

The analyses were repeated, excluding both cereal grain and chaff from the analysis:

Analysis	Variables	Transformation
DA–5	weeds (>10%)	square root
DA–6	weeds (> 5%)	square root
DA–7	weeds (>10%)	octave scale
DA–8	weeds (> 5%)	octave scale

The results are given in Table 10.5.; only the results of Analysis DA–5 are presented in detail (Figures 10.7 and 10.8). The value of Wilk's Lambda is still very low (0.10) and the percentage of correctly reclassified samples remains 100% (Figure 10.7). The variables causing most of the variation are (Figure 10.8): *Polygonum lapathifolium/persicaria* and *Chenopodium album* (associated with Group A), and *Sieglingia decumbens* and *Montia fontana* (associated with Group B).

10.1.5 Conclusion

The results of the multivariate analyses indicate that the prehistoric data set divides into two separate subsets of data, Group A samples characterized by the presence of emmer wheat and weed species such as *Chenopodium album*, *Polygonum lapathifolium/persicaria*, *Stellaria media*, *Atriplex* sp., and small grasses (including *Poa annua*). The Group B samples are characterized by the presence of spelt wheat associated with weeds such as *Sieglingia decumbens*, *Montia fontana*, *Bromus mollis/secalinus*, and *Galium aparine*. In the next section the ecological differences between these different weed associations will be analysed, using the autecological approach as discussed in Chapter 9 above.

10.2 Autecological analysis of the weed assemblages

10.2.1 Introduction

In this section the ecological behaviour of the two weed assemblages will be analysed, looking at the behaviour of the species in relation to climatic, edaphic, biotic, and anthropogenic factors. In Chapter 9 the importance of using local autecological studies was emphasized. Unfortunately, there are, at present, no studies available which provide autecological information for all British species. Fitter (1978) contains information for the flowering plants, but leaves out the Gramineae and the Cyperaceae. Plants of these two families represent a large proportion (*ca.* 30 per cent) of the weed assemblages in the present

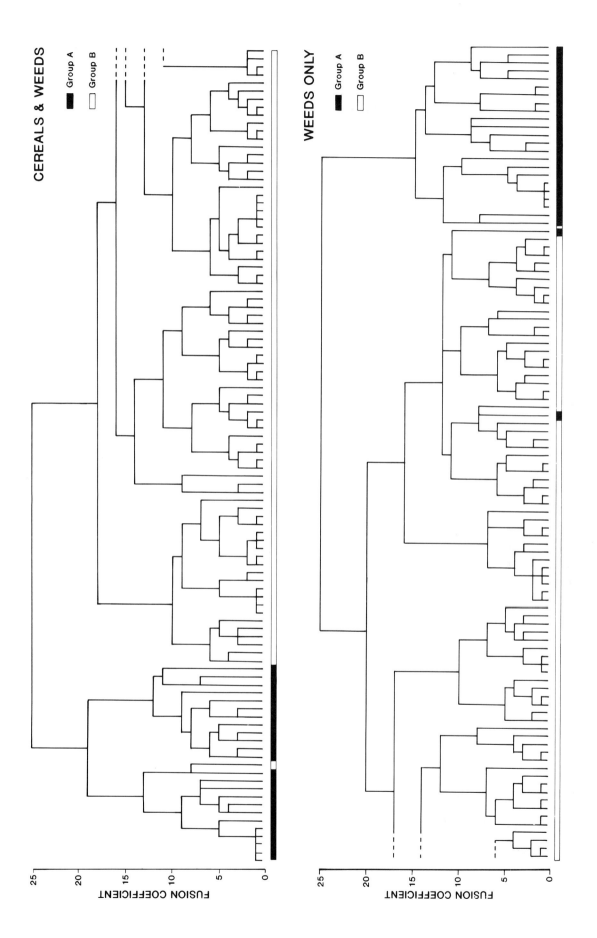

Figure 10.4 Cluster analysis of the prehistoric assemblages, using both cereals and weed species (top), and weed species only (below).

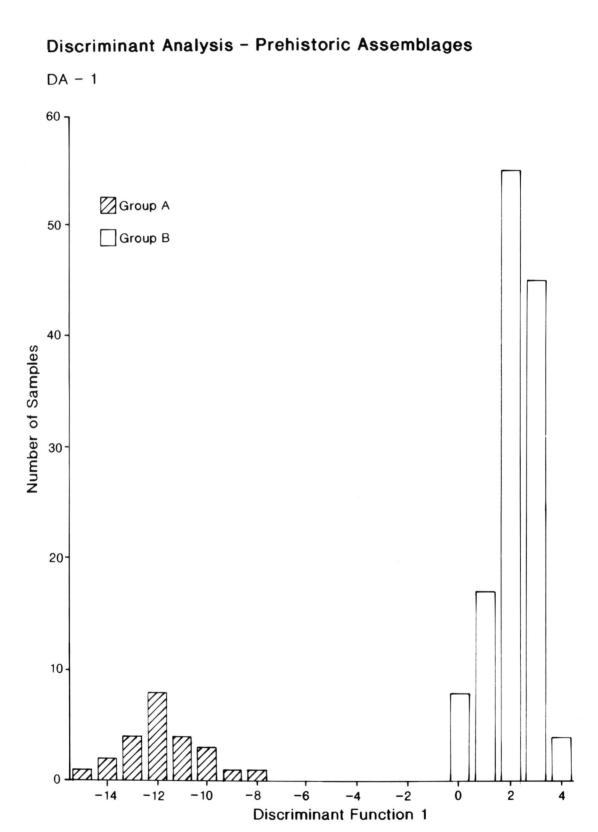

Figure 10.5   Discriminant analysis of the prehistoric assemblages, using both cereals and weed species.

Discriminant Analysis - Prehistoric Assemblages

DA - 1

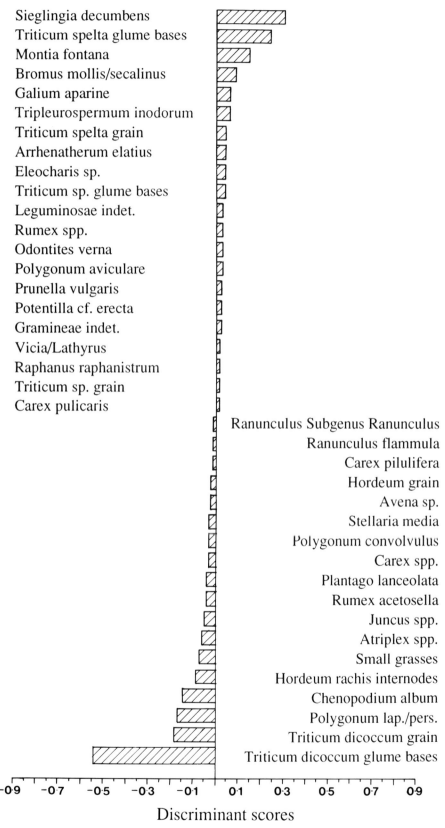

Discriminant scores

Figure 10.6   Discriminant analysis of the prehistoric assemblages, using both cereals and weed species. Group A is associated with negative, Group B with positive discriminant scores.

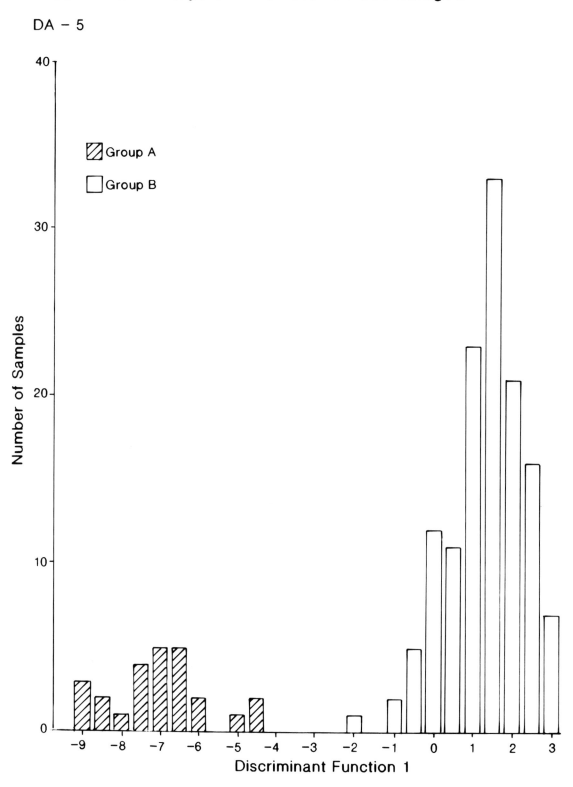

Figure 10.7   Discriminant analysis of the prehistoric assemblages, using weed species only.

Discriminant Analysis - Prehistoric Assemblages

DA - 5

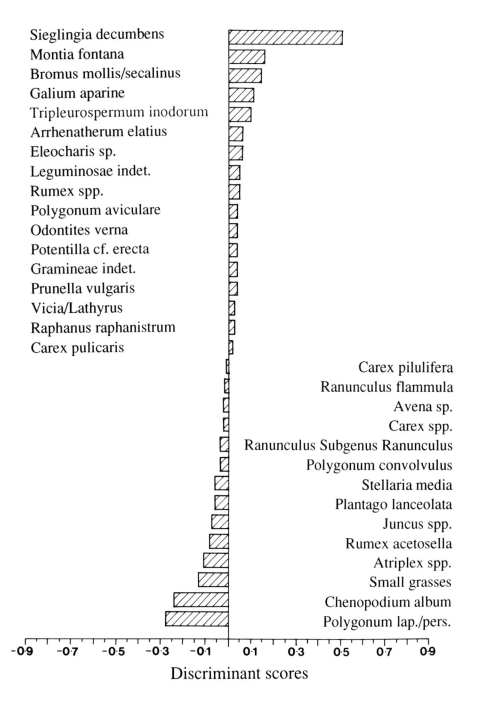

Figure 10.8   Discriminant analysis of the prehistoric assemblages, using weed species only. Group A is associated with negative, Group B with positive discriminant scores.

study. An analysis of the data which excluded these species would be misleading. *The Biological Flora of the British Isles* (published in the *Journal of Ecology* since 1941) does not yet cover all the weed species occurring in the present data set. The most complete autecological work to date is that by Ellenberg (1979), but this is based on observations made in Central Europe. The following procedure has been followed here: for each of the weed species occurring in the present data set for which the information was available, the information from Fitter (1978), *The Biological Flora of the British Isles*, and from Clapham *et al.* (1962) has been compared with that provided by Ellenberg (1979). While there are minor differences (Fitter often gives a slightly broader ecological amplitude to species than Ellenberg), there is very good agreement between the different studies regarding the tendency of the plants to occur in either dry, well-drained or wet/ acid, weakly acid or neutral/ and poor, intermediate or rich soils. This suggests that Ellenberg's data can be used for the present study. Furthermore, the autecological information provided by Runhaar *et al.* (1987), based on the Dutch flora which is more similar to the British flora than that from Central Europe, will also be used, to provide an additional control on the appropriateness of using Ellenberg's figures here.

As there was so little difference between the four analyses carried out each time in the previous section (using different numbers of weed species, and different transformations) in this section only the first analysis will be discussed each time. In all cases the abundance figures of the species are used.

10.2.2 Climatic factors

The reaction of the plant species to climatic factors such as light, temperature and continentality have been described by Ellenberg (1979) (see also Table 9.1). His indicator figure for 'light' refers to the occurrence of a species in relation to relative light intensity during summer time and ranges from L1 (full shadow plant) to L9 (full light plant). His 'temperature' figure refers to the distribution of plants according to latitudinal zones and altitudinal belts and ranges from T1 (cold climate, i.e. boreal, arctic or alpine) to T9 (very warm, mediterranean climate). The 'continentality' figure refers to the distribution of plants according to the degree of continentality of the general climate with special emphasis on minimum and maximum temperature. The continentality figure ranges from K1 (euoceanic, reaching Central Europe only in the extreme west) to K9 (eucontinental, scarcely reaching Central Europe) (Ellenberg 1979).

In order to analyse whether the difference between the two weed assemblages can be explained by differences in the behaviour of the plants with regard to these climatic factors, a discriminant analysis was

carried out, using the indicator values for each species and each climatic factor as the variables (see also Table 10.6), and using Groups A and B as the groups to be discriminated:

Analysis	Variables	Transformation
DA–9	L6, L7, L8, Lx T4, T5, T6, Tx K2, K3, K4, K5, K6, Kx (weeds>10%)	square root

The results of the analysis are given in Figures 10.9 and 10.10. The percentage of correctly reclassified samples is 100% (Figure 10.9) and Wilk's Lambda is 0.17, suggesting the separation of the two groups using these variables is good. Group A is characterized by species which are indifferent to light (Figure 10.10), while Group B is characterized by species which prefer a fair amount of light (L7 + L8). There is no difference in the temperature requirements of the two groups, half of all species being indifferent to this factor (Tx). The figures for continentality suggest that the Group B species are slightly more oceanic than those of Group A (Figure 10.10).

To test how robust the difference in climatic factors was for the two groups, a cluster analysis was run, using the same variables:

Analysis	Variables	Transformation
CA–9	L6, L7, L8, Lx T4, T5, T6, Tx K2, K3, K4, K5, K6, Kx (weeds>10%)	square root

The results of the cluster analysis are illustrated in Figure 10.11. Four samples (three from Group A: Murton 623 and 630, Chester House 117, and one from Group B: Rock Caste 69) were classified in the 'wrong' group. More importantly, the samples from Group A were clustered with some of those from Group B, before being joined to the remaining samples of Group B (Figure 10.11). Thus, using these climatic variables, the two groups are not as dissimilar as the discriminant analysis suggested. While there clearly are some differences between the two groups in their behaviour with regard to these climatic factors (mainly in relation to the amount of light required), these differences are not very great and do not represent the main cause of variation, indicating that there is no environmental gradient underlying the difference between the two groups.

10.2.3 Edaphic factors

The behaviour of plant species in relation to edaphic or soil factors has also been studied by Ellenberg (1979) (see also Table 9.1). He gave each plant an indicator value for F (= moisture; the occurrence in relation to soil moisture or water level), R (= reaction; occurrence in relation to soil acidity),

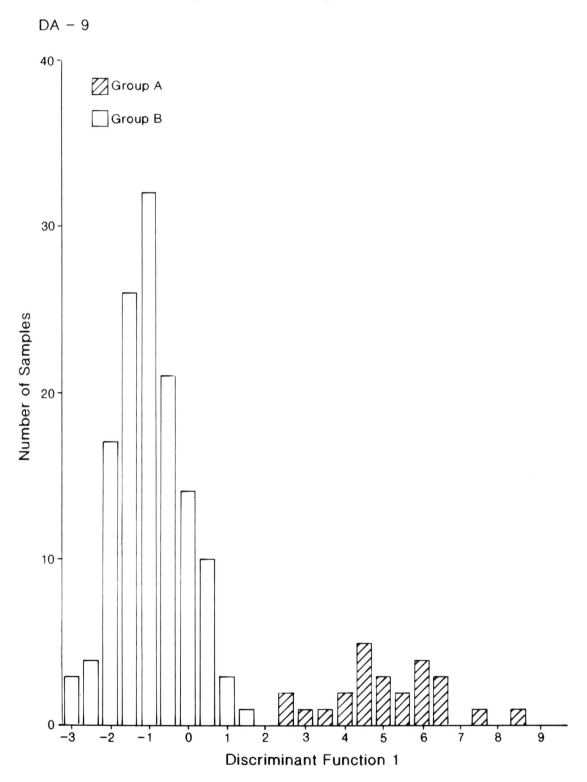

Figure 10.9   Discriminant analysis of the prehistoric assemblages, using Ellenberg's (1979) indicator values for climatic factors.

Discriminant Analysis - Ellenberg's Climatic Factors

DA - 9

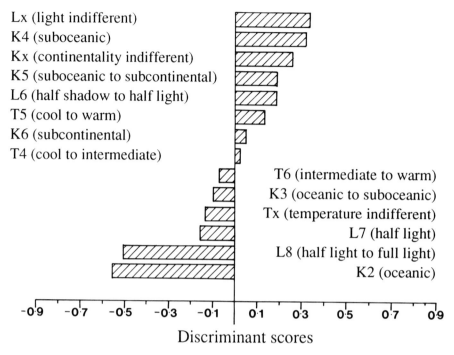

Discriminant scores

Figure 10.10   Discriminant analysis of the prehistoric assemblages, using Ellenberg's (1979) indicator values for climatic factors. Group A is associated with positive, Group B with negative discriminant scores.

and N (= nitrogen; occurrence in relation to ammonia and nitrate supply). The indicator values for each species for each of these edaphic factors have been used in a discriminant analysis (see also Table 10.7):

Analysis	Variables	Transformation
DA–10	F3, F4, F5, F6, F7, F8, F9, F10, Fx; R2, R3, R4, R6, R7, Rx N2, N4, N5, N6, N7, N8, Nx (weeds>10%)	square root

The results of the analysis are presented in Figure 10.12. The percentage of correctly reclassified samples is 100% and Wilk's Lambda is low (0.15). Group A is characterized by a high value for N7 (soils rich in mineral nitrogen), while Group B is characterized by high values for Fx (indifferent to soil moisture), R3 (acid soils), and N2 (soils poor in nitrogen).

In order to take account of the fact that Fitter (1978) gave several of the species a wider ecological amplitude than Ellenberg (see section 10.2.1 above), the analysis was repeated, grouping the indicator values into slightly broader categories (see also Table 10.8):

Analysis	Variables	Transformation
DA–11	F3+4, F5+6, F7+8, F9+10, Fx, R2+3, R4+5+6, R7+8, Rx, N2+3, N4+5+6, N7+8, Nx (weeds>10%)	square root

The results were very similar to the previous analysis and are illustrated in Figures 10.13 and 10.14. Again, the percentage of correctly reclassified samples is 100% (Figure 10.13), and Wilk's Lambda is low (0.16). Group A is characterized by very fertile, well-drained soils, while Group B is characterized by poor, rather acid soils (Figure 10.14).

To test the robustness of these results a cluster analysis was carried out:

Analysis	Variables	Transformation
CA–10	F3+4, F5+6, F7+8, F9+10, Fx, R2+3, R4+5+6, R7+8, Rx, N2+3, N4+5+6, N7+8, Nx (weeds>10%)	square root

The results of the cluster analysis (Figure 10.11) confirm the results of the discriminant analysis: the samples from Group A and B form two separate clusters which only join right at the top of the

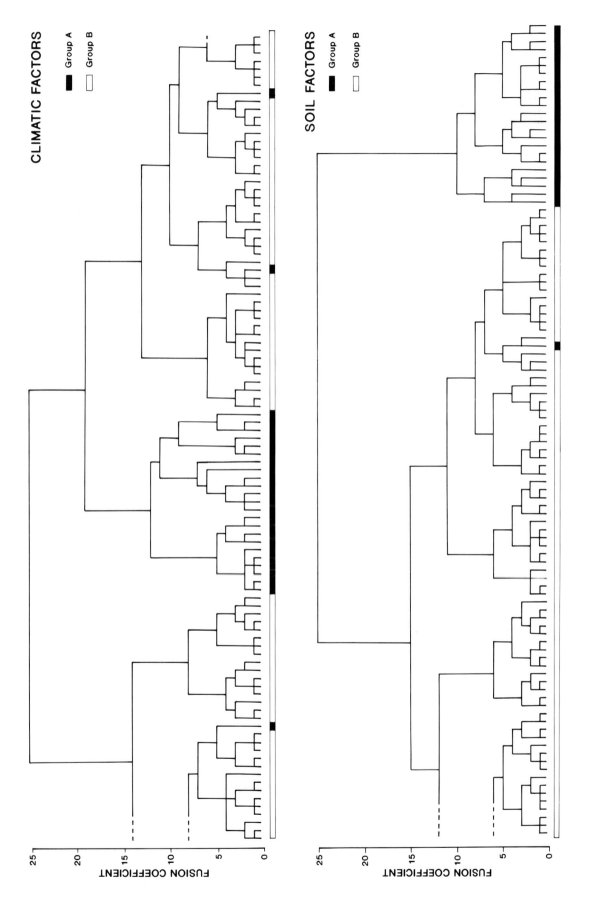

Figure 10.11   Cluster analysis of the prehistoric assemblages, using Ellenberg's (1979) climatic factors (top), and Ellenberg's (1979) edaphic factors (below).

Discriminant Analysis - Ellenberg's Edaphic Factors

DA - 10

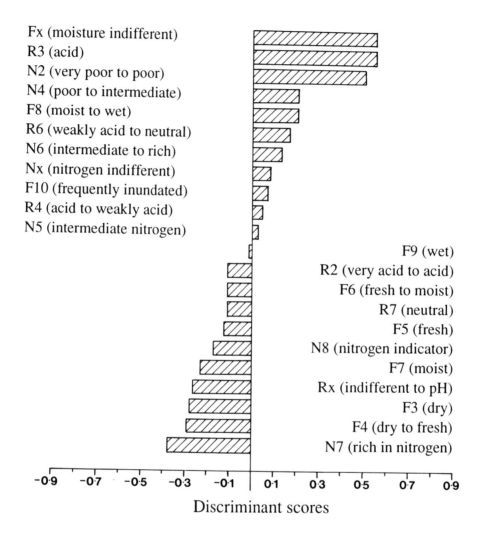

Fx (moisture indifferent)
R3 (acid)
N2 (very poor to poor)
N4 (poor to intermediate)
F8 (moist to wet)
R6 (weakly acid to neutral)
N6 (intermediate to rich)
Nx (nitrogen indifferent)
F10 (frequently inundated)
R4 (acid to weakly acid)
N5 (intermediate nitrogen)

F9 (wet)
R2 (very acid to acid)
F6 (fresh to moist)
R7 (neutral)
F5 (fresh)
N8 (nitrogen indicator)
F7 (moist)
Rx (indifferent to pH)
F3 (dry)
F4 (dry to fresh)
N7 (rich in nitrogen)

-0·9 -0·7 -0·5 -0·3 -0·1 0·1 0·3 0·5 0·7 0·9

Discriminant scores

Figure 10.12   Discriminant analysis of the prehistoric assemblages, using Ellenberg's (1979) indicator values for edaphic factors. Group A is associated with negative, Group B with positive discriminant scores.

dendrogram at fusion coefficient 25. Only two samples (from Group A) were classified 'wrongly' (Murton 630 and Chester House 117).

As Ellenberg's study is based on the flora of Central Europe, a similar discriminant analysis was carried out, using the data by Runhaar *et al.* (1987), which is based on a study of the Dutch flora, to test whether any differences in the results occur. Runhaar *et al.* (1987) gives information for each species

regarding a number of biotic and abiotic characteristics, such as vegetation structure, stage of succession, moisture regime, nutrient availability etc. (see also section 9.2.2 and Table 9.2). Here only the information regarding the edaphic factors is used (i.e. moisture regime, nutrient availability, and pH). The codes used in Runhaar *et al.* (1987) consist of two figures, one (the first figure) for the moisture regime (aquatic/wet/moist/dry), and one (the second figure)

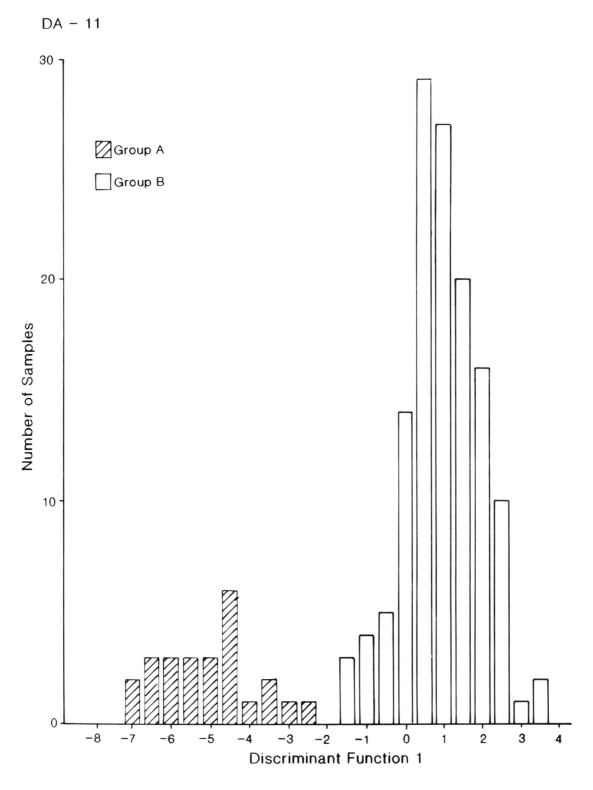

Figure 10.13   Discriminant analysis of the prehistoric assemblages, using Ellenberg's (1979) edaphic factors, having combined the indicator values into broader groups.

Discriminant Analysis - Ellenberg's Edaphic Factors

DA - 11

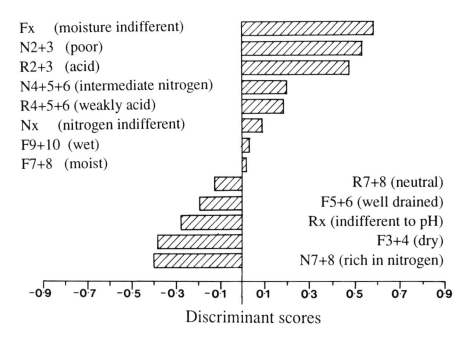

Figures on chart:

Fx (moisture indifferent)
N2+3 (poor)
R2+3 (acid)
N4+5+6 (intermediate nitrogen)
R4+5+6 (weakly acid)
Nx (nitrogen indifferent)
F9+10 (wet)
F7+8 (moist)

R7+8 (neutral)
F5+6 (well drained)
Rx (indifferent to pH)
F3+4 (dry)
N7+8 (rich in nitrogen)

-0.9 -0.7 -0.5 -0.3 -0.1 | 0.1 0.3 0.5 0.7 0.9

Discriminant scores

Figure 10.14   Discriminant analysis of the prehistoric assemblages, using Ellenberg's (1979) edaphic factors, having combined the indicator values into broader groups. Group A is associated with negative, Group B with positive discriminant scores.

for nutrient availability and acidity (low/moderate/high nutrient availability and acid/moderately acid/basic/ neutral). The information regarding nutrient availability and pH has been combined into one figure as these two factors are closely related. In contrast to Ellenberg (1979), but in agreement with Fitter (1978), an allowance is made for the ecological amplitude of a species, by classifying it into as many categories as is necessary to explain two thirds of the occurrences of the plant species in The Netherlands (Runhaar *et al.* 1987). The classification of the weed species according to this study is given in Table 10.9. A discriminant analysis was carried out using these categories as variables:

Analysis	Variables	Transformation
DA–12	edaphic categories according to Runhaar *et al.* 1987, see Table 10.9	square root

The results are very similar to those where Ellenberg's indicator values were used, see Figures 10.15 and 10.16. The percentage of correctly classified samples is 98.08%, i.e. two samples from

Group A (Murton 630 and Chester House 117) were classified as belonging to Group B (Figure 10.15), and Wilk's Lambda is 0.24, not as low as in the previous analyses but still giving a reasonably good separation. Although the discrimination between the two groups is slightly less strong, perhaps because broader edaphic categories were used, it identifies the same variable as causing the main source of variation, i.e. the nutrient and pH status of the soils. Group A is characterized by very fertile soils, while Group B is characterized by poor, rather acid soils. Again, differences in soil moisture do not appear to form a determinant factor.

To summarize, the difference in the weed assemblages between Groups A and B appears to be related to differences in the soil conditions, and especially the amount of nitrogen available in the soil, and, to a lesser extent, the pH level of the soils (the two are, of course, linked). Group A is characterized by weed species which demonstrate a great preference for well-drained, very fertile soil conditions, while Group B is characterized by species which tend to occur largely on poor, rather acid soils, or are indifferent to soil conditions.

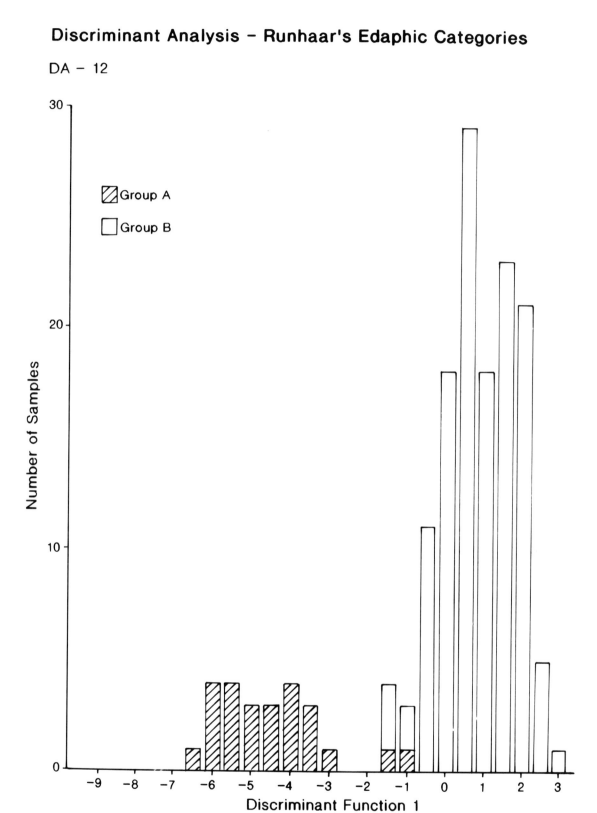

Figure 10.15   Discriminant analysis of the prehistoric assemblages, using Runhaar's (1987) edaphic categories.

10.2.4 Biotic factors

As we have seen in Chapter 9, it has often been suggested that the type of crop could influence the weed flora in the fields, either by the amount and type of shade cast by its leaves, or by its growth requirements and consequent cultivation regime. The crops present in Group A are emmer wheat and six-row barley (with some spelt wheat), those in Group B spelt wheat and six-row barley (no emmer wheat). As the multivariate analysis has identified the presence or absence of emmer wheat versus spelt wheat as the main source of variation, as far as the crop plants are concerned, the possible differences between these two crops are explored below.

10.2.4.1 Emmer and spelt wheat: stand height and shade

Very little information is available regarding either emmer or spelt wheat. As these species have not been grown in Europe since the early medieval period, with the exception of a few isolated regions, and parts of southern Germany where spelt is still grown today (Körber-Grohne 1987), no detailed studies of the ecology of these crops are available. The particular varieties grown at the experimental farm at Butser, Hampshire, southern England, show little difference in stand heights, though spelt wheat is generally some 10 cm taller than emmer (Reynolds, 1987, 1988, 1989). In order to assess the variation in performance levels between emmer and spelt wheat on different soils and in different climatic zones of Britain, a small-scale experiment was initiated in 1987 by the writer, growing both crops (the same forms as at Butser) at 22 plots across the country (Van der Veen 1989a, 1989b). The results from the first year indicate that spelt wheat is, in general, *ca.* 15 cm taller than emmer (Van der Veen 1989b). Care must be taken with this type of information, however, as there are very many different 'varieties' of emmer and spelt, which may vary in their growth habits.

No information is available regarding the type of shade cast by the leaves. In the present writer's experience, the leaves of spelt are slightly larger and broader than those of emmer, but emmer plants tiller more, so that there may be little difference, but this needs further investigation.

10.2.4.2 Emmer and spelt wheat: autumn versus spring sowing

It was suggested by Percival (1921) that the majority of emmer varieties are not very frost-resistant and are, therefore, spring-sown, while spelt wheat is one of the hardiest cereals, and is usually autumn-sown. This suggestion has become generally accepted in the literature (e.g. M. Jones 1981, Gregg 1988), but has met with strong criticism from Hillman (1981a). Hillman pointed out that all cereals were initially likely to have been autumn-sown as their wild ancestors were species germinating in the autumn (Hillman 1981a), a point also made by Willerding (1988). It is not known when spring-sown varieties were first developed. Hillman stresses that farmers in general only grow spring crops if they are forced to, that is either when weather conditions are so severe that autumn-sown crops would not survive the winter, when the soil is subject to winter waterlogging, or when time limitations in the autumn require a spreading of the burden of ploughing and sowing over two seasons. In the latter case the crop sown in spring is either a minor crop or a crop well suited to spring sowing (e.g. pulses) (Hillman 1981a). The reason that Percival suggested that emmer was a spring-sown crop stems, according to Hillman, from the fact that by the beginning of the twentieth century, when Percival did his research, emmer had ceased to be an important crop, and was consequently rare and only grown as a minor, spring-sown crop (Hillman 1981a).

The advantage of autumn-sown crops over spring-sown ones is that of higher yields. The development of the cereal ear is largely controlled by day-length: the temperate cereals (wheat, barley, rye, oat) are long-day plants which change from the vegetative to the flowering stage when a certain day-length is reached in spring. A change in the time of sowing, therefore, does not greatly alter the time of flowering or harvest. Late sowing will tend to give a low yield at about the normal harvest time, owing to a reduction in the total photosynthesis, rather than a normal yield at a later date (Gill and Vear 1980). High yielding spring cereals are a rather modern phenomenon (Hillman 1981a).

There is no *a priori* reason why crops have to be spring-sown in the north east of England. The climate of the region is characterized by a cold, dry spring and cool summers (Shirlaw 1966), not by a severe winter. With the exception of Hallshill (at 230 m O.D.) all the sites are located in a lowland position, on the coastal plain (below 200 m O.D.), where the weather is modified by the sea. In this region the risk of frost damage is less great than in parts of Central Europe.

As mentioned in Chapter 9, several authors have tried to assess the presence of autumn and spring sowing by looking at the proportions of Secalietea (mostly winter annuals) versus Chenopodietea (mostly summer annuals) in an assemblage. The disadvantage of using phytosociological Classes (or Orders) has already been discussed in Chapter 9, but the figures are presented here to illustrate this point (see opposite page).

Phytosociological Classes or Orders

| | Ellenberg (1979) | | Silverside (1977) | |
	Chenopodietea	Secalietea	Polygono-Chenopodietalia	Centauretalia
Hallshill	32.3%	6.1%	16.0%	8.2%)
Murton	33.9%	1.4%	5.6%	1.1%) A
Dod Law	56.9%	3.1%	3.5%	1.5%)
Chester House	4.7%	9.4%	3.2%	6.3%)
Thorpe Thewles	6.7%	12.9%	0.8%	1.8%)
Stanwick	5.6%	18.8%	1.0%	0.4%) B
Rock Castle	17.4%	6.1%	0.2%	2.6%)

Preferred germination time

| | Annuals | | | Perennials |
	spring	*both*	*autumn*	
Hallshill	40.2%	29.9%	-	19.6%)
Murton	28.7%	31.4%	-	32.3%) A
Dod Law	59.2%	21.2%	-	16.8%)
Chester House	12.6%	14.0%	-	60.2%)
Thorpe Thewles	6.8%	19.7%	-	50.0%)
Stanwick	5.3%	30.2%	-	55.4%) B
Rock Castle	11.4%	31.4%	-	39.2%)

Using Ellenberg's classification the Chenopodietea are clearly much better represented in Group A than in Group B, but when we use Silverside's classification, the contrast is considerably less marked. In neither group are the Secalietea (Centauretalia) well represented. In both cases, but especially so when using Silverside's classification, the proportion of Chenopodietea (Polygono-Chenopodietalia) and Secalietea (Centauretalia) together only represents a small part of the overall weed assemblage, pointing to the fact that the results may not be representative (i.e. do not explain the variation in the assemblages).

As mentioned in Chapter 9 above, Silverside (1977) found that, in contrast to Central Europe, these two syntaxa did not segregate according to sowing time in Britain. Most weed species show two peaks of germination, one in the autumn, and one more substantial one in the spring. In Britain the maritime influence on the climate extends the period over which weed species may germinate and populations may contain substantial ranges (Silverside 1977). Consequently, there is a blurring of the boundaries between these two Orders, with many species occuring in both, and very few species being restricted to only one of them. Silverside (1977) has, in fact, suggested that the difference between the two Orders might, at least in Britain, have more to do with competition for nitrogen than with the time of sowing, a suggestion also put forward by G. Jones (1983, forthcoming) and M. Jones (1988). Thus, the low proportions for the Secalietea (Centauretalia) in Group A relative to

Group B may be related to the fact that the soils in Group A were more fertile. In fertile soils species of the Secalietea (Centauretalia) cannot compete well with species of the Chenopodietea (Polygono-Chenopodietalia) (Bannink *et al.* 1974). Note the slightly anomalous position of Chester House in Group A, a point I will come back to in Chapter 11.

The scarcity of species of the Secalietea (Centauretalia) in both groups may be related to the fact that many species of this Class (Order) are thermophilous and are, consequently, poorly represented in northern Europe, and show a southern and eastern distribution in Britain (Holzner 1978, Silverside 1977; see also section 9.4.2 above). This highlights the problem of using these phytosociological syntaxa in interpreting agricultural practices such as spring and autumn sowing.

Furthermore, the results suggest that these two Classes or Orders of weed species do not really explain the nature of the weed assemblages in total. To analyse this aspect further, a discriminant analysis was run, using Ellenberg's phytosociological Classes as variables (see Table 10.10):

Analysis	Variables	Transformation
DA–13	phytosociological Orders according to Ellenberg(1979) see Table 10.10	square root

The results (Figure 10.17) indicate that while the Chenopodietea are more closely associated with Group A, the main variation between the two groups is caused by different representations of the Nardo-Callunetea (Group B) and Bidentetea (Group A),

Discriminant Analysis - Runhaar's Edaphic Categories

DA - 12

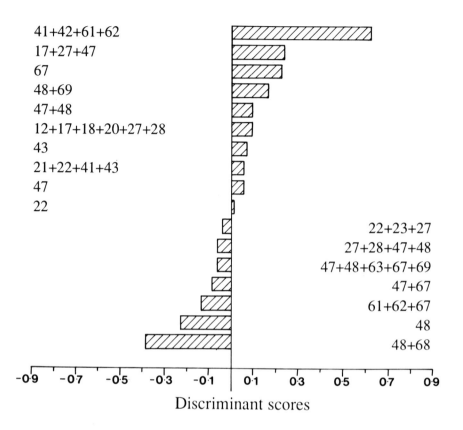

First Figure = Moisture Regime

1 = aquatic
2 = wet
4 = moist
6 = dry

Second Figure = Nutrient Availability and pH

1 = low nutrient availability, acid
2 = low nutrient availability, moderately acid to neutral
3 = low nutrient availability, basic
4 = low nutrient availability
7 = moderate nutrient availability
8 = high nutrient availability
9 = moderate to high nutrient availability

Figure 10.16   Discriminant analysis of the prehistoric assemblages, using Runhaar's (1987) edaphic categories. Group A is associated with negative, Group B with positive discriminant scores.

which emphasizes the fact that these archaeological weed communities are quite different from modern ones. Past weed communities cannot be characterized satisfactorily by the present-day segetal and ruderal communities. As discussed in Chapters 6 and 9 above, the writer believes that the regular occurrence of weed species from phytosociological Classes other than the Secalietea and Chenopodietea should be taken as pointing to the fact that past weed communities did not closely resemble modern ones, rather than to the fact that carbonized seed

assemblages consist of residues of several different activities. The application of the phytosociological approach to the present data set has been used here to highlight the difference between past and present weed communities.

Using the autecological approach, the question of spring and autumn sowing can be analysed by looking at the preferred germination time of the individual weed species. Unfortunately, neither Clapham *et al.* (1962), nor Fitter (1978) give information regarding germination time. Here the

Discriminant Analysis - Ellenberg's Phytosociological Classes

DA - 13

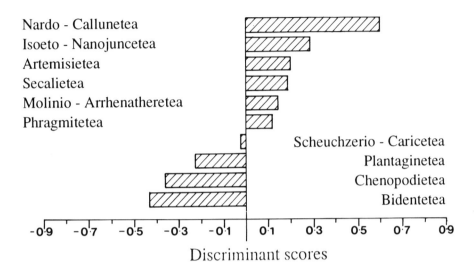

Figure 10.17   Discriminant analysis of the prehistoric assemblages, using Ellenberg's (1979) phytosociological Classes. Group A is associated with negative, Group B with positive discriminant scores.

data provided by the Geigi Weed Tables (Häfliger and Brun-Hool 1968–1977) have been used (see page 130 and also Table 10.11).

These figures indicate that Group A is characterized by a fairly high proportion of species which germinate in spring and a slightly lower proportion of species which can germinate in both spring and autumn (note the anomalous position of Chester House, which I will come back to in Chapter 11). Group B is characterized by much lower proportions of species germinating in spring, but the proportions of species germinating in both spring and autumn are similar to those in Group A. Neither group contains species which germinate exclusively in the autumn.

These figures must, however, be treated with great care. First of all, the information regarding the preferred germination time should relate to local studies, as Silverside (1977) has suggested that, in Britain, the period of germination might be influenced by climatic conditions. Insufficient local information was available, however. Secondly, there is some conflicting evidence in the literature regarding the germination time, which may perhaps be related to local variations in climate. For example,

both *Polygonum convolvulus* and *Avena fatua* are listed as germinating in spring in the *Geigi Weed Tables*, but Brenchley and Warington (1930) give both spring and autumn for *Polygonum convolvulus*, and Harper (1977) gives both spring and autumn for *Avena fatua*. Furthermore, *Polygonum convolvulus* and *Avena fatua* are listed by Ellenberg (1979) as belonging to the Secalietea, and *Avena fatua* is listed by Silverside (1977) as belonging to the Centauretalia. The information regarding *Galium aparine* is more complex. M. Jones (1981) has characterized this species as an indicator of autumn sowing, as it germinates in the autumn. Reynolds, however, describes the germination characteristics of this plant as 'showing a major peak at the end of March/beginning of April with a minor peak in late October' (Reynolds 1981, 112). He does, however, go on to say that the plant does not survive soil cultivation in spring associated with the practice of spring sowing. In autumn-sown fields considerable numbers of the plants manage to escape this destruction. Thus, while *Galium aparine* may represent the practice of autumn sowing, this is not due to the fact that the plant germinates exclusively in the autumn, but due to its inability to survive cultivation in spring.

A third reason for treating the figures regarding the preferred germination time with great care, is that of dependant variables. The Group A assemblages are characterized by nitrogen-loving species (section 10.2.3 above). These are mainly annual species and mainly species which germinate in spring. Thus, the high proportion of spring germinating species in Group A may be a function of the fact that Group A contains a high proportion of nitrogen-loving annuals, something already referred to above. These factors are, in fact, all closely related, an aspect discussed below, in section 10.2.6.

Thus, the data provided by the preferred time of germination of the weed species associated with Groups A and B do not provide clear-cut information regarding the practice of spring or autumn sowing. While neither Group A nor Group B contains any species which germinate exclusively in the autumn, suggesting an absence of the practice of autumn sowing, the presence of *Galium aparine* in one Group A site and all Group B sites, does suggest that autumn sowing was practised. The much greater proportion of species germinating in spring in the Group A assemblages may point to the practice of spring sowing at the Group A sites, but these high proportions may equally be a function of the soil nutrient status, see section 10.2.6 below.

10.2.4.3 Barley

Six-row barley was present in both groups. As we saw in Chapter 7, several samples, more in Group A than in Group B, contained some early processing waste of barley, in the form of barley rachis internodes, as well as fine-sieving residues. The presence of these rachis internodes does not form a very important source of variation in the samples, however. In the principal components analysis (PCA–1, Figures 10.1, 10.2, and 10.3) the rachis internodes did not score high on the first axis, and scored low on the second axis. On the third axis (Figure 10.3) both grains and rachis internodes of barley scored high positive, while wheat grains scored negatively, suggesting that the remains of wheat and barley came in as separate crops. In order to analyse this aspect more closely for each of the two groups, principal component analyses were carried out on the samples from Groups A and B separately:

Analysis	Variables	Transformation
PCA–5	grain/chaff/weeds (weeds>10%) samples Group A	square root
PCA–6	grain/chaff/weeds (weeds>10%) samples Group B	square root

The results are presented in Figures 10.18 and 10.19. Barley rachis internodes and grains have high positive loadings on the first axis of the analysis for Group A (Figure 10.18), while emmer grains and glume bases have high negative loadings, which suggests that the two crops were grown separately. In the principal component analysis of Group B samples, a similar pattern occurred, but here the separation between wheat and barley was registered on the second axis (Figure 10.19), the first axis separating weeds of different soil conditions. In both groups barley is associated with weeds typical of rather poor soil conditions (e.g. *Carex pilulifera*, *Rumex acetosella*, *Potentilla* cf. *erecta*, Leguminosae indet.). This probably means that barley was grown as a separate crop, and not as a mixture with wheat. It also indicates that the husbandry practices for the two crops differed, especially in Group A, where barley is associated with weed species indicative of poor soil conditions and perennials, suggestive of little soil disturbance.

10.2.4.4 Summary

To summarize, there is too little information available regarding the three crops present in the samples, emmer wheat, spelt wheat and six-row barley, to allow a detailed analysis of the influence of the crops on the composition of the weed flora in the fields. There is some evidence that emmer wheat is associated with weeds characteristic of spring-sown crops, but the evidence is not conclusive, and may be caused by other factors, such as the availability of nitrogen. There is little evidence for the practice of autumn sowing, with the possible exception of the presence of *Galium aparine* in Group B and one site of Group A. The evidence does, however, suggest that wheat and barley were grown as separate crops. Overall, the information available is insufficient to allow definitive conclusions to be drawn.

10.2.5 Anthropogenic factors

The influence of human actions is regarded as one of the most important ecological factors in the formation of weed communities (Chapter 9), and two particular actions are considered here, i.e. the harvesting method, which influences which weeds are found in the assemblage, and the amount of soil disturbance (depth of cultivation), which influences the weed composition in the field. Other actions have already been considered indirectly above, e.g. manuring under edaphic factors and time of sowing under biotic factors.

10.2.5.1 Harvesting methods

Cereal ears can be harvested by plucking the ears, by reaping (cutting) the ears and the straw separately, and by reaping the ears and straw simultaneously

Principal Components Analysis - Group A Assemblages

PCA - 5

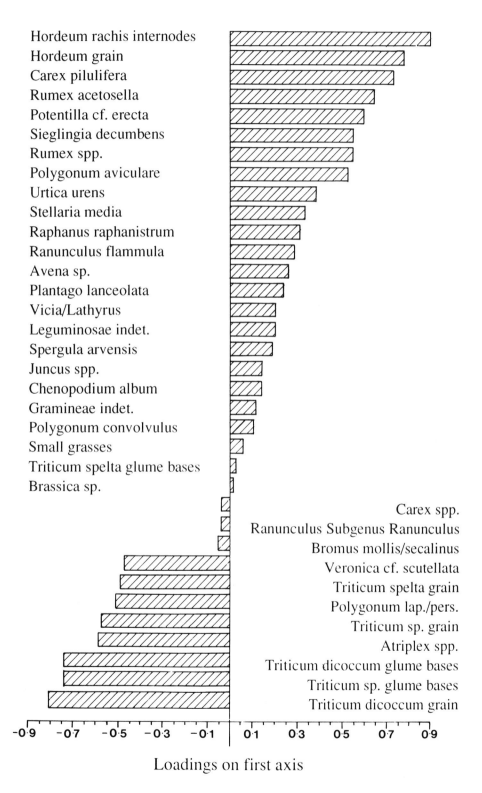

Loadings on first axis

Figure 10.18   Principal components analysis of the Group A assemblages only, using both cereals and weed species. Factor loadings on the first axis.

Principal Components Analysis - Group B Assemblages

PCA - 6

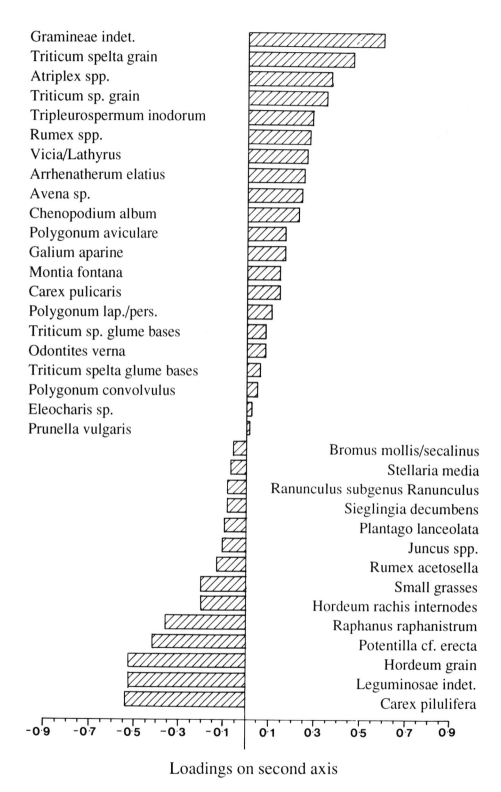

Loadings on second axis

Figure 10.19   Principal components analysis of the Group B assemblages only, using both cereals and weed species. Factor loadings on the second axis.

(Hillman 1981a). Reynolds (1981) has suggested that the plucking (hand picking) of the ears is much easier than cutting them with a sickle. The presence of the basal rachis nodes of both wheat and barley in the samples, however, indicates that handpicking was not practised by the inhabitants of the settlements studied here (Hillman 1981). The separate cutting of ears and straw does not appear to have been practised either, as the samples contain a mixture of weed seeds and glume bases. If the ears were cut separately, the fine-sieve residues would contain very few weed seeds (Hillman 1981a). The sample composition suggests that the harvesting method used was that of cutting the straw and ear together.

10.2.5.2 Harvesting height

When the straw and ears are harvested simultaneously, the straw can be cut close to the ground, or half-way up the stem. The presence of low growing weed species (such as *Rumex acetosella* and *Plantago lanceolata*) points to low reaping, the presence of tall growing species (such as *Avena fatua*, *Rumex* spp., *Chenopodium album*) combined with an absence of low growing species points to reaping high on the straw.

The weed species present in the samples have been classified into nine categories, according to their maximum flowering height (following Clapham *et al.* 1962), and these have been used as the variables in a discriminant analysis (see Table 10.12):

Analysis	Variables	Transformation
DA–14	H1 (1–30 cm),	square root
	H2 (31–40 cm),	
	H3 (41–50 cm),	
	H4 (51–60 cm),	
	H5 (71–80 cm),	
	H6 (81–90 cm),	
	H7 (91–100cm),	
	H8 (101–120cm),	
	H9 (121–200cm)	
	(weeds>10%)	

The results are illustrated in Figure 10.20. Wilk's Lambda for this analysis is 0.19, and the percentage of correctly classified samples is 98.72%, with two samples from Group A (Hallshill 23C and Chester House 117) classified 'wrongly', i.e. classified with Group B. The results cannot be interpreted in terms of differences in harvesting height. Both groups contain low growing and tall growing species, suggesting that the cereals were harvested low down in both groups.

These results were compared with a cluster analysis using the same variables:

Analysis	Variables	Transformation
CA–11	H1 (1–30 cm),	square root
	H2 (31–40 cm),	
	H3 (41–50 cm),	
	H4 (51–60 cm),	
	H5 (71–80 cm),	

H6 (81–90 cm),
H7 (91–100 cm),
H8 (101–120cm),
H9 (121–200cm)
(weeds>10%)

In the cluster analysis (results not illustrated) eight samples of Group A (i.e. 32%) were 'wrongly' classified, suggesting that the difference between the two groups is not based on a difference in the harvesting height.

10.2.5.3 Tillage method

As mentioned in Chapter 9 above, the method of tillage will influence the weed communities growing in the arable fields. Major changes in the weed flora are thought to have taken place with the introduction of the mouldboard plough, first recorded in the later Roman period and probably in widespread use during the medieval period (Behre 1981b, Hillman 1981a). Unlike the earlier ploughs or ards the mouldboard plough not only cuts the soil, but turns it over. This destroys many of the perennial species either by damaging their root systems, or by inversion and burial. It favours annual weed species, as these can recover more quickly from such damage. The ard plough in use during the prehistoric period may have caused much less soil disturbance and may not have destroyed the roots and rhizomes of perennial plants such as grasses, plantain etc. to the same extent. The suggested detrimental influence of deep cultivation on perennial weeds is confirmed by recent studies of the effect of tillage on the weed flora of fields sown with spring barley. Pollard and Cussans (1976) found that annual species like *Polygonum aviculare*, *P. convolvulus*, *P. lapathifolium*, *P. persicaria*, and *Raphanus raphanistrum* increased with cultivation (i.e. mouldboard ploughing, inversion and disturbance down to 22 cm). Annual grasses such as *Avena fatua*, were favoured by reduced cultivation (disturbance down to 16 or 8 cm only), while direct drilling (i.e. no cultivation and no disturbance) favoured perennial species like *Rumex* spp., *Agropyron repens*, *Taraxum officionale*, etc. R. Jones (1966) also noted the effect of seed-bed preparation on the weed flora, with annual species being increased by normal cultivation (deep ploughing), while this caused a decrease in perennial grasses.

Thus, different methods of soil cultivation may be identified by looking at the life form of the weed species in the two weed assemblages. A large number of annuals points to the presence of a high degree of soil disturbance, large numbers of perennials to low levels of soil disturbance. However, certain perennial weeds are favoured by soil disturbance, as they are able to grow from small pieces of their fragmented rhizomes (Hillman 1981a). *Ranunculus repens* and *Arrhenatherum elatius* ssp. *bulbosum* are such species, and in the calculation below these two

Life Form

	Annuals	Annuals which can overwinter	Perennials	
Hallshill	46.4%	25.8%	17.5%)	
Murton	29.9%	30.4%	32.2%)	A
Dod Law	61.1%	19.5%	16.5%)	
Chester House	54.0%	10.8%	21.9%)	
Thorpe Thewles	21.8%	18.8%	49.2%)	
Stanwick	25.5%	11.7%	55.1%)	B
Rock Castle	24.3%	20.3%	38.8%)	

species have been grouped with the annuals. Consequently, the percentage figures for the perennials given below differ from those in section 10.2.4.2 above. There were no biennials in the assemblages, but a few annual species could overwinter (*Stellaria media*, *Montia fontana* and *Poa annua*, the latter a species in the category 'small grasses'). These have been listed separately below.

The figures show a clear difference between the two groups. Group A has twice as many annuals as Group B, while Group B has twice as many perennials as Group A. The figures suggest that on Group A sites the amount of soil disturbance was quite high, while in Group B it was considerably lower, allowing the growth of many perennials.

Thus, while there does not appear to be any difference between the two groups of weed assemblages as far as method and height of harvesting is concerned, there is a marked difference in the dominant life form of the weed assemblages of the two groups, suggesting that different tillage methods had been practised.

10.2.6 Conclusion

The autecology of the weed species in the two groups suggests that there are two main differences between the two groups, the soil preferences and tillage practice.

Species strongly associated with Group A are *Chenopodium album*, *Polygonum lapathifolium*, *Polygonum persicaria*, *Atriplex* spp., small grasses (including *Poa annua*) and *Stellaria media*. These are all annual species which are known to increase in abundance with the application of farmyard manure or mineral fertilizers. All are able to survive a great deal of disturbance. *Poa annua* is noted for its ability to survive after being uprooted (Hutchinson and Seymour 1982). All these species also occur in Group B samples, but in much smaller numbers.

Species strongly associated with Group B are *Sieglingia decumbens*, a tufted perennial today growing in poor, acid grassland, *Montia fontana*, a species found on slightly acid soils which are seasonally waterlogged, *Bromus mollis/secalinus*, an annual grass indifferent to soil conditions and a common arable weed. Other species, rarely present in Group A samples, but common in Group B are *Arrhenatherum elatius*, *Galium aparine*, *Tripleurospermum inodorum* and *Prunella vulgaris*. *Arrhenatherum* and *Prunella* are perennials, *Galium* and *Tripleurospermum* are annuals. *Arrhenatherum*, *Galium* and *Tripleurospermum* are species not characteristic of poor soils, but require moderate levels of nutrients. *Galium aparine* does not tolerate much disturbance, however. It is today often found in shallowly ploughed, non-inverted soils, its germinated seeds being unable to penetrate a solid soil depth of more than 4 cm, which makes it vulnerable to burial (Froud-Williams and Chancellor 1982, M. Jones 1984b). Reynolds (1981) has also pointed to the inability of *Galium aparine* to survive cultivation. No information is available specific to *Tripleurospermum*, but there is some concerning the closely related species *Anthemis arvensis* and *Anthemis cotula*. All three species can produce vigorous new shoots from the undamaged lower parts of the plant after the upper stems have been cut off during the harvest of the cereal crop (Kay 1971a and 1971b), but both *Anthemis arvensis* and *A. cotula* are unable to regenerate if they have been completely buried after ploughing, which is possibly why they are found to be more abundant along the edges of fields, rather than in the centre of the fields, as ploughing may be less thorough there (Kay 1971a and 1971b). *Tripleurospermum* may well behave similarly. *Arrhenatherum*, once established in a field, is difficult to remove, as it can regenerate from the bulbous stem tubers which are dispersed by ploughing (Pfitzenmeyer 1962). The fact that this species only occurred in small numbers may also point to low levels of soil disturbance.

The presence of these three species suggests that the level of soil disturbance is at least as important a factor in the separation between the two groups as the levels of soil nitrogen. Their presence would also suggest that the soils were not all as poor in nitrogen as the presence of *Sieglingia* would suggest. While *Sieglingia decumbens*, the most common weed species in Group B, is nowadays associated with poor soils, this may be due to competition, in that

Discriminant Analysis - Maximum Flowering Height

DA - 14

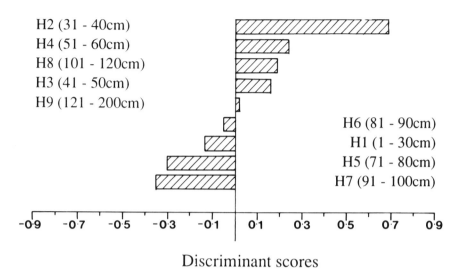

Discriminant scores

Figure 10.20   Discriminant analysis of the prehistoric assemblages, using the maximum flowering height (after Clapham *et al.* 1962). Group A is associated with negative, Group B with positive discriminant scores.

Sieglingia may not be able to compete with other species on very fertile soils, where tall growing species such as *Chenopodium album*, *Atriplex* spp., *Polygonum lapathifolium* and *P. persicaria* can easily outcompete the shorter, light-demanding *Sieglingia* plants. It may be significant here that the climatic factors (section 10.3.2 above) indicated a slight preference for open, light conditions in Group B, and an indifference to light in Group A, and that the information regarding the maximum flowering height of the weed species pointed to the presence of many tall growing species in Group A (section 10.2.5.2). The conditions in the fields of Group B may well have been less dense, and more open than in Group A, allowing *Sieglingia* plants to develop.

Thus, while *Sieglingia* is an indicator of poor soils, it does not necessarily 'like' or 'need' poor soils (see Chapter 9, section 9.5.2 above). In fact, the small leguminous weeds (*Vicia/Lathyrus* and indeterminate Leguminosae), identified by M. Jones (1984b) as indicative of soil exhaustion, are not very prominent in the assemblages, which also points to the fact that the soils, while not very fertile, were not disastrously low in nitrogen.

Not all the fields in Group A were very fertile, however. The results of the principal component analysis of the Group A samples separately (PCA–5 and Figure 10.18) suggest that barley was associated with species characteristic of rather poor soils, such as *Carex pilulifera*, *Rumex acetosella*, *Potentilla cf. erecta*, and *Sieglingia decumbens*. While the data for Group B are slightly less clear cut, the principal component analysis of the Group B samples (PCA–6 and Figure 10.19) also suggest a correlation between barley and poor soil indicators, such as *Carex pilulifera*, Leguminosae, *Potentilla cf. erecta*, and *Raphanus raphanistrum*.

The availability of mineral nitrogen and the method of tillage, both identified as the two main factors causing the difference between the two groups, should not, in fact, be seen as separate environmental factors; the two are intricately linked. Intensive working of the soil improves the mineralization of the organic matter in the soil, increasing the levels of nitrate nitrogen (Bannink *et al.* 1974, King 1966). This is caused by the fact that through tillage the soil becomes better aerated and this allows the growth of the bacteria *Nitrosomonas* and *Nitrobacter*, which are responsible for the conversion of organic matter into nitrate, which is the form in which plants can take up nitrogen. The process of converting organic matter into nitrate, known as nitrification and ammonification, is, under most conditions, associated with tillage of the soil (King 1966). Furthermore, intensive soil working often goes hand-in-hand with manuring.

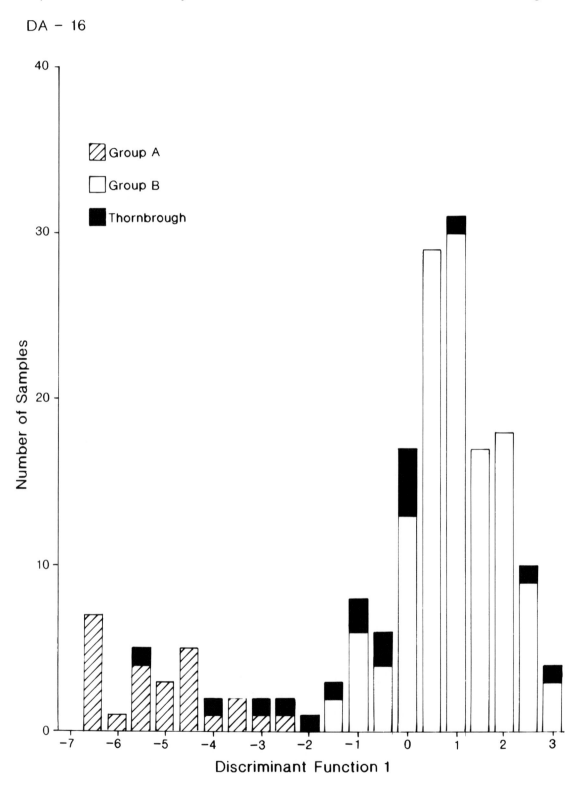

Figure 10.21   Discriminant analysis using Ellenberg's edaphic factors and the life form of the weed species as the variables, and the Group A and Group B assemblages as the control groups to which the samples from Roman period Thornbrough are compared.

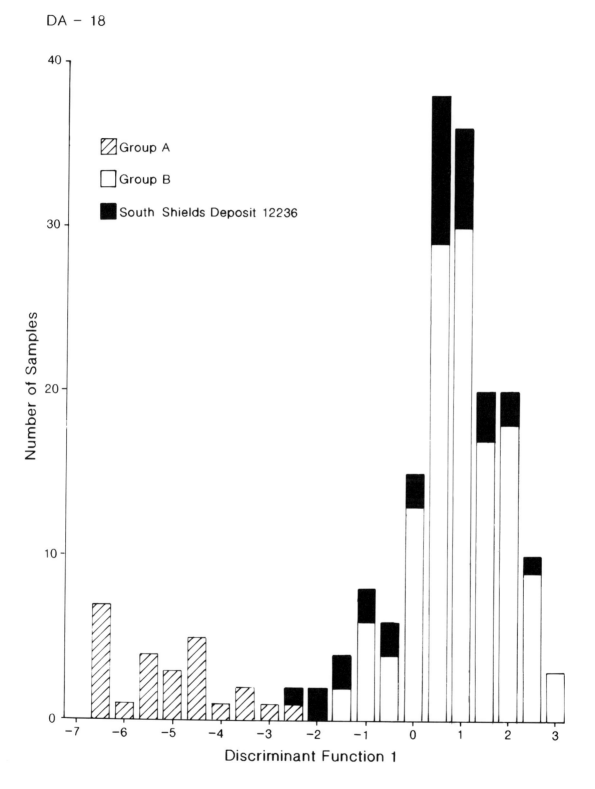

Discriminant Analysis – Prehistoric and Roman Assemblages

DA – 18

Figure 10.22 Discriminant analysis using Ellenberg's edaphic factors and the life form of the weed species as the variables, and the Group A and Group B assemblages as the control groups to which the samples from Roman South Shields, deposit 12236, are compared.

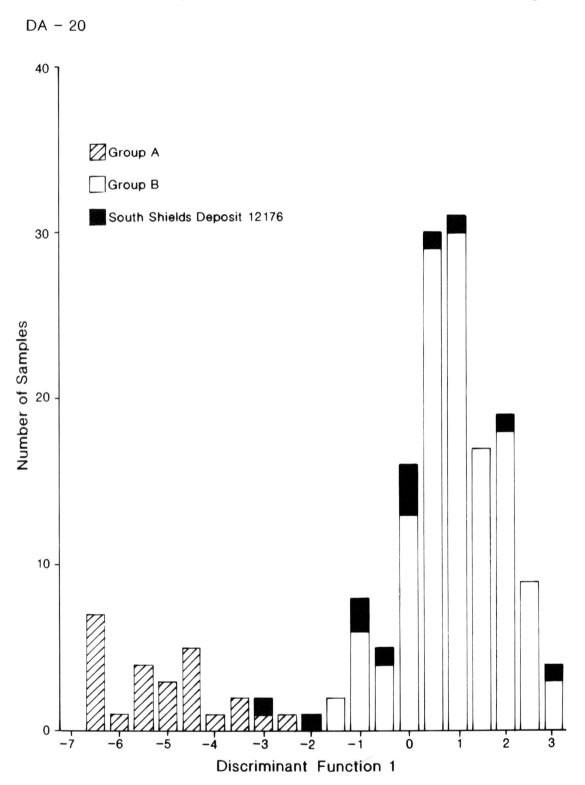

Figure 10.23   Discriminant analysis using Ellenberg's edaphic factors and the life form of the weed species as the variables, and the Group A and Group B assemblages as the control groups to which the samples from Roman South Shields, deposit 12176, are compared.

This link between the type of weed flora and the anthropogenic factor was emphasized by Bannink *et al.* (1974) in a detailed study of the correlations between arable weed vegetation, soil fertility and moisture regime in The Netherlands. They found that the type of weed flora present in the fields was not determined by the time of sowing, the type of crop, or the inherent soil conditions, but largely by the actions of the farmer herself/himself, in the form of the amount of soil disturbance (deep or shallow cultivation), manuring and weeding (Bannink *et al.* 1974). Thus, the differences between the weed assemblages of Groups A and B can best be interpreted as differences in the crop husbandry regimes of the two groups of sites. This aspect will be explored further in Chapter 11.

10.3 The Roman assemblages

It was concluded in Chapter 7 that no direct comparisons could be made between the prehistoric and the Roman assemblages, as they consist of different remains, i.e. the samples represent different crop processing groups. The prehistoric assemblages consist of fine-sieving by-products, while the Roman samples contain largely fine-sieving products. These are, however, closely related crop processing groups, i.e. they both come at the end of the sequence, so that an indirect comparison is possible. To gain some impression of the extent to which the Roman assemblages resemble either Group A or Group B, a discriminant analysis was performed, using the assemblages of Groups A and B as the control groups, to which the Roman samples were compared. As the difference between the assemblages of Group A and B was not related to differences in crop processing groups, it was felt that such an analysis could be successfully performed.

The variables used in the discriminant analysis were not the actual weed species themselves. The Roman assemblages contained a number of species (e.g. *Agrostemma githago*, *Bromus sterilis*, and *Vicia hirsuta*) which were not present in the prehistoric samples, so that a discriminant analysis using the actual weed species would automatically have classified the Roman samples as very different from

the prehistoric ones. Furthermore, we are interested in the ecological information that the weed species represent, rather than in the species themselves. Thus, the variables used in the discriminant analysis were Ellenberg's indicator values for the soil factors moisture, pH, and nitrogen, and the life form of the weeds (annual/perennial).

Analysis	Variables	Transformation
DA–15	F3+4, F5+6, F7+8, F9+10, Fx, R2+3, R4+5+6, R7+8, Rx N2+3, N4+5+6, N7+8, Nx (weeds>10%) – Thornbrough	square root
DA–16	F3+4, F5+6, F7+8, F9+10, Fx, R2+3, R4+5+6, R7+8, Rx N2+3, N4+5+6, N7+8, Nx Annuals, Perennials (weeds>10%) – Thornbrough	square root
DA–17	as DA–15, but for South Shields, deposit 12236	
DA–18	as DA–16, but for South Shields, deposit 12236	
DA–19	as DA–15, but for South Shields, deposit 12176	
DA–20	as DA–16, but for South Shields, deposit 12176	

As the two Roman sites are quite different in nature, i.e. Thornbrough is a 'native' site and South Shields is a Roman fort, the analysis was carried out separately for the two sites. As the two granary contexts from South Shields varied slightly in composition, they were also analysed separately.

The results are summarized in Table 10.13. Figures 10.21, 10.22, and 10.23 give a graphical display of the results. The results clearly indicate that the Roman assemblages have a greater affinity with Group B than with Group A. When using Ellenberg's edaphic factors the percentages of Roman samples classified with Group B lie at 71% (Thornbrough), 91% (South Shields deposit 12236), and 83% (South Shields deposit 12176). When the variable of life form is added, these figures increase to 77%, 94%, and 92% respectively. Thus, the assemblages from these two Roman sites are likely to have been produced by crop husbandry regimes similar to those at the Group B sites. The implications of this are explored further in the next chapter.

11. The Results in their Wider Archaeological Context

We have seen in the previous chapter that the prehistoric assemblages fall into two distinct groups, Group A and Group B, on the basis of the following characteristics:

	Group A	Group B
Crop type	emmer, barley, and some spelt	spelt, barley, and no emmer
Soil conditions	high in nitrogen	low in nitrogen
Soil disturbance	high	low

It was suggested in the previous chapter that these differences may be best interpreted as differences in the crop husbandry regimes at the two groups of sites. In the first section of this chapter these crop husbandry regimes will be analysed further, looking at the information provided by both the crops and the weed species present.

In the second section of this chapter we will look at possible reasons for the presence of these two crop husbandry regimes side by side (it was already established in Chapter 5 that the assemblages of the two groups of sites were contemporary). Both environmental and cultural factors will be considered. The third section will discuss the results from the region in their wider national and international contexts.

Finally, in the last section the Roman assemblages will be discussed. It was established in the previous chapter that the Roman assemblages showed more similarities with those of Group B than with those of Group A, and the implications of this fact will be considered in the light of the question of the grain supply of the Roman army.

11.1 Crop husbandry practices

11.1.1 Emmer versus spelt

Very little is known about the ecological behaviour of emmer and spelt wheat, but Percival (1921) provides the following information:

> 'Emmer will grow on soils which are too light to yield a good crop of spelt. It is more or less immune to rust fungi and suited to warm, dry climates; the majority of forms are rapid-growing spring varieties, with erect caespitose habit and little power of resisting frost. One late variety with black glumes, however, is more hardy, and is a typical winter form whose young shoots lie close to the surface of the ground'

> (Percival 1921, 188).

About spelt Percival comments:

> 'Spelt will grow wherever bread wheat will thrive; it yields best on good land, but will succeed on soils which are too dry and light for the commonly cultivated varieties of *T. vulgare* [= bread wheat] . . . It is one of the hardiest cereals, being rarely affected by frosts which destroy other wheats, and grows at all elevations, from about 300 to 3000 feet above sea-level in Germany and Switzerland ... Other points in its favour are its greater resistance [than bread wheat] to smut, bunt, and rust fungi, and freedom from the attacks of birds . . . The majority of the kinds of Dinkel [= spelt] are winter forms, though a few less hardy spring varieties are cultivated; Emmer (*T. dicoccum*), however, takes the place of *T. spelta* where a spring 'spelt' is grown'

> (Percival 1921, 326).

Körber-Grohne (1987) gives very similar information regarding the ecological behaviour of spelt (which is still grown in parts of southern Germany today), stressing that spelt wheat is not very demanding of soil conditions, fertility or soil cultivation. By the end of the last century there were still five land races of both emmer and spelt cultivated in Germany: four of the emmer races were spring-sown varieties, one was a winter crop; four spelt races were autumn-sown crops, and one was a spring-sown variety (Körber-Grohne 1987).

This information has given rise to the commonly held notion in the archaeological and archaeobotanical literature that emmer wheat is a spring-sown crop well-suited to light soils, while spelt wheat is an autumn-sown crop more tolerant of the heavier soils (Applebaum 1954, 1972, M. Jones 1981, Gregg 1988). Hillman (1981a and Davies and Hillman 1988) has warned against this tendency to regard these ancient wheats as physiologically uniform and has suggested that this simplistic stereotyping of their ecological behaviour can be misleading. For example, an experiment on the effect of flooding on the growth and yield of populations of tetraploid and hexaploid wheats has indicated that there are significant differences between and within species, with one form of emmer being less affected than the hexaploids (Davies and Hillman 1988). It is clear that a great deal more experimental research needs to be done before we can begin to understand the full ecological range of the different forms of emmer and spelt wheat. In the mean time, however, the available evidence suggests that spelt can be grown on a wider range of soils than emmer, although this does not rule out the existence of emmer forms tolerant of heavy or wet soils, and that spelt wheat is hardier than emmer wheat, although, again, this does not rule out the existence of some hardy forms of emmer.

The question of the preferred time of sowing (i.e. autumn versus spring sowing) remains a vexed issue (see also section 10.2.4.2 above). As already discussed in Chapter 10, all cereals are initially likely to have been autumn-sown, as their wild ancestors were species germinating in the autumn (Hillman 1981a). It is not known when spring-sown varieties were first developed. Their development probably originated from a wish to spread the workload of soil preparation and sowing over a longer period of time and/or in response to the severe winters in parts of Europe. At the experimental farm at Butser, Hampshire, spring and autumn-sown varieties of both emmer and spelt wheat are grown, and at the Plant Breeding Institute at Cambridge both species were autumn-sown (T. Miller, pers. comm.). In a small-scale national wheat growing experiment being carried out at present, both species are autumn-sown and the results of the last three years indicate that emmer wheat (at least this particular form) does survive the winter in the north east of England (Van der Veen 1989b and forthcoming e), but as the last few winters have been uncharacteristically mild, and as only one form of emmer is used in the experiment, the matter remains an open question.

Little information is available regarding the yield potential of both crops. At the experimental farm at Butser, Hampshire, the yields of both crops over the last 17 years have averaged out at *ca.* 2 tonnes per hectare on unmanured plots and 5 tonnes per hectare on manured plots, with no appreciable difference between emmer and spelt (P. Reynolds, pers. comm. and 1987, 1988, and 1989). Percival mentions a yield for emmer of 25 to 50 bushels per acre (= 1.8 to 3.6 tonnes per hectare) and for spelt, on the better soils, of 40 to 80 hectolitres per hectare (1600 to 3200 lbs. per acre), and an average for winter spelt in a good season of *ca.* 40 hectolitres per hectare (= 1.8 tonnes per hectare) (Percival 1921). These yields of *ca.* 1.8 to 3.6 tonnes per hectare are remarkably similar to those obtained at Butser. However, Percival's actual words when describing the yield of emmer are: 'between 25 and 50 bushels of 'spelt' grain' (Percival 1921, 188), which suggests that he means spikelets, rather than dehusked grains. This would mean that the yield of emmer was *ca.* 25 per cent lower than that of spelt, as the chaff takes up *ca.* 25 per cent of the spikelet.

The soils at the experimental farm at Butser consist of a thin redzina over Middle Chalk and these are very susceptible to drought (Reynolds 1987). The present evidence suggests that while spelt wheat can tolerate a wide range of soils, it yields best on the heavier soils. The fact that spelt produces a good yield on the very thin and light soils at Butser corroborates its tolerance of the lighter soils, but this would also suggest that on heavier soils spelt might do even better and would probably out-yield emmer. The national wheat experiment was set up to assess the difference in yield potential between emmer and spelt on different soils and in different climatic zones of Britain (Van der Veen 1989a, 1989b). So far, the results from only one year are available. In that year spelt gave a higher yield than emmer on 11 out of the 12 plots for which information was available (Van der Veen 1989b). These results are, of course, not representative until a sequence of years giving the same results is available, but they do indicate that spelt wheat (at least this particular form of spelt) can out-yield emmer wheat (at least this particular form of emmer).

It must be said, however, that it is difficult to determine any reliable yield figures for these ancient crops, as the yield will vary with the variety grown, soil conditions, and climate, as well as the cultivation regime. In some situations it may be judged expedient to invest the effort required to realize high yields on a specific plot of land, while in other situations it may be judged preferable to realize lower yields over larger areas (M. Jones 1981).

Obviously, large scale experiments growing emmer and spelt on a range of different soils, in a number of different climatic regions, and adopting different cultivation regimes, are urgently required to improve our knowledge of the ecological behaviour of these ancient crops. The present evidence, however, suggests that spelt wheat had a number of advantages over emmer wheat, which may have been the cause of it replacing emmer during the first millennium BC. Spelt appears to tolerate a wider range of soil conditions, is hardier than emmer, and is resistant to a range of different pests and diseases. It may also, in certain circumstances, have been higher yielding than emmer, but this remains an uncertain factor. It has been suggested that the replacement of emmer by spelt during the first millennium BC, at the same time as a range of other crops were being introduced, took place in response to an expansion of agriculture requiring both the continuation of cultivation on deteriorating soils as well as the expansion onto hitherto marginal lands, at a moment in time when the climate was deteriorating (M. Jones 1981). The evidence from the carbonized seed assemblages studied here indicates that this replacement of emmer by spelt wheat, and by implication the expansion of agriculture, only occurred in part of the present study region.

11.1.2 Methods of cultivation

The different levels of available nitrogen in the soils of the two groups of sites suggest that at the Group A sites the soil fertility was kept high, either by the regular replacement of nutrients, i.e. manuring, or by demanding little of the soil, i.e. letting the land lie fallow for long periods of time. At the Group B sites the soil fertility was not kept up and nitrogen levels were comparatively low. This suggests the fields were intensively cropped without adequate replacement of the nutrients.

Furthermore, the Group A assemblages are characterized by large numbers of annual weeds, those of Group B by perennials as well as annuals. High numbers of annuals point to high levels of soil disturbance (annuals can recover better from soil disturbance), while high numbers of perennials point to low levels of soil disturbance, which allows the growth of these plants.

The intricate relationship between intensive soil working and high levels of nitrogen has already been discussed in Section 10.2.6 above. Not only does the intensive working of the soil often go hand-in-hand with manuring, but intensive soil working also increases the mineralization of the organic matter in the soil by increasing the growth of certain soil bacteria.

It is tempting to interpret the Group A assemblages as the product of a horticultural type of cultivation, with the fields being dug and hoed regularly, and those of Group B as the product of plough cultivation, with the fields being ploughed once or twice before sowing, but no form of soil disturbance implemented afterwards. The possible existence of these two types of soil cultivation (spade versus plough) in north-east England has been speculated upon by Gates (1982a, 1983; see below), but for the sites under study there is no direct evidence for the use of either implement. There is, of course, no need to postulate the use of two different agricultural implements. An ard plough can create as much soil disturbance as a spade if enough effort is exerted, i.e. by ploughing the field several times in opposite directions combined with tilting the ard to throw a furrow (Payne 1949, Hillman and Fowler, pers. comm.). This method of ploughing obviously requires much more time and energy than ploughing the field once or twice. The difference between the two groups can, therefore, be better described as representing different levels of energy spent per unit of land, than by inferences regarding the implement used.

Extensive evidence for field systems has been recovered from the upland areas of Northumberland (Gates 1982a, 1983, Topping 1989a, 1989b). Gates (1982a, 1983) has tentatively suggested that the

difference in size between fields associated with Bronze Age unenclosed settlements and Roman period enclosed settlements in Northumberland might be related to a difference in spade versus plough cultivation. A survey of cord rig cultivation in Northumberland and the Borders, broadly dated to the Iron Age, has also identified a basic division in plot size (Topping 1989a, 1989b). The presence of ard marks underneath some of these cord rig plots has, however, been taken as tentative evidence that cord rig cultivation was associated with the ard plough rather than the spade, even though the actual mechanics by which the ridges were constructed are still not understood (Topping 1989a, 1989b). It will be interesting to study charred seed assemblages from sites associated with different types and sizes of field systems, to see whether they do coincide with different crop and weed species. With the exception of Hallshill, none of the sites studied here was associated with (recognizable) field systems. Hallshill, a Bronze Age unenclosed settlement, had a plot of *ca.* 0.6 hectare, which is larger than the average for these sites. The seed assemblage from this site was a Group A assemblage.

It was suggested above that the soil fertility at the Group A sites was kept up either by manuring or by leaving the fields to lie fallow to allow a restoration of the nitrogen levels. This last option does, in fact, not look very likely in the light of the fact that the assemblage is characterized by annuals. The practice of fallow in between periods of cultivation tends to introduce perennials, especially perennial grassland species, into the arable fields (Behre 1981b, Van Haaster forthcoming, Wasylikowa 1981). The very low occurrence of these plants in the Group A assemblages seems to suggest that the soil fertility was kept up through regular manuring, rather than periods of fallow.

The evidence from the principal components analysis of the Group A and B assemblages separately (section 10.2.4.3 above and Figures 10.18 and 10.19) indicates that we should interpret the differences between the two assemblages as reflecting the cultivation regimes of the wheat crops only (emmer and spelt). Barley was clearly cultivated separately, and especially in Group A, treated differently. The weeds associated with barley in Group A are suggestive of the cultivation regime of Group B, i.e. low levels of nitrogen and low levels of soil disturbance. This suggests that barley was receiving considerably lower levels of energy input, which may be a reflection of its different status as a crop.

To summarize, the evidence from the weed assemblages suggests the existence of two distinct crop husbandry regimes. At the Group A sites the emmer fields were intensively worked and the soil fertility was kept up, probably through regular manuring. This points to a high level of energy input per unit of land, which suggests that the cultivated area was relatively small. The barley fields at these sites were, however, treated differently; they received much less manure and much lower levels of cultivation, probably reflecting the different status of the crop. At the Group B sites, the fields (both spelt and barley) were much less intensively worked than the emmer fields of Group A, allowing the growth of perennial weeds, and the soil fertility was not kept up, i.e. there was no regular replacement of nutrients. The presence of many grasses and other perennial weed species may also suggest that some pasture land may have been ploughed up and converted to arable and/or that the land was allowed to lie fallow, allowing such species to establish themselves. The Group A assemblage probably represents small-scale subsistence agriculture, while the Group B assemblage is suggestive of arable expansion.

11.1.3 Summary

Thus, both the crop plants and the weed species point to the presence of intensive, small-scale agriculture at the Group A sites, and larger-scale, arable expansion at the Group B sites. The occurrences of naked barley as a trace in the Group A assemblages and bread/club wheat in one sample of a Group B assemblage are interesting in this respect, as they confirm the slightly 'traditional' character of the Group A assemblage and the 'innovative' character of the Group B assemblage. Thus, both the crops and the weed species point in the same direction, i.e. small-scale subsistence agriculture in one part of the region, and larger-scale arable production and expansion in another.

11.2 Intra-regional variability

In this section we will look at the possible reasons for the presence of arable expansion at some sites and the continuation of small-scale, intensive practices at others. The sites belonging to Group A (Hallshill, Murton, Dod Law, and Chester House) are all located north of the River Tyne, those of Group B (Thorpe Thewles, Stanwick, and Rock Castle) are all located south of the Tyne, in the Tees lowlands (Figure 4.1). This marked divergence in geographical distribution of the two groups of sites will be analysed, and consideration will be given to both environmental and cultural factors which may underlie this distribution.

11.2.1 Environmental factors

11.2.1.1 Climate

The principal factor determining the climate of the region is relief (see also Chapter 2). The region can be divided into an upland and a lowland zone and any variation in the climate is east-west rather than north-south. The western upland zone has higher levels of rainfall and lower average temperatures than the eastern coastal zone. With the exception of Hallshill, the sites all lie in the lowland, coastal plain. Hallshill is different from the other sites not merely in its location, but also in the date of its occupation (Bronze Age rather than late Iron Age), and, consequently, in the climatic conditions prevailing at that time (see Chapter 2). For these reasons Hallshill has been omitted from the discussions below.

The climatic conditions during the late Iron Age were probably not too dissimilar to those of today (see Chapter 2), which means that the present-day figures may have some relevance. The present-day mean annual rainfall figures for the two groups of sites are more or less identical and cannot explain the variation, nor can differences in altitude, as both groups of sites have a similar range (see opposite page).

Regional differences in temperature relevant to agriculture are usually measured by the start and length of the growing season and the accumulated temperatures above the minimum for plant growth (Coppock 1971). As the temperature at which plants start to grow varies from species to species, the length of the growing season is a rather general measure. Conventionally 6 °C is taken as the threshold temperature, representing the onset of grass growth (Coppock 1971). The length of the growing season is of limited relevance to cereal crops, as these are harvested up to three months before the growing season ends (Coppock 1971). The onset of the growing season may be a more relevant factor. An alternative measure is that of the number of day degrees, which expresses the relationship between the length of the growing season and the range of temperatures above the minimum (the rate of growth increases with temperature) (Coppock 1971). This measure is given as the cumulative total degrees by which the daily temperature exceeds the minimum for plant growth between January and June (in this case above 0 °C, after Jarvis *et al.* 1984), expressed as the number of day degrees centigrade (see opposite page).

Environmental Factors

Mean annual rainfall (after Jarvis *et al.* 1984)

Group A		*Group B*	
Murton	600–700 mm	Thorpe Thewles	600–700 mm
Dod Law	*ca.* 700 mm	Stanwick	*ca.* 700 mm
Chester House	600–700 mm	Rock Castle	*ca.* 700 mm

Altitude

Group A		*Group B*	
Murton	90 m O.D.	Thorpe Thewles	60 m O.D.
Dod Law	182 m O.D.	Stanwick	100 m O.D.
Chester House	41 m O.D.	Rock Castle	190 m O.D.

Length of the growing season (after Smith 1984)

Group A		*Group B*	
Murton	237 days	Thorpe Thewles	248 days
Dod Law	238 days	Stanwick	248 days
Chester House	238 days	Rock Castle	248 days

Onset of the growing season (after Smith 1984)

Group A		*Group B*	
Murton	April 2	Thorpe Thewles	April 2
Dod Law	April 4	Stanwick	April 2
Chester House	April 4	Rock Castle	April 2

No. of day degrees above 0 °C (after Jarvis *et al.* 1984)

Group A		*Group B*	
Murton	1250–1350	Thorpe Thewles	1250–1350
Dod Law	1150–1250	Stanwick	1250–1350
Chester House	1350–1400	Rock Castle	1250–1350

Soil type (after Mackney 1974)

	Surface-Water Gley soils	Brown soils	Other
Group A			
Murton	60 %	40 %	-
Dod Law	20 %	65 %	15 %
Chester House	70 %	10 %	20 %
Group B			
Thorpe Thewles	100 %	-	
Stanwick	65 %	35 %	-
Rock Castle	30 %	70 %	-

Mean accumulated maximum potential soil moisture deficit, PSMD (after Jarvis *et al.* 1984, Figure 7)

Group A		*Group B*	
Murton	100–125 mm	Thorpe Thewles	125–150 mm
Dod Law	100–125 mm	Stanwick	100–125 mm
Chester House	*ca.* 100 mm	Rock Castle	100–125 mm

Mean accumulated potential soil moisture deficit for mid July, adjusted for winter wheat, MD (after Jarvis *et al.* 1984, Figure 8)

Group A		*Group B*	
Murton	75–100 mm	Thorpe Thewles	75–100 mm
Dod Law	50–75 mm	Stanwick	75–100 mm
Chester House	50–75 mm	Rock Castle	75–100 mm

Median field capacity period (days) (after Jarvis *et al.* 1984, Figure 9):

Group A		*Group B*	
Murton	150–175 days	Thorpe Thewles	125–150 days
Dod Law	175–200 days	Stanwick	175–200 days
Chester House	*ca.* 150 days	Rock Castle	175–200 days

These figures indicate that there are today no appreciable differences in climate between the two parts of the region. Assuming these figures have some relevance to the climate in the late Iron Age, they suggest that the difference between the two groups of sites cannot be explained by differences in the climatic conditions at these sites.

11.2.1.2 Soils

The present-day soils within a 1 km radius of the sites are listed in Table 11.1. The soils around each site are largely loams and clays and, broadly speaking, consist of Surface-Water Gley Soils and Brown Soils, the latter being the more permeable and more easily worked category of the two. The relative proportions of these two broad categories of soils are given on the previous page.

Soil moisture is an important factor in determining the suitability of a soil for agricultural purposes. Both soil droughtiness and soil wetness can cause problems. Soil droughtiness is measured using the concept of potential soil moisture deficit (PSMD). This is calculated as the difference between potential transpiration and rainfall during the growing season (Jarvis *et al.* 1984). The maximum PSMD is most appropriate when studying perennial crops like grasses. As cereal crops stop growing before the maximum deficit has developed, the figures are often adjusted for particular crops. Here the MD figures are also listed, representing the mean PSMD figures for mid-July adjusted for the average growth pattern of winter wheat (Jarvis *et al.* 1984). Finally the concept of field capacity is used to assess the wetness of the soils. Field capacity is calculated from rainfall and potential transpiration, using a standard soil water abstraction model for short rooted crops (Jarvis *et al.* 1984). The duration of field capacity determined in this way varies broadly with annual rainfall and provides a measure of accessibility to land, provided that allowance is made for soil properties. Well-drained, coarse-textured soils can usually be cultivated during the field capacity period, but slowly permeable soils would be damaged (Jarvis *et al.* 1984), (see previous page).

These figures do not seem to point to any significant differences between the two groups as far as soil type and soil moisture are concerned. We must, therefore, conclude that neither climatic nor soil conditions, so far as we can assess them from present-day conditions, can explain much of the differences in crop type and weed assemblage between the two groups of sites. Furthermore, the autecological information of the weed species (using Ellenberg's climatic factors, see Chapter 10, section 10.2.2) indicated that there was no climatic gradient in the weed assemblages.

The ecological analysis of the weed assemblages pointed to the availability of nitrogen as the main soil factor causing the discrimination. While the capacity to hold nitrogen will vary with soil type, nitrogen is a nutrient easily leached from soils under arable cultivation, and as such the availability of nitrogen is largely the result of human activities, i.e. a function of the intensity with which the fields are cropped and the frequency with which they are manured. Thus, there appear to have been no immediate environmental constraints which forced the people north of the Tyne to restrict their agriculture to small-scale, intensive cultivation. In the next section some cultural factors are considered.

11.2.2 Cultural factors

11.2.2.1 Settlement type

As has already been discussed in Chapter 2 above, later Iron Age settlement in Northumberland, the area between the Tweed and the Tyne, is characterized by defended sites, i.e. small hillforts, while the area between the Tyne and Tees is characterized by undefended sites, i.e. open settlements and, more commonly, sub-rectangular enclosures. These sub-rectangular enclosures are normally sited in the coastal lowlands on good quality soils (Haselgrove 1982, Vyner 1988), which may suggest an emphasis on arable farming. The location of the hillforts does not appear to respect the requirements for arable farming, and may point to an emphasis on animal husbandry (Gates 1982a, 1983). The location of hillforts in East Yorkshire has been interpreted as taking advantage of upland pasture and downslope arable areas (Manby 1980).

Two of the Group B sites, Thorpe Thewles and Rock Castle, are sub-rectangular enclosures, although Thorpe Thewles became an open, unenclosed settlement in Phase 3. The occupation at Stanwick probably started as an open settlement not dissimilar to Thorpe Thewles in Phase 3, later became enclosed, and only became defended in the last phase of its occupation. As the Group B assemblages are indicative of arable expansion, they would appear to confirm the association of sub-rectangular enclosure settlements (and unenclosed settlements) with arable farming.

The Group A sites consist of a mixture of different types of settlements. Hallshill is a small, late Bronze Age, unenclosed settlement, Murton and Dod Law are both defended settlements, and Chester House is an undefended site (enclosed or unenclosed). The charred seed assemblages clearly indicate that all four sites were actively involved in arable production, but also suggest, especially when compared with Group B, that the crop husbandry practice was intensive and probably small-scale. The similarities between the late Bronze Age assemblage of Hallshill and those of the late Iron Age assemblages of Murton, Dod Law, and Chester House suggest a stable arable system in this part of the region, with little change during the first millennium BC.

The plant remains from Chester House originate from an undefended settlement, either enclosed or open. Unfortunately, the archaeological remains were not conclusive on the latter aspect (see Chapters 4 and 5, sections 4.4 and 5.2.4 above). If the above-mentioned difference in function between defended sites and undefended sites holds true, then Chester House ought to have an assemblage different from those of Murton and Dod Law. It is remarkable in this respect that the analyses discussed in Chapter 10 did frequently classify the sample from Chester House with the Group B sites (only one sample from this site was large enough to be included in the analysis), rather than with the Group A sites. The Chester House assemblage belongs to Group A in that it contains emmer and only small quantities of spelt, but the weed assemblages are more characteristic of Group B. While the seed assemblage from Chester House is extremely small, and from a statistical point of view, therefore, unreliable, it does suggest that there may well have been a difference of emphasis on arable farming between these two types of settlement. As a speculation, Chester House might be suspected of representing a 'pioneer' site in the process of moving from a Group A to a Group B type of cultivation regime, while holding on to the more 'traditional' suite of crops.

11.2.2.2 Social structure

The settlements north of the Tyne, but especially in the northern part of Northumberland, are characterized by hillforts, often quite small ones, with elaborate defences. They are often located on the upland fringes above the Rivers Tweed, Till and Coquet and their tributaries. From this location both the valley bottoms and the upland grazing areas might have been exploited. The presence of fortifications at these sites may point to an element of competition for control of land (Higham 1986). The small internal area and the absence of a water supply suggest that they were intended to deter sudden or casual raids by neighbours, rather than to function as points of resistance during lengthy military campaigns (Higham 1986). The finds of metal work in this region consist largely of items of display, ornamented weaponry and horse trappings, suggesting the presence of a warrior aristocracy. This preoccupation with defensive structures and weaponry may point to the existence of a society without strong centralization, consisting of competing aristocrats (Higham 1986). Cunliffe

(1983) sees this emphasis on defence as evidence of social stress. The inferred deterioration of the climate during the first half of the first millennium BC must have made cereal growing at high altitudes difficult, if not impossible. In such a situation, an adjustment to the economy might have been required and, in the absence of a decline in population (for which there is no evidence), pressure on land, causing social stress, may have resulted. The large number of fortified homesteads built during the later part of the first millennium BC may represent a physical manifestation of the social dislocation that Cunliffe (1983) believes must have taken place. In this type of society, arable farming may never have developed beyond its subsistence base, either due to a shortage of land, or, in the case of large-scale unrest, because the risks might have been too high. It may have been more expedient in such a situation to combine small-scale subsistence arable farming with a greater reliance on animal husbandry, as animals can be moved in times of danger. Unfortunately, the faunal assemblages from the region are too limited to allow any interpretation regarding the nature of animal husbandry (see also Table 2.3).

The area between the Tyne and the Tees is characterized by sub-rectangular enclosures; defended settlements are more or less absent, with the notable exception of Stanwick. Finds of weaponry and horse trappings are also rare, again with the exception of the famous hoard of metalwork (consisting of as many as four sets of harness and chariot fittings) found close to Stanwick (Haselgrove 1990). This may point to the existence of a greater degree of political authority and centralization compared with northern Northumberland. According to Higham (1986) the few defended sites may have administered larger areas, which may have allowed a greater degree of local stability and security. The scale of the earthworks at Stanwick does certainly suggest the existence of a political authority capable of mobilizing the necessary labour as well as of generating the surplus to 'finance' the operation (Haselgrove 1982). It has been suggested that the status of the site was based on its diplomatic relationship with Rome, in the period after the conquest of southern Britain, but before the military occupation of the North (Haselgrove 1990). The presence of a remarkable collection of Roman fine wares dated to the period between *ca.* AD 40–70 at Stanwick also points to the presence of a ruling élite (Haselgrove 1984). Finds of much smaller quantities of these fine wares at sites such as Thorpe Thewles are suggestive of the existence of a redistribution network around Stanwick (Haselgrove 1982), although some of the pottery at Thorpe Thewles dates from after the demise of Stanwick as the regional capital. In principle, arable farming could have developed more easily in this type of society, and an expansion of arable agriculture may have been encouraged to raise a surplus, both to support the ruling élite and to provide the necessary exchange goods in return for Roman luxuries. However, the expansion of arable agriculture in the Tees lowlands started as early as 300 cal BC, i.e. well before Stanwick had become the residence of the local élite, and well before diplomatic and/or trading contacts were established with the Romans in southern Britain. This indicates that other factors, such as population growth, an absence of pressure on land and, consequently, an absence of social stress, contributed to the onset of arable expansion.

The fact that the seed assemblage from Stanwick, the seat of the ruling élite, is very similar to those of Thorpe Thewles and Rock Castle, both 'ordinary' farmsteads, does need an explanation. First of all, the settlement is very extensive and the excavations which have taken place cover only a minute part of the site. This means that while the samples are representative of the excavated area, they cannot be regarded as representative of the site as a whole. Secondly, so far only a small proportion of the samples has been analysed (see Figure 4.11); more will be studied by the writer in future years. Thirdly, the occupation on the site started off as an open settlement not dissimilar to Thorpe Thewles in Phase 3. The present evidence suggests that the site only acquired its special nature during the mid to late first century AD, when the fortifications were built and large numbers of Roman imports reached the site. By the end of the first century AD the north of Britain was conquered by the Romans and the occupation at Stanwick ceased. The time period during which the site functioned as a central place may have been very short indeed. The present seed assemblage is unlikely to represent this brief period adequately.

Thus, the archaeological evidence seems to point to the possibility of the existence of two different 'types' of society in the two parts of the region. These do, in fact, coincide with parts of the territories of the two tribal groupings of the region, the Votadini (in Northumberland) and the Brigantes (in the Tees lowlands; see also Chapter 2 and Figure 2.7). It was Queen Cartimandua of the Brigantes, thought to have lived at Stanwick, who entered into a treaty with Rome by AD 51, and this diplomatic contact may have encouraged a further expansion of arable production. The Votadini do not appear to have had trade links with Rome during this early period. On present evidence the distribution of early Roman fine wares (i.e. AD 40–70) is restricted to the area south of the Tyne.

The River Tyne is usually taken to form the frontier between the territories of the Votadini and the Brigantes (Frere 1978, Maxwell 1980). However,

the find of a statuette of the goddess Brigantia at Birrens, Dumfries, has been interpreted by some as evidence that the territory of the Brigantes extended northwards beyond the Tyne-Solway gap, at least in the North West. (Hartley and Fitts 1988, Rivet and Smith 1981, Salway 1981). Higham (1986) has suggested that this may also have been the case in the North East, with the Brigantes occupying both banks of the Tyne. Jobey (1966) already pointed to the dissimilarity between settlements in northern and southern Northumberland, and Higham (1986) regards the poor-grade lands of Rothbury Forest from Otterburn to Alnwick as the natural boundary between the two areas. This would mean that the area roughly south of the River Coquet might have had more affinities with the Brigantes than with the Votadini. It is interesting in this respect that the site at Chester House, already identified as representing a slight anomaly within Group A, lies just south of this river and might, therefore, have been within the sphere of influence of the Brigantes. The present evidence is, of course, quite insufficient to resolve this issue; further seed assemblages from sites located between the Rivers Coquet and Tyne may well throw light on this matter.

11.3 The results in their wider context

During the first millennium BC major changes took place in the nature of arable production in Britain (see also Chapter 6). Emmer wheat was replaced by spelt wheat, naked barley by hulled barley, and several new crops were introduced (e.g. Celtic bean, oats, rye). Initially it was thought that these developments did not take place in the Highland Zone of Britain (M. Jones 1981), but the information from new regional studies, such as the present one, indicates that the situation is more complex than that. A considerable degree of regional variation has been recognized in both the time when and the degree to which these new practices were adopted (M. Jones 1984a).

In the Upper Thames Valley, emmer wheat goes into decline and naked barley disappears during the first half of the first millennium BC. The new crops in this region are spelt wheat and bread/club wheat (M. Jones 1984a). In Wessex the pattern is a little different in that emmer and naked barley retain a significant presence (although emmer is always less common than spelt), while the new crops include the pulses and brassicas (Green 1981, M. Jones 1984a). The evidence from Wales is difficult to assess at the moment as so little of the data has been published. Hillman has indicated, however, that by the late first millennium BC regional differences do occur (Hillman, forthcoming a). Assemblages from various hillforts contain both emmer and spelt, as well as barley. At one hillfort (Breiddin) emmer appears to predominate, while at the late Iron Age and Romano-British farmstead at Cefn Graeanog spelt wheat was the dominant wheat crop (Hillman 1981a, and forthcoming a and b). It is interesting to note that at this last site, spelt wheat was accompanied by seeds of heath grass, *Sieglingia decumbens*, as at the Group B assemblages from north-east England.

The information from Scotland is still very sparse. There are relatively few excavated settlements from which botanical evidence is available, and their distribution is uneven. Taking this into account, the present evidence suggests that six-row barley was the most important crop throughout the prehistoric and Roman period (Boyd 1988). Emmer wheat is common on Bronze Age sites, but only a few records of this species are known from Iron Age and Roman settlements. Spelt wheat and bread/club wheat have, so far, not been found on prehistoric or Romano-British sites, but both are known from Roman forts such as Bearsden, Castlecary, Rough Castle (Dickson 1989), and Newstead (Van der Veen, forthcoming f). Oats (both *Avena sativa* and *Avena strigosa*) have been found in small quantities on Iron Age and Roman period sites (Boyd 1988).

The present new evidence from north-east England fits in well with this pattern. Part of the region continued to grow emmer wheat as part of an intensive, small-scale arable strategy, while another part switched to spelt wheat as part of a strategy of arable expansion. The timing of these developments varies across the country, but broadly speaking the southern and central parts of the country appear to have been the earliest to adopt these changes (by *ca.* 500 BC), while in Wales and north-east England these changes (on present evidence) may have occurred a few centuries later. In Scotland they appear never to have occurred, although the evidence is really insufficient to assess the situation accurately. The pollen evidence for large-scale forest clearance mirrors this picture, in that the south and east of the country were deforested early in the first millennium BC, while the west and north of the country were cleared progressively later (Turner 1981a, 1981b). In north-east England, the large-scale deforestation of the landscape did not occur until the first century BC (with the area around Bishop Middleham on the Durham Plateau as the only exception; see Chapter 2). The landscape in north-west England and Scotland was not cleared until the Roman period (Turner 1981a, 1981b).

The variation across the country in the timing of the adoption of these changes in arable farming has been explained in various ways. In the past, differences in the environmental conditions in the various regions of Britain were usually seen to have caused the pattern (Faechem 1973, Fowler 1983, Fox 1932, Piggott 1958, Wheeler 1954) and the division of Britain into a Highland and a Lowland zone derives from this viewpoint. Hillman has suggested that the 'continued cultivation of outmoded crops could have been caused either out of agricultural or culinary conservatism, or because the first wave of new crops yielded poorly under local climatic and edaphic conditions' (Hillman 1981b). M. Jones (1981, 1984a) has suggested that there must have been an increase in the scale of arable production in response to increasing population pressure and declining soil fertility, requiring the diversification of soils and crops under cultivation. He has postulated that the regional divergence in the rate of adoption of these new practices may possibly be related to differences in the social circumstances of the various groups of farmers, for example in access to wealth, capital, technical information etc. (M. Jones 1984a).

The weed assemblages associated with the emmer and spelt crops from north-east England do corroborate M. Jones' suggestion that the switch to spelt wheat should be interpreted as representing arable expansion, and the associated archaeological evidence from this region does suggest that socio-economic circumstances do, indeed, appear to influence the choice of arable strategy. The emmer wheat assemblages from the North East (Group A) have been interpreted as representing intensive, small-scale subsistence farming, the spelt assemblages as representing larger-scale arable production and expansion. It has been suggested above that differences in site function and site location and differences in the social structure of society may lie at the root of this divergence. Why these supposed differences in social structure might have existed and to what extent, if at all, they may have arisen due to certain environmental constraints, are questions well outside the scope of the present study.

These changes in crop production are not only restricted to Britain. The switch from emmer to spelt wheat has also been recorded in southern Germany (especially in the lower Rhineland and Baden-Württemberg) during the late Iron Age and Roman period (Körber-Grohne 1979, Knörzer 1971b and 1984a). In northern Germany and in the northern part of The Netherlands, however, there are virtually no records of spelt wheat. Here emmer wheat remained the principal wheat crop, although barley would appear to have been more important than wheat in this area (Willerding 1979, Van Zeist 1968 and 1981).

Willerding (1979) has pointed to both geographical and cultural differences between northern and southern Germany in trying to explain the pattern: the northern region was thought to be less favourable for crop production, and the area was inhabited by Germanic tribes, in contrast to the southern region which was more favourable to arable agriculture and was inhabited by Celtic tribes. He has also noted the fact that the autecology of the weed species associated with the crops throughout the prehistoric period in Central Europe points to the fact that the fields were well-drained and high in nitrogen. The indicator values for the edaphic factors (moisture, pH, and nitrogen) are similar to those of the Group A assemblages from north-east England (Willerding 1981). In contrast, in the lower Rhineland, Knörzer has noted that during the late Iron Age and Roman period it is the assemblages containing spelt wheat which are the assemblages also containing high numbers of weed species indicative of poor soil conditions, which he related to an increase in the scale of cultivation and a consequent degradation of the soil conditions (Knörzer 1964 and 1984a).

Thus, the evidence from Germany parallels the pattern found in Britain: some areas of the country switch to spelt wheat as part of a strategy of arable expansion. As in Britain, both environmental and social factors have been suggested as lying at the root of the regional variation in the adoption of this development. The evidence from the present study would suggest that social and economic factors may play a larger role than environmental ones in causing the divergence.

11.4 Grain supply to the Roman army in north-east England

Charred seed assemblages dated to the Roman period are still very rare in the study region, making it difficult to draw any conclusions regarding the origin of the grain supplied to the Roman army. A few points can be made, however. First, the composition of the seed assemblage from the third century AD granary of the Roman fort at South Shields indicates that at least part of the grain was supplied by a local source. The presence of seeds of *Sieglingia decumbens* in association with spelt wheat is very characteristic for the Group B assemblages. This particular combination of species has, so far, only been recorded in north-east England and parts of Wales (see section 10.3 above; Hillman 1981a and forthcoming b). While it cannot be ruled out that the grain was supplied from Wales, the presence of a nearby local source makes the local supply of the grain much more likely, especially as the Group B assemblages were interpreted as representing arable expansion.

The granary deposit at South Shields did, however, also contain substantial quantities of bread/club wheat. Bread/club wheat has, so far, been recorded in only one context on one late Iron Age site in the

North East, i.e. at Rock Castle (a Group B site). There are at least a hundred years between the occupation of Rock Castle and the burning down of the granary. The find of bread/club wheat at Rock Castle may represent the introduction of this new crop to the region. It is quite possible that in the subsequent hundred years this crop became widely cultivated. The presence of a few grains of bread/club wheat at the Romano-British settlement at Catcote, Cleveland (Huntley 1989), points at least to a continuation of the cultivation of this species since the late Iron Age in the region. The bread/club wheat at South Shields could, therefore, also have been supplied by a local farmer, but we lack the evidence to demonstrate this at the moment. Future charred seed assemblages from Romano-British sites in the region may clarify this point.

The role of bread/club wheat as a crop in Britain is still poorly understood. The glume wheats (emmer and spelt) are replaced by free-threshing cereals such as bread/club wheat and rye sometime during the first millennium AD. The data base of charred seed assemblages from Romano-British settlements is still very patchy, so that this development cannot be traced in any detail. In parts of central-southern Britain bread/club wheat appears to have become an important crop as early as the late Iron Age (e.g. at Barton Court Farm, Oxfordshire and Bierton, Buckinghamshire; M. Jones 1984a, 1986), but most of the present evidence points to the Saxon period as the time when bread/club wheat gained importance as a crop. Bread/club wheat has, however, been found on a number of Roman sites, such as Castlecary, Rough Castle (Dickson 1989), Newstead (Van der Veen, forthcoming f), Caerleon (Helbaek 1964), London (Straker 1984), Rocester (Moffett 1989), and Colchester (Murphy 1984), but in most of these cases the number of grains was very small. As such, the find from South Shields is apparently unusual, but this may simply reflect the fact that large grain assemblages from Roman sites in Britain are still relatively rare.

There is some evidence from the South Shields granary which could be interpreted as pointing to long-distance importation of grain. A number of bones of the garden dormouse, *Eliomys quercinus*, and of the yellow-necked mouse, *Apodemus flavicollis*, were found in the granary deposit (Younger 1986). The fact that this species of dormouse is not native to Britain and that the distribution of the yellow-necked mouse today is restricted to southern England, might be taken to mean that these species represent accidental imports, i.e. were brought in with imported grain (Younger 1986). Alternative explanations for their presence can, however, also be offered. The present distribution of the yellow-necked mouse appears to be a relic distribution, suggesting that the species was previously much more widely distributed (T. O'Connor, pers. comm.). There is, in fact, a pre-1911 record of the species from Riding Mill, Northumberland (Corbet 1971). Furthermore, the yellow-necked mouse is not a grain pest, but lives in all types of woodland, hedgerows, and gardens (Dobroruka 1988). The presence of this species, therefore, does not, in itself, support an argument for grain importation. As far as the dormouse is concerned, it is known that the Romans did eat dormice (Zeuner 1963). They were kept in enclosures or earthenware jars to be fattened up for the tables. Although it is normally assumed that these references refer to the edible dormouse (*Glis glis*), there is no reason to rule out *Eliomys quercinus*, as discussed in Younger (1986). Only one other example of the garden dormouse is known from Britain, from the Roman deposits at Tanner Row, York (O'Connor 1989). Here the bones were found associated with other food remains and the find is interpreted as representing an imported luxury food item (O'Connor 1989). The garden dormouse is also an unlikely candidate for an accidental import. It is not a grain pest, but lives in woodland, orchards and gardens (Dobroruka 1988); it will eat grain, but only if nothing else is available (T. O'Connor, pers. comm.). Thus, while the presence of dormice in the granary could, in principle, point to the long-distance importation of grain, the supporting evidence is far from conclusive.

The other Roman-period seed assemblage studied here is that from Thornbrough, a native settlement just south of Hadrian's Wall. The assemblage was dominated by spelt and barley remains and the ecology of the associated weed species was similar to those of the Group B sites. The composition of the samples suggested that most of the grain could have been brought in, rather than have been locally produced. The samples contained the first record for rye in the region and the remains of rye did point to local production. The fact that the assemblage is very similar to those of Group B would suggest that little change occurred between the late Iron Age and the later Roman period as far as arable production was concerned. One free-threshing cereal is present, i.e. rye, but only in very small quantities. No bread/club wheat was found. Obviously, carbonized seed assemblages from many more native sites are required before we can assess the situation accurately, but there is some other evidence that the economic development of the region might have stagnated during the Roman period. The construction of villas during this period has been regarded as an expression of rural wealth and status (Hingley 1989). The virtual absence of villas from the North may be a reflection of the actual economic poverty of the area,

although the possibility that capital was invested in a different way in the North needs to be explored further (Hingley 1989). Millett (1990) has suggested that the very heavy presence of the Roman army in the region may have prevented the emergence of a civil authority amongst the tribal élite, which may have had a negative effect on the socio-economic development of the region. The end of the occupation at Thorpe Thewles and Rock Castle in the first half of the second century AD may also be relevant here.

To conclude, the evidence from the granary at South Shields indicates that the Roman army relied, at least in part, on a local supply of grain. This may well have been supplemented by grain imports from further afield. After all, the demand for grain will have varied enormously, depending on the political situation and the numbers of soldiers stationed along the wall at any one time (see also Chapter 2). The construction of 22 granaries in advance of the Severan campaign into Scotland in the early third century is an example of such a temporary increase in demand. Investigations of major charred seed assemblages from many more Roman forts and native sites are required before we can draw more detailed conclusions on these issues.

11.5 Conclusion

The prehistoric seed assemblages studied here fall into two distinct groups, Group A and Group B. On the basis of both the crops and the weed species present, the assemblages of Group A have been interpreted as representing small-scale subsistence production, and those of Group B as representing larger-scale production or arable expansion. The sites of Group A (Hallshill, Murton, Dod Law, and Chester House) lie to the north of the River Tyne, while those of Group B lie to the south of the Tyne, in the Tees lowlands.

It has not been possible to explain this marked difference in the geographical distribution of the two groups of sites by intra-regional variation in climate or soil type. While arable farming in parts of the Highland Zone, especially in the upland parts, must have suffered many environmental constraints, the differences in the scale of arable production found in the prehistoric assemblages from these lowland sites in north-east England cannot be explained in this way.

These differences do, however, appear to relate to differences in settlement type and location, with the small hillforts associated with small-scale subsistence agriculture and the sub-rectangular enclosures associated with larger-scale production. This may be connected with the location of these two types of sites. Sub-rectangular enclosures are usually located in a lowland position on good arable land, while the hillforts, in general, take up positions from which both upland and lowland resources can be exploited. Furthermore, the differences in the distribution of these types of site and the distribution of the metalwork (weaponry and horse trappings) have been interpreted as reflecting differences in the social structure of society, perhaps with a more unstable society north of the Tyne, possibly consisting of competing aristocrats, and a more stable, centralized society in the area south of the Tyne. These areas coincide with parts of the territories of the Votadini (north of the Tyne) and the Brigantes (south of the Tyne).

The presence of the small hillforts is, in fact, restricted to the most northern part of the region. The settlements between the Rivers Coquet and Tyne are not dissimilar to those south of the Tyne, i.e. they largely consist of sub-rectangular enclosures rather than defended sites. The assemblage from Chester House, located just south of the River Coquet, is, unfortunately, too small to be reliable from a statistical point of view, but it may suggest two things. Firstly, it appears to confirm a difference in the scale of arable production between defended and undefended settlements, with the weed assemblage (but not the type of crop) from Chester House suggesting some arable expansion. Secondly, the fact that the Chester House assemblage still contains more emmer than spelt wheat points to differences between the areas north and south of the Tyne, which may reflect both cultural and socio-economic differences. Obviously, larger seed assemblages from many more sites are required before we can test these hypotheses, an issue I will return to in Chapter 12 below.

These differences in the scale of arable production and the development of arable expansion have been recognized in other parts of the country and in other parts of Europe. During the late Iron Age and Roman period we witness an increase in the scale of arable production in response to a number of different, interrelated developments, such as an increase in population, social stratification and the emergence of civil authority amongst the tribal élite, economic specialization and the development of a market economy, and contact with or the incorporation into the Roman Empire. These developments did not occur synchronously across Europe, but varied according to local circumstances, and the changes in the scale of arable production consequently also varied in time and space.

12. Summary and Conclusion

It has been the aim of the present study to analyse and interpret recently collected archaeobotanical data from north-east England, a lowland area within the Highland Zone of Britain, in order to improve our understanding of the role of arable farming in this region, and to assess the extent to which the increase in the scale of arable farming, as witnessed in parts of the Lowland Zone of Britain, took place in this region. The data used in this study are carbonized seed assemblages collected from seven prehistoric and two Roman period sites located in this region.

Although the data base for this study, consisting of 325 samples and *ca.* 89,000 seeds, is one of the largest collections of carbonized seeds from the Highland Zone, it still cannot be regarded as necessarily representative of the whole of the north east of England during the period under study. The assemblages from the Group A sites are particularly small, and, consequently, their interpretation is particularly tentative. The interpretations and conclusions drawn in this study should, therefore, be taken as hypotheses, which need testing against further data sets.

The results of the archaeobotanical analysis of the present data set indicate that, contrary to earlier held beliefs, spelt wheat was introduced into the north of England as early as into the southern part of the country. The three radio-carbon dates on the spelt wheat from Hallshill cluster around the very early first millennium BC. So far, the seed assemblage of only one Bronze Age site has been analysed (Hallshill). It will be necessary to collect seed assemblages from several more such sites before we can assess how widespread the cultivation of spelt wheat was during this period. The results from pollen analysis have suggested that the area of the Durham Plateau witnessed an early expansion of agriculture, with evidence for large scale deforestation being dated to the Bronze Age. It would be interesting to collect seed assemblages from one or more sites on the Durham Plateau and dated to this period, to allow a comparison between the archaeobotanical and palynological results.

The evidence from Thorpe Thewles, Stanwick, and Rock Castle indicates that spelt wheat had replaced emmer wheat as the principal wheat crop in the Tees lowlands by *ca.* 300 cal BC. The radio-carbon dates on the seed assemblages from these three sites demonstrate that this development can be regarded as an indigenous one, rather than as a consequence of contact with the Romans. At present no information is available from sites dated to the mid first millennium BC. In order to establish exactly when this replacement of emmer took place, we need to analyse seed assemblages dated to this earlier period. The palisaded enclosure underlying the sub-rectangular enclosure at Rock Castle would form an obvious choice for excavation.

The dominance of emmer wheat over spelt wheat in the assemblages from Murton, Dod Law, and Chester House, all located north of the Tyne, demonstrates that the replacement of emmer by spelt wheat did not occur across the whole region. While spelt wheat is present in the assemblages of all three sites, it is secondary in numerical importance, suggesting that emmer wheat had remained the principal wheat crop here. The data sets from these three sites are very small, however, and substantial seed assemblages from a number of sites in this part of the region are needed to test the representativeness of the present assemblages. The similarity of these three late Iron Age assemblages with that from the late Bronze Age site of Hallshill appears to suggest that there was a remarkable stability in the nature of arable farming in this part of the region during the first millennium BC. Seed assemblages from sites located in this northern part of the region and dated to the middle part of the first millennium BC are needed to explore this aspect further.

The first record of a free-threshing wheat in the region is that of bread/club wheat at Rock Castle, dated by radio-carbon assays to the first century cal AD. The near absence of substantial seed assemblages from native sites dated to the Roman period prevents a detailed assessment of the importance of this crop during this period, but the presence of a few grains of this species at Catcote, Cleveland (Huntley 1989), suggests at least a continuation of the cultivation of this crop in the region. It cannot yet be established where the bread/club wheat present in the granary at the Roman fort at South Shields originated from, but the spelt wheat from the same granary deposit could be identified as a product of local cultivation through its association with weed seeds of *Sieglingia decumbens*. At Thornbrough another free-threshing cereal, rye, was present, though only as a minor component. Seed assemblages from many more Roman period sites are needed before we can establish whether bread/club wheat and rye became extensively grown during this period, and to provide further evidence on the extent to which the local farmers supplied the Roman army with the necessary grain.

The data have been analysed further using three multivariate statistical techniques. The prehistoric assemblages were analysed using Principal Components Analysis, Cluster Analysis, and

Discriminant Analysis. Two methods of data transformation were used (square root and the octave scale), as well as two different cut-off points for the exclusion of rare weed species (10 and 5 per cent). The results of the analyses consistently pointed to a division of the data into two separate groups, independent of the type of analysis, transformation, or cut-off point used, a clear indication that the data are fairly robust. The two groups have been named Group A and Group B, and the division between these two groups is based on differences in both the cereal crops (i.e. the presence of emmer versus spelt wheat) and the weed species present (i.e. the presence of species like *Chenopodium album*, *Polygonum lapathifolium/ persicaria*, and *Atriplex* spp. versus *Sieglingia decumbens*, *Montia fontana*, and *Bromus mollis/ secalinus*).

The differences between the two groups were explored further using the autecological data available for the weed species. This indicated that the main difference between the two weed assemblages lay in the soil preferences and life form of the weed species. These differences have been interpreted as pointing to differences in soil conditions and tillage practice at the two groups of sites. Group A sites were characterized by the presence of emmer wheat, fertile soil conditions (i.e. the presence of a large proportion of nitrogen-loving species), and a high degree of soil disturbance (i.e. a high proportion of annuals), while Group B was characterized by the presence of spelt wheat, poor soil conditions (i.e. a large proportion of species indicative of low levels of nitrogen), and relatively low levels of soil disturbance (i.e. the presence of many perennials).

The Roman assemblages could not be analysed directly with those of the prehistoric sites as they represented different stages in the processing of crops. To gain some impression of the extent to which the Roman assemblages resembled either Group A or Group B, a discriminant analysis was performed using the Group A and B assemblages as the control groups to which the Roman assemblages were compared. The results clearly indicated that the Roman assemblages had a greater affinity with Group B than with Group A.

The differences between the two groups of prehistoric assemblages have been interpreted as differences in crop husbandry regimes, with Group A sites representing small-scale, intensive, subsistence cultivation, and Group B sites representing larger-scale cultivation, reflecting an expansion of arable farming. The radio-carbon dates on the seed assemblages demonstrated that these two groups of assemblages were contemporary, so that these differences could not be 'explained' by variation through time (though there might be a similar time trend within each group).

The two groups of sites were located in two different parts of the region, with Group A sites (Murton, Dod Law, and Chester House) located between the Rivers Tweed and Tyne, and Group B sites (Thorpe Thewles, Stanwick, and Rock Castle) in the Tees lowlands. This marked geographical difference between the two groups of sites was analysed, looking at both environmental and cultural factors which might have contributed to the divergence. It has not been possible to explain the differences by intra-regional variation in environmental factors, such as the climatic conditions or soil types present at the individual sites.

While the present information from pollen diagrams does not show any evidence for major differences in the start of large-scale deforestation in the two parts of the region, the number of diagrams with radio-carbon dates for this period is very small, and no diagrams are available from the Northumberland coastal plain. Further radio-carbon dated pollen diagrams are needed from the lowland areas of the region to establish whether it is possible to recognize these different crop husbandry regimes, resulting in different arable weed assemblages, in the palynological record.

While the intra-regional variation in the scale of arable farming could not be linked to variation in the environmental conditions at the two groups of sites, it did, however, appear to be related to differences in settlement type and location, and these two factors appear to be connected to cultural and socio-economic differences in the two parts of the region. Differences in both the tribal affinity and the socio-economic structure of society have been tentatively identified. The area between the Tweed and the Tyne was occupied by the Votadini, and the settlement and artefact evidence from this part of the region has been interpreted as reflecting social stress and an absence of a strong social élite. The Tees lowlands were inhabited by the Brigantes, and the archaeological evidence from this part of the region has been interpreted as pointing to a greater degree of political authority, centralization, and stability compared with the area north of the Tyne. These differences in the socio-economic structure of society may partly explain the recognized differences in the husbandry regimes of arable production in the two parts of the region. Unfortunately, the present evidence from animal bone studies from these sites, and from the region as a whole, is far too limited to allow an assessment of the extent to which there was also a difference in the type of animal husbandry between the two groups of sites or parts of the region.

Thus, the results of the present study have indicated that arable farming played an important role in the economy of the late Iron Age people of this part of the Highland Zone, that Piggott's

Stanwick-type economy consisted of arable expansion rather than pastoralism, and that the Roman army in the region was, at least partly, supplied by local farmers. The evidence also suggests, however, that within the region substantial differences existed in the husbandry regimes of arable farming. In the area between the Tweed and the Tyne emmer wheat maintained its importance as a crop into the late Iron Age, although some spelt was also cultivated. The evidence from the arable weeds suggests that the type of cultivation was intensive and small-scale. In the Tees lowlands spelt wheat had completely replaced emmer by *ca.* 300 cal BC and the arable weeds point to an extensive type of cultivation, suggestive of arable expansion. In both areas barley was cultivated as well, but as a separate crop.

These results could, of course, be used simply to redraw the dividing line between the Highland and Lowland Zones of the country from the Exe to the Tyne, instead of to the Tees, but doing this would mean a continuation of the simplistic use of broad generalizations and environmental determinism, without adding any understanding to the issues at stake. The evidence from this study suggests that the differences in the scale of arable farming in this region cannot be explained by differences in environmental conditions between the two areas, but has pointed, instead, to socio-economic and cultural differences as possible underlying factors. What is not understood at the moment is why these socio-economic and cultural differences occurred, and to what extent these were influenced by certain broader environmental factors.

The present study has indicated that arable expansion did take place in a part of a lowland area within the Highland Zone (in fact, in precisely that area which gave its name to the Highland Zone economy) during the late Iron Age. In order to improve our understanding of this development and its underlying factors, further information is required. A number of suggestions have been put forward above for further work in the North East. Similar studies are, of course, necessary from other lowland areas within the Highland Zone of Britain, before we can accurately assess the representativeness of the present results.

159

Appendix

Table 2.1   Location of pollen diagrams in the region.

```
Site                    Grid Ref.        Altitude          Reference

Upland area
Threepwood Moss         NT 425 515       290 m O.D.        Mannion 1979
Blackpool Moss          NT 517 289       250 m O.D.        Butler in press
The Dod                 NT 473 060       215 m O.D.        Shennan and Innes 1986
Camp Hill Moss          NU 100 263       205 m O.D.        Davies and Turner 1979
Broad Moss              NT 963 215       390 m O.D.        Davies and Turner 1979
Steng Moss              NY 965 913       305 m O.D.        Davies and Turner 1979
Coom Rigg Moss          NY 690 790       300 m O.D.        Chapman 1964
Muckle Moss             NY 798 669       200 m O.D.        Pearson 1960
Fellend Moss            NY 679 658       200 m O.D.        Davies and Turner 1979
Steward Shield          NY 983 438       390 m O.D.        Roberts et al. 1973
Bollihope Bog           NY 980 356       345 m O.D.        Roberts et al. 1973
Valley Bog              NY 763 331       549 m O.D.        Chambers 1978
Cow Green               NY 810 300       550 m O.D.        Turner et al. 1973
Red Sike Moss           NY 820 287       490 m O.D.        Turner et al. 1973
Simy Folds              NY 763 549       350 m O.D.        Donaldson 1983

Coastal Plain
Linton Loch             NT 793 254        92 m O.D.        Mannion 1978
Cranberry Bog           NZ 232 545        90 m O.D.        Turner & Kershaw 1973
Hallowell Moss          NZ 251 439        75 m O.D.        Donaldson & Turner 1977
Hutton Henry            NZ 410 350       137 m O.D.        Bartley et al. 1976
Thorpe Bulmer           NZ 453 354       100 m O.D.        Bartley et al. 1976
Bishop Middleham        NZ 324 304        76 m O.D.        Bartley et al. 1976
Nunstainton             NZ 320 295        76 m O.D.        Bartley et al. 1976
Mordon Carr             NZ 321 253        80 m O.D.        Bartley et al. 1976
Neasham Fen             NZ 332 116        40 m O.D.        Bartley et al. 1976
```

Table 2.2   Calibration of radio-carbon dates dating the start of large-scale forest clearance (Stuiver and Pearson 1986)

```
                              1 sigma                        2 sigma

Steng Moss
Q-1520                 52 cal BC -  86 cal AD        160 cal BC - 176 cal AD

Fellend Moss
SRR-876                 2 cal AD - 110 cal AD         88 cal BC - 132 cal AD

Steward Shield Meadow
GaK-3/033             350 cal BC -  70 cal AD        390 cal BC - 190 cal AD

Bollihope Bog
GaK-3/031             140 cal AD - 420 cal AD         80 cal AD - 540 cal AD

Valley Bog
SRR-88               372 cal BC - 204 cal BC        394 cal BC - 124 cal BC
SRR-89               360 cal BC - 186 cal BC        376 cal BC - 116 cal BC

Hallowell Moss
SRR-415               42 cal BC - 118 cal AD        114 cal BC - 220 cal AD

Hutton Henry
SRR-600               84 cal AD - 240 cal AD         12 cal AD - 340 cal AD

Thorpe Bulmer
SRR-404              168 cal BC -  10 cal BC        350 cal BC -  70 cal AD
```

Table 2.3   Faunal assemblages from the region (after Haselgrove 1982 with modifications)

		cattle	pig	sheep/ goat	horse	other	sample size
1.	Doubstead	+	–	+	+	–	<50
2.	Kennel Hall	30	2	3	–	bird	437
3.	Hartburn	+	–	+	+	bird	<50
4.	Burradon	+	+	+	–	–	32
5.	Tynemouth	–	+	+	–	bird	<50
6.	Coxhoe	47%	4%	32%	17%	dog, deer	164
7.	Catcote ('68)	46%	9%	40%	5%	deer	735
8.	Catcote ('89)	51	7	14	4	fish	342
9.	Thorpe Thewles	53%	13%	24%	10%	dog, cat deer, bird	8000
10.	Stanwick ('54)	40%	16%	18%	13%	dog, deer	large
11.	Stanwick ('84)	50%	21%	18%	4%	dog, deer	28MNI

```
References:        1. Rackham 1982a, 2. Rackham 1978, 3. Hodgson 1973, 4. Hodgson 1970, 5. Hodgson 1967,
                   6. Rackham 1982b, 7. Hodgson 1968, 8. Gidney 1989, 9. Rackham 1987, 10. Wheeler 1954,
                   11. Haselgrove 1984.
```

Table 3.1   List of species present in the samples.

BOTANICAL NAME	ENGLISH NAME
CEREALS	
Triticum dicoccum (Schrank.) Schubl.	emmer wheat
Triticum spelta L.	spelt wheat
Triticum aestivo-compactum Schiem	bread wheat
Hordeum vulgare L.	six-row barley
Secale cereale L.	rye
WEEDS	
Ranunculus acris L.	meadow buttercup
Ranunculus repens L.	creeping buttercup
Ranunculus Subgenus Ranunculus	
Ranunculus flammula L.	lesser spearwort
Papaver argemone L.	prickley-headed poppy
Papaver rhoeas/dubium L.	field/long-headed poppy
Papaver sp.	poppy
Raphanus raphanistrum L.	wild radish
Brassica sp.	
Cruciferae indet.	
Viola Subgenus Melanium	pansy
Montia fontana, ssp. chondro-sperma (Fenzl) Walters	blinks
Stellaria media (L.) Vill.	chickweed
Stellaria palustris Retz.	marsh stitchwort
Spergula arvensis L.	corn spurrey
Agrostemma githago L.	corn cockle
Caryophyllaceae indet.	
Chenopodium album L.	fat hen
Chenopodium sp.	goosefoot
Atriplex spp.	orache
Chenopodiaceae indet.	
Malva sylvestris L.	common mallow
Malva sp.	mallow
Linum catharticum L.	purging flax
Vicia hirsuta (L.) S. F. Gray	hairy tare
Vicia/Lathyrus	vetch/pea
Leguminosae indet. (small)	small-seeded legumes
Aphanes arvensis agg.	parsley piert
Potentilla cf. erecta (L.) Räusch	common tormentil
Heracleum spondylium L.	cow parsnip
Conium maculatum L.	hemlock
Anthriscus caucalis Bieb.	bur chervil
Polygonum aviculare agg.	knotgrass
Polygonum convolvulus L. (= Fallopia convolvulus (L.) A. Löve) (= Bilderdykia convolvulus (L.) Dumort.)	black bindweed
Polygonum lapathifolium L.	pale persicaria
Polygonum persicaria L.	red shank, persicaria
Polygonum lap/pers	persicaria
Polygonum sp.	
Rumex acetosella agg.	sheep's sorrel
Rumex spp.	dock
Polygonaceae indet.	
Urtica dioica L.	stinging nettle
Urtica urens L.	small nettle
Solanum nigrum L.	black nightshade
Hyoscyamus niger L.	henbane
Odontites verna (Bell.) Dum.	red bartsia
Veronica arvensis L.	wall speedwell
Veronica cf. scutellata L.	marsh speedwell
Rhinanthus sp.	yellow-rattle
Verbascum sp.	mullein
Ajuga reptans L.	bugle

BOTANICAL NAME	ENGLISH NAME
Lamium album/purpureum L.	white/red dead nettle
Stachys arvensis (L.) L.	field woundwort
Mentha arvensis/aquatica L.	mint
Galeopsis sp.	hemp-nettle
Prunella vulgaris L.	self heal
Plantago lanceolata L.	ribwort plantain
Plantago major L.	great plantain
Galium aparine L.	goosegrass
Galium palustre L.	marsh bedstraw
Sherardia arvensis L.	field madder
Valerianella dentata (L.) Poll.	lamb's lettuce
Tripleurospermum inodorum (L.) Schultz Bip.	scentless mayweed
Lapsana communis L.	nipplewort
Sonchus asper (L.) Hill	sow-thistle
Hypochoeris glabra/radicata L.	cat's ear
Centaurea cf. cyanus L.	cornflower
Compositae indet.	
Avena sp.	oat
Bromus mollis/secalinus	bromegrass
Bromus sterilis L. (=Anisantha sterilis (L.) Nevski)	barren brome
Sieglingia decumbens (L.)Bernh. (= Danthonia decumbens (L.) DC.)	heath grass
small grasses (including Poa annua)	
Gramineae indet.	grasses
Arrhenatherum elatius, ssp. bulbosum (Willd.) Spenn.	onion couch
rhizomes Gramineae indet.	
Juncus squarrosus L.	heath rush
Juncus sp., capsule	rush
Eleocharis sp.	spike-rush
Isolepis setacea (L.) R. Br.	bristle scirpus
Carex pilulifera L.	pill-headed sedge
Carex pulicaris L.	flea sedge
Carex spp.	sedge
OTHER	
Linum cf. usitatissimum L.	flax
Corylus avellana L.	hazelnut
Crataegus cf. monogyna Jacq.	hawthorn
Prunus spinosa L.	sloe
Sambucus nigra L.	elderberry
tree buds indet.	
Rosa sp.	rose
Rubus fruticosus agg.	blackberry
Rubus sp.	blackberry/raspberry
Thelycrania sanguinea (L.) Fourr. (=Cornus sanguinea L.)	dogwood
Calluna vulgaris (L.) Hull, leafshoots	heather
Calluna vulgaris (L.) Hull, flowers	heather
Erica sp., flowers	heather
Vaccinium myrtillus L.	bilberry
Empetrum nigrum L.	crowberry
Pteridium aquilinium (L.) Kuhn, fronds	bracken
Lycopus europaeus L.	gipsy-wort
Viola Subgenus Viola	violet
Caltha palustris L.	marsh marigold
Menyanthes trifoliata L.	bogbean
Potamogeton spp.	pondweed

Table 3.2 List of abbreviations used in the data tables.

ABBREVIATION	BOTANICAL NAME
CEREALS	
Trit dico	Triticum dicoccum (Schrank.) Schübl.
Trit spel	Triticum spelta L.
Trit aest	Triticum aestivo-compactum Schiem.
Trit sp.	Triticum sp.
Hord vulg	Hordeum vulgare L.
Seca cere	Secale cereale L.
Cere inde	Cerealia indet.
coleopti.	detached coleoptiles
CHAFF	
glum dico	glume bases Triticum dicoccum
glum spel	glume bases Triticum spelta
glum inde	glume bases Triticum sp.
glumes	glume fragments Triticum sp.
rach brit	rachis internodes of a brittle rachis wheat
rach aest	rachis internodes Triticum aestivum
base Trit	basal nodes Triticum sp.
rach Hord	rachis internodes Hordeum vulgare
base Hord	basal nodes Hordeum sp.
flor Avef	floret bases Avena fatua
flor Aven	floret bases Avena sp.
awns Aven	awn fragments Avena sp.
culm node	culm nodes cereals/large grasses
awns Trit	awn fragments Triticum sp.
lemm Hord	lemma fragments Hordeum sp.
chaf inde	chaff fragments indet.
rach Seca	rachis internodes Secale cereale
WEEDS	
Ranu acri	Ranunculus acris L.
Ranu repe	Ranunculus repens L.
Ranu Ranu	Ranunculus Subgenus Ranunculus
Ranu flam	Ranunculus flammola L.
Papa arge	Papaver argemone L.
Papa rh/d	Papaver rhoeas/dubium L.
Papa sp.	Papaver sp.
Raph raph	Raphanus raphanistrum L.
Bras sp.	Brassica sp.
Crucif.	Cruciferae indet.
Viol Mela	Viola Subgenus Melanium (DC.) Hegi
Mont font	Montia fontana, ssp. chondrosperma (Fenzl) Walters
Stel medi	Stellaria media (L.) Vill.
Stel palu	Stellaria palustris Retz.
Sper arve	Spergula arvensis L.
Agro gith	Agrostemma githago L.
Caryoph.	Caryophyllaceae indet.
Chen albu	Chenopodium album L.
Chen sp.	Chenopodium sp.
Atri spp.	Atriplex spp
Chenop.	Chenopodiaceae indet.
Malv sylv	Malva sylvestris L.
Malv sp.	Malva sp.
Linu cath	Linum catharticum L.
Vici hirs	Vicia hirsuta (L.) S. F. Gray
Vici Lath	Vicia/Lathyrus
Legumin.	Leguminosae indet. (small)
Apha arve	Aphanes arvensis agg.
Pote erec	Potentilla cf. erecta (L.) Räusch
Hera spon	Heracleum spondylium L.
Coni macu	Conium maculatum L.
Anth cauc	Anthriscus caucalis Bieb.
Poly avic	Polygonum aviculare agg.
Poly conv	Polygonum convolvulus L.
Poly lapa	Polygonum lapathifolium L.
Poly pers	Polygonum persicaria L.
Poly l/p	Polygonum lapathifolium/persicaria
Poly sp.	Polygonum sp.

ABBREVIATION	BOTANICAL NAME
Rume acet	Rumex acetosella agg.
Rume spp.	Rumex spp.
Polygon.	Polygonaceae indet.
Urti dioc	Urtica dioica L.
Urti uren	Urtica urens L.
Sola nigr	Solanum nigrum L.
Hyos nige	Hyoscyamus niger L.
Odon vern	Odontites verna (Bell.) Dum.
Vero arve	Veronica arvensis L.
Vero scut	Veronica cf. scutellata L.
Rhin sp.	Rhinanthus sp.
Verb sp.	Verbascum sp.
Ajug rept	Ajuga reptans L.
Lami a/p	Lamium album/purpureum L.
Stac arve	Stachys arvensis (L.) L.
Ment a/a	Mentha arvensis/aquatica L.
Gale sp.	Galeopsis sp.
Prun vulg	Prunella vulgaris L.
Plan lanc	Plantago lanceolata L.
Plan majo	Plantago major L.
Gali apar	Galium aparine L.
Gali palu	Galium palustre L.
Sher arve	Sherardia arvensis L.
Vale dent	Valerianella dentata (L.) Poll.
Trip inod	Tripleurospermum inodorum (L.) Schultz Bip.
Laps comm	Lapsana communis L.
Sonc aspe	Sonchus asper (L.) Hill
Hypo g/r	Hypocheris glabra/radicata L.
Cent cyan	Centaurea cf. cyanus L.
Compos.	Compositae indet.
Aven sp.	Avena sp.
Brom m/s	Bromus mollis/secalinus
Brom ster	Bromus sterilis L.
Sieg decu	Sieglingia decumbens (L.)Bernh.
smal gras	small grasses (including Poa annua)
Gramin.	Gramineae indet.
Arrh elat	Arrhenatherum elatius, ssp. bulbosum (Willd.) Spenn.
rhiz Gram	rhizomes Gramineae indet.
Junc squa	Juncus squarrosus L.
Junc sp.	Juncus sp., capsule
Eleo sp.	Eleocharis sp.
Isol seta	Isolepis setacea (L.) R. Br.
Care pilu	Carex pillulifera L.
Care puli	Carex pulicaris L.
Care spp.	Carex spp.
OTHER	
Linu usit	Linum cf. usitatissimum L.
Cory avel	Corylus avellana L.
Crat mono	Crataegus cf. monogyna Jacq.
Prun spin	Prunus spinosa L.
Samb nigr	Sambucus nigra L.
tree buds	tree buds indet.
Rosa sp.	Rosa sp.
Rubu frut	Rubus fruticosus agg.
Rubu sp.	Rubus sp.
Thel sang	Thelycrania sanguinea (L.) Fourr.
Call leaf	Calluna vulgaris (L.) Hull, leafshoots
Call flow	Calluna vulgaris (L.) Hull, flowers
Eric flow	Erica sp. flowers
Vacc myrt	Vaccinium myrtillus L.
Empe nigr	Empetrum nigrum L.
Pter aqui	Pteridium aquiliniu. (L.) Kuhn, fronds
Lyco euro	Lycopus europaeus L.
Viol Viol	Viola Subgenus Viola
Calt palu	Caltha palustris L.
Meny trif	Menyanthes trifoliata L.
Pota spp.	Potamogeton spp.

Table 3.3   Relative proportions of weed species present in >10% of the samples.

SITE:	HH86	MT83	DL85	CH85	TT81	SW85	RC87	TH84	SS84
NO. OF SAMPLES:	21	10	12	14	127	32	23	23	63
NO. OF LITRES:	252	68	303	890	3,556	431	598	635	892
WEEDS									
Ranu acri	.00	.00	.00	.00	.01	.00	.00	.00	.00
Ranu repe	.00	.00	.13	.00	.04	.04	.03	.20	.00
Ranu Ranu	2.05	.21	.00	37.12	.29	.28	.36	.41	.00
Ranu flam	.00	.43	.13	.00	.19	.00	.10	.20	.06
Raph raph	.00	.21	.26	.00	.06	.48	.13	1.42	1.91
Mont font	.00	.00	.00	.00	13.31	1.37	1.30	.20	.13
Stel medi	2.05	10.26	1.47	.00	.29	.76	3.45	.20	.25
Chen albu	14.36	13.03	28.01	.76	2.73	3.02	2.02	3.25	.06
Chen sp.	1.03	6.84	24.74	1.52	1.09	.80	4.62	1.02	.38
Atri spp.	6.67	.85	.45	.76	.91	.20	1.71	.61	.32
Chenop.	.00	.00	.00	1.52	.50	.16	.10	.00	.00
Vici Lath	.00	.43	.51	2.27	.24	.24	3.73	.61	1.53
Legumin.	3.08	1.50	.96	6.06	2.40	6.24	10.34	1.63	1.27
Pote erec	.00	.85	.71	.00	.50	1.61	1.10	.61	.13
Poly avic	.00	1.92	.77	.00	.52	.40	1.71	1.63	.00
Poly conv	.51	.43	.45	.76	.18	.28	.28	1.02	.06
Poly lapa	6.15	2.56	.96	.76	.32	.12	.03	.61	.06
Poly pers	6.15	2.35	1.28	.76	.24	.16	.03	.41	.00
Poly l/p	3.59	.43	.90	1.52	.17	.16	.13	.81	.00
Rume acet	.51	.64	1.47	2.27	.13	1.57	.43	1.22	.83
Rume spp.	2.05	1.28	1.92	.00	3.39	12.11	.94	.61	8.21
Odon vern	.00	.00	.00	.00	.17	.12	.03	.00	.06
Prun vulg	.00	.21	.00	.00	.15	.08	.08	.20	.00
Plan lanc	1.54	1.28	1.67	.76	.38	.52	.79	1.63	.38
Gali apar	.00	.00	.06	.00	.55	.89	2.02	1.02	1.59
Trip inod	.00	.21	.00	.00	1.35	.52	5.46	.41	.06
Aven sp.	1.54	.21	.96	.65	.08	.08	.92	1.02	10.31
Brom m/s	4.10	.64	1.28	3.03	11.84	17.87	4.72	47.76	44.75
Sieg decu	.51	2.99	.71	11.36	35.22	30.14	28.55	4.47	16.23
smal gras	23.59	20.09	17.88	10.61	5.14	9.50	15.37	6.71	1.85
Gramin.	7.18	5.56	1.41	4.55	7.58	1.21	2.40	10.98	1.08
Arrh elat	.00	.00	.06	.00	.45	.00	.08	1.63	.00
Eleo sp.	.00	.00	.00	.00	.18	.20	.10	.41	.19
Junc sp.	.51	.00	.83	1.52	.01	.12	.54	.20	.00
Care pilu	.00	1.50	1.41	.76	.23	2.09	.54	.41	.64
Care puli	.00	2.78	.00	.00	.43	.00	.00	.00	.13
Care spp.	12.82	20.30	8.53	6.06	8.15	6.64	5.87	6.50	7.51

Table 3.4   Number of seeds per 1 litre of sieved sediment of weed species present in >10% of the samples.

SITE:	HH86	MT83	DL85	CH85	TT81	SW85	RC87	TH84	SS84
NO. OF SAMPLES:	21	10	12	14	127	32	23	23	63
NO. OF LITRES:	252	68	308	890	3,556	431	598	635	892
WEEDS									
Ranu acri	.00	.00	.00	.00	.00	.00	.00	.00	.00
Ranu repe	.00	.00	.01	.00	.00	.00	.00	.00	.00
Ranu Ranu	.02	.01	.00	.06	.01	.02	.02	.00	.00
Ranu flam	.00	.03	.01	.00	.01	.00	.01	.00	.00
Raph raph	.00	.01	.01	.00	.00	.03	.01	.01	.03
Mont font	.00	.00	.00	.00	.51	.08	.09	.00	.00
Stel medi	.02	.71	.07	.00	.01	.04	.23	.00	.00
Chen albu	.11	.90	1.42	.00	.10	.17	.13	.03	.00
Chen sp.	.01	.47	1.25	.00	.04	.05	.30	.01	.01
Atri spp.	.05	.06	.02	.00	.03	.01	.11	.00	.01
Chenop.	.00	.00	.00	.00	.02	.01	.01	.00	.00
Vici Lath	.00	.03	.03	.00	.01	.01	.24	.00	.03
Legumin.	.02	.10	.05	.01	.09	.36	.68	.01	.02
Pote erec	.00	.06	.04	.00	.02	.09	.07	.00	.00
Poly avic	.00	.13	.04	.00	.02	.02	.11	.01	.00
Poly conv	.00	.03	.02	.00	.01	.02	.02	.01	.00
Poly lapa	.05	.18	.05	.00	.01	.01	.00	.00	.00
Poly pers	.05	.16	.06	.00	.01	.01	.00	.00	.00
Poly l/p	.03	.03	.05	.00	.01	.01	.01	.01	.00
Rume acet	.00	.04	.07	.00	.00	.09	.03	.01	.01
Rume spp.	.02	.09	.10	.00	.13	.70	.06	.00	.14
Odon vern	.00	.00	.00	.00	.01	.01	.00	.00	.00
Prun vulg	.00	.01	.00	.00	.01	.00	.01	.00	.00
Plan lanc	.01	.09	.08	.00	.01	.03	.05	.01	.01
Gali apar	.00	.00	.00	.00	.02	.05	.13	.01	.03
Trip inod	.00	.01	.00	.00	.05	.03	.36	.00	.00
Aven sp.	.01	.01	.05	.01	.02	.00	.06	.01	.18
Brom m/s	.03	.04	.06	.00	.45	1.03	.31	.37	.79
Sieg decu	.00	.21	.04	.02	1.35	1.74	1.87	.03	.29
smal gras	.18	1.38	.91	.02	.20	.55	1.01	.05	.03
Gramin.	.06	.38	.07	.01	.29	.07	.16	.09	.02
Arrh elat	.00	.00	.00	.00	.02	.00	.01	.01	.00
Eleo sp.	.00	.00	.00	.00	.01	.01	.01	.00	.00
Junc sp.	.00	.00	.04	.00	.00	.01	.04	.00	.00
Care pilu	.00	.10	.07	.00	.01	.12	.04	.00	.01
Care puli	.00	.19	.00	.00	.02	.00	.00	.00	.00
Care spp.	.10	1.40	.43	.01	.31	.38	.38	.05	.13

164

Table 3.5   Coefficient of skewness for the distribution of the variables present in >10 % of the samples in the prehistoric assemblages.

COEFFICIENT OF SKEWNESS

	TRANSFORMATION		
	%	SQRT%	OCTAVE SCALE%
GRAIN			
Triticum dicoccum	4.4	3.6	3.2
Triticum spelta	3.7	0.7	0.1
Triticum sp.	1.0	-0.1	-0.7
Hordeum vulgare	0.6	-0.7	-2.3
CHAFF			
glume bases Triticum dicoccum	2.9	2.2	1.9
glume bases Triticum spelta	-0.2	-1.1	-2.0
glume bases Triticum sp.	0.2	-1.4	-3.1
rachis internodes Hordeum	2.9	1.2	0.3
WEEDS			
Ranunculus Subgenus Ranunculus	6.4	2.1	1.5
Ranunculus flammula	5.6	3.5	3.8
Raphanus raphanistrum	3.6	2.7	2.8
Stellaria media	3.9	1.6	1.1
Montia fontana	3.2	1.1	-0.1
Chenopodium album	2.8	1.2	-0.4
Atriplex spp.	6.4	2.2	1.2
Vicia/Lathyrus	3.7	2.0	1.7
Leguminosae indet.	2.7	0.6	-0.3
Potentilla cf. erecta	5.7	1.7	1.3
Polygonum aviculare	2.0	1.0	1.0
Polygonum convolvulus	6.1	2.6	2.5
Polygonum lapathifolium/persicaria	3.5	1.8	1.0
Rumex acetosella	4.5	2.4	2.1
Rumex spp.	2.4	0.7	0.2
Odontites verna	5.3	3.5	3.6
Prunella vulgaris	7.0	3.6	4.0
Plantago lanceolata	4.7	1.7	1.2
Galium aparine	1.9	0.9	0.9
Tripleurospermum inodorum	2.5	1.2	0.9
Avena sp.	5.1	1.9	1.7
Bromus mollis/secalinus	2.3	0.3	-1.0
Sieglingia decumbens	-0.2	-1.1	-1.9
small grasses	1.9	0.4	-1.2
Arrhenatherum elatius	4.1	2.1	2.0
Gramineae indet.	1.5	-0.1	-0.7
Eleocharis sp.	3.9	2.3	2.4
Carex pilulifera	5.6	2.0	1.5
Carex pulicaris	4.8	3.2	3.3
Carex spp.	1.9	0.1	-1.6
Juncus spp.	5.8	2.8	2.5

Table 4.1   Sample contexts from Hallshill.

```
CONTEXT DESCRIPTION                        VOLUME IN LITRES

16         fill of posthole                      2
17         fill of posthole                      2
18         fill of posthole                     10
19         fill of posthole                      2
20         fill of posthole                      2
21         fill of posthole                     11
24         fill of posthole                      8
26         fill of posthole                     12
27         fill of posthole                     10
28         fill of posthole                     12
30         fill of posthole                     13

 8         area of burning (hearth?)            70
23A        fill of pit                          10
23B        fill of pit                          10
23C        fill of pit                          10
23D        fill of pit                          10
23E        fill of pit                          10
23F        fill of pit                          10
23G        fill of pit                           4
25L        lower fill of pit                    17
25U        upper fill of pit                    17

TOTAL VOLUME OF SIEVED SEDIMENT IN LITRES:     252
```

Table 4.2a   Carbonized seeds from Hallshill.

CONTEXTS:	16	17	18	19	20	21	24	26	27	28	30	sub-total
VOL. IN LITRES:	2	2	10	2	2	11	8	12	10	12	13	84
GRAIN												
Trit dico	.	.	1	.	.	2	.	1	.	.	.	4
Trit spel	.	.	.	.	.	.	.	.	.	.	.	.
Trit sp.	.	.	2	1	.	.	.	.	.	.	.	3
Hord vulg	.	1	2	.	.	1	.	.	.	.	2	6
Cere inde	.	.	.	.	.	2	1	2	.	3	2	10
CHAFF												
glum dico	1	.	91	.	.	21	8	81	54	27	27	310
glum spel	.	.	4	.	.	.	1	.	.	.	7	12
glum inde	.	.	44	.	.	9	5	23	20	11	43	155
rach brit	.	.	4	.	.	.	1	4	11	1	9	30
base Trit	.	.	.	.	.	.	.	.	2	.	1	3
rach Hord	.	.	9	1	.	4	.	1	2	1	3	21
awns Aven	.	.	.	.	.	.	.	.	2	.	.	2
culm node	.	.	.	.	.	.	.	.	.	.	.	.
WEEDS												
Ranu Ranu	.	.	.	.	.	.	.	.	1	.	.	1
Bras sp.	.	.	.	.	.	.	.	1	.	.	.	1
Stel medi	.	.	.	.	.	1	.	.	1	.	.	2
Sper arve	.	.	.	.	.	.	.	.	.	.	.	.
Atri spp.	.	.	3	.	.	.	.	.	.	.	1	4
Chen albu	1	.	2	.	.	.	.	2	2	.	1	8
Chen sp.	.	.	1	.	.	.	.	.	.	.	.	1
Legumin.	.	.	.	.	.	.	.	1	.	1	.	2
Poly conv	.	.	.	.	.	.	.	.	.	.	.	.
Poly lapa	.	.	1	.	.	.	2	.	1	.	1	5
Poly pers	.	.	1	.	.	.	.	.	.	.	1	2
Poly l/p	.	.	.	.	.	.	.	.	.	.	.	.
Rume acet	.	.	.	.	.	.	.	.	1	.	.	1
Rume spp.	.	.	.	.	.	.	.	.	.	.	.	.
Vero scut	.	.	.	.	.	.	.	.	.	.	.	.
Ajug rept	.	.	.	.	.	.	.	.	.	1	.	1
Stac arve	.	.	.	.	.	.	.	.	.	.	.	.
Plan majo	.	.	.	.	.	.	.	.	.	.	.	.
Plan lanc	.	.	.	.	.	.	.	.	.	.	.	.
Brom m/s	.	.	.	.	.	.	1	.	.	.	1	2
Aven sp.	1	.	.	.	.	.	.	.	.	.	.	1
Sieg decu	.	.	.	.	.	1	.	.	.	.	.	1
smal gras	.	.	.	.	.	1	.	1	7	.	2	11
Gramin.	.	.	.	.	.	.	1	.	.	.	.	1
rhiz Gram	.	.	.	.	1	.	.	.	.	.	.	1
Care spp.	.	.	.	.	2	.	.	1	.	.	1	4
Junc sp.	.	.	.	.	.	.	.	.	.	.	.	.
OTHER												
Linu usit	.	.	1	.	.	.	.	.	.	.	.	1
Rubu sp.	.	.	.	.	.	.	.	.	1	.	.	1
Cory avel	.	.	1	.	.	.	4	2	1	2	3	13
Lyco euro	.	.	.	.	.	.	.	.	3	.	.	3
Call leaf	.	.	.	.	.	.	.	.	.	.	.	.
Eric flow	.	.	.	.	.	.	.	.	.	.	.	.
Pter aqui	.	.	.	.	.	.	.	.	.	.	1	1
INDET.	.	.	1	2	.	2	1	.	2	1	4	13
TOTAL	3	1	168	4	3	44	25	120	111	48	110	637

Table 4.3 Sample contexts from Murton.

CONTEXT	DESCRIPTION	VOLUME IN LITRES
617	fill of palisade trench P1	1
623	from floor area of timber-built house T9, occupation earth sealed by paved floor of stone-built house S7	20
624	from earth incorporated into wall of stone-built house S7, probably derived from context 623	20
625	from earth beneath enclosure wall and over filled innner ditch	20
630	from fill of post hole of timber-built house T9, sealed by paved floor of stone-built house S7	1.5
631	from fill of unlined pit sealed by paved floor	1
633	fill of posthole of timber-built house T9, sealed by paved floor of stone-built house S7	1.5
636	from fill of clay lined pit in floor area of timber-built house T9, sealed by paved floor of stone-built house S7	1
638	from fill of unlined pit sealed by paved floor of stone-built house S7	1
639	from fill of clay lined pit, probably associated with timber-built house T3, sealed by paved floor of stone-built house S1	1
	TOTAL VOLUME OF SIEVED SEDIMENT IN LITRES:	68

Table 4.2b Carbonized seeds from Hallshill.

CONTEXTS:	8	23A	23B	23C	23D	23E	23F	23G	25L*	250*	TOTAL
VOL. IN LITRES:	70	10	10	10	10	10	10	4	17	17	252
GRAIN											
Trit dico	18	7	4	2	5	4	1	.	23	36	104
Trit spel	2	.	.	1	.	1	1	.	6	18	29
Trit sp.	33	.	6	4	2	1	1	.	16	57	123
Hord vulg	6	4	6	.	5	2	1	1	4	5	40
Cere inde	29	5	7	5	7	10	3	2	36	63	177
CHAFF											
glum dico	5	86	38	70	98	58	90	12	133	157	1057
glum spel	3	3	1	3	3	3	.	5	3	5	35
glum inde	.	100	51	58	139	81	98	22	117	176	997
rach brit	2	9	5	8	23	11	7	5	18	24	142
base Trit	.	.	.	2	3	.	1	.	.	.	6
rach Hord	.	6	.	2	1	4	6	.	4	2	48
awns Aven	.	1	.	1	1	1	1	1	.	.	7
culm node	3	3	1	.	.	1	.	.	.	.	6
WEEDS											
Ranu Ranu	.	1	.	.	.	.	.	.	.	.	4
Bras sp.	.	.	.	.	.	2	.	.	.	.	4
Stel medi	.	.	.	1	.	1	.	.	.	1	4
Sper arve	.	.	.	.	.	.	.	1	1	1	13
Atri spp.	1	1	2	2	2	1	2	.	.	.	28
Chen albu	.	.	5	2	5	2	4	.	2	1	28
Chen sp.	.	.	.	2	2	.	1	.	.	.	6
Legumin.	.	.	.	.	1	.	.	.	.	.	2
Poly conv	1	1	1	1	1	2	1	.	1	1	12
Poly lapa	.	.	2	1	1	3	1	2	1	1	12
Poly pers	.	.	1	3	3	1	1	.	.	1	7
Poly l/p	.	.	.	.	.	.	.	.	.	.	1
Rume acet	.	.	.	.	1	.	.	.	.	.	1
Rume spp.	2	1	1	.	1	3	2	1	1	.	4
Vero scut	.	.	.	3	4	.	.	.	.	.	12
Ajug rept	.	.	1	.	1	3	.	.	.	.	3
Stac arve	.	.	.	.	1	2	.	.	1	.	1
Plan majo	1	1	.	1	1	2	.	.	.	.	3
Plan lanc	.	.	.	.	2	2	.	1	.	.	3
Brom m/s	1	.	.	.	.	.	.	1	.	.	8
Aven sp.	1	1	.	.	.	.	.	1	.	.	3
Sieg decu	.	.	.	.	.	.	.	.	.	.	1
smal gras	4	4	7	.	3	7	2	1	5	6	46
Gramin.	1	1	1	.	5	.	1	2	2	1	14
rhiz Gram	.	.	.	.	1	.	.	.	.	.	1
Care spp.	4	4	.	1	4	5	2	.	2	2	25
Junc sp.	1	1	.	.	.	.	.	.	.	.	1
OTHER											
Linu usit	1	.	.	.	.	.	.	1	1	.	2
Rubu sp.	.	.	.	.	.	.	.	1	.	.	3
Cory avel	.	5	5	.	9	8	.	1	2	1	44
Lyco euro	.	.	.	.	.	.	.	1	.	.	3
Call leaf	.	.	.	.	1	.	.	.	.	.	2
Eric flow	.	.	.	.	1	.	1	.	.	.	1
Pter aqui	7	7	3	7	6	7	8	5	.	.	44
INDET.	4	4	2	1	5	7	3	.	.	2	37
TOTAL	100	254	150	174	346	230	240	57	377	559	3124

KEY: * = only 25 per cent of this sample analysed.

167

Table 4.4   Carbonized seeds from Murton.

CONTEXTS:	617	623	624	625	630	631	633	636	638	639	TOTAL
VOL. IN LITRES:	1	20	20	20	1.5	1	1.5	1	1	1	68
GRAIN											
Trit dico	.	1	.	1	.	.	.	.	.	.	2
Trit spel	.	.	.	2	.	.	.	.	.	.	2
Trit sp.	.	.	1	5	.	.	.	.	.	.	6
Hord vulg	.	34	49	28	14	1	3	4	5	3	141
Cere inde	.	12	20	31	3	.	2	1	3	2	74
CHAFF											
glum dico	2	6	7	21	.	.	.	.	1	.	37
glum spel	.	.	2	16	.	.	.	.	.	.	18
glum Trit	.	4	4	16	.	.	.	.	.	.	24
glumes	.	.	.	1	.	.	.	.	.	.	1
rach brit	.	2	2	8	1	.	.	1	.	1	15
rach Hord	.	39	28	12	6	1	1	6	2	9	104
base Hord	.	.	.	.	.	.	.	2	.	.	2
chaf inde	.	.	.	2	.	.	.	.	.	.	2
WEEDS											
Ranu flam	.	.	1	1	.	.	.	.	.	.	2
Ranu Ranu	.	.	1	.	.	.	.	.	.	.	1
Raph raph	.	.	.	.	.	.	.	.	.	1	1
Stel medi	.	1	46	1	.	.	.	.	.	.	48
Sper arve	.	1	.	.	.	.	.	.	.	.	1
Chen albu	.	4	18	28	.	.	9	2	.	.	61
Chen sp.	.	2	5	24	.	.	.	.	.	1	32
Atri spp.	.	.	.	4	.	.	.	.	.	.	4
Vici Lath	.	.	1	1	.	.	.	.	.	.	2
Legumin.	.	1	2	2	2	.	.	.	.	.	7
Pote erec	.	1	3	.	.	.	.	.	.	.	4
Poly avic	.	.	7	1	.	.	1	.	.	.	9
Poly conv	.	.	1	1	.	.	.	.	.	.	2
Poly lapa	1	1	5	5	.	.	.	.	.	.	12
Poly pers	.	2	6	2	.	.	1	.	.	.	11
Poly l/p	.	.	.	2	.	.	.	.	.	.	2
Poly sp.	.	1	.	.	.	.	.	.	.	.	1
Rume spp.	.	1	.	1	4	.	.	.	.	.	6
Rume acet	.	1	.	.	2	.	.	.	.	.	3
Urti uren	.	.	9	5	.	.	1	.	.	.	15
Gale sp.	.	.	1	.	.	.	.	.	.	.	1
Prun vulg	.	.	1	.	.	.	.	.	.	.	1
Stac arve	.	.	.	1	.	.	.	.	.	.	1
Plan lanc	.	1	4	1	.	.	.	.	.	.	6
Trip inod	.	1	.	.	.	.	.	.	.	.	1
Brom m/s	.	.	2	.	1	.	.	.	.	.	3
Aven sp.	.	.	1	.	.	.	.	.	.	.	1
Sieg decu	.	1	2	9	2	.	.	.	.	.	14
smal gras	.	6	76	5	4	.	1	.	.	2	94
Gramin.	.	11	7	4	2	.	.	.	2	.	26
Care pilu	.	1	1	.	4	1	.	.	.	.	7
Care puli	.	1	12	.	.	.	.	.	.	.	13
Care spp.	.	3	77	11	2	.	2	.	.	.	95
OTHER											
Cory avel	.	.	1	.	1	.	.	.	.	.	2
Call leaf	.	5	2	.	1	.	.	.	.	.	8
Empe nigr	.	1	.	.	.	.	.	.	.	.	1
INDET.	1	5	30	9	4	1	.	.	.	.	50
TOTAL	4	149	436	261	53	4	21	16	13	19	976

Table 4.5   Sample contexts from Dod Law.

CONTEXT	DESCRIPTION	VOLUME IN LITRES
AREA A		
45/ 8	ground surface underneath inner rampart	29
45/ 9	ground surface underneath inner rampart	18
40/10	rubbish deposit in between the inner and outer rampart	25
40/11	rubbish deposit in between the inner and outer rampart	30
38/ 7	rubbish deposit in between the inner and outer rampart	30
25/ 3	rubbish deposit in between the inner and outer rampart	29
25/ 5	rubbish deposit in between the inner and outer rampart	18
30/ 6	rubbish deposit accumulated against the outer rampart	30
AREA B		
8/ 2	outer rampart deposit	23
8/12	above context 51	29
51/13	lense within make-up of outer rampart	22
AREA C		
24/ 1	charcoal rich deposit outside hut circles in extra-mural settlement	25
TOTAL VOLUME OF SIEVED SEDIMENT IN LITRES:		308

Table 4.6   Carbonized seeds from Dod Law.

CONTEXT:	45	45	40	40	38	25	25	30	8	8	51*	24	TOTAL
SAMPLE NO.:	8	9	10	11	7	3	5	6	2	12	13*	1	
VOL. IN LITRES:	29	18	25	30	30	29	18	30	23	29	22	25	308
GRAIN													
Trit dico	.	.	1	2	1	.	3	3	.	.	.	.	10
Trit spel	.	.	.	1	10	1	4	1	.	.	.	.	17
Trit sp.	2	.	7	4	4	9	6	9	.	.	.	.	41
Hord vulg	9	5	104	332	226	104	150	84	.	8	14	4	1040
Cere inde	.	3	50	304	192	35	127	81	.	8	15	4	819
CHAFF													
glum dico	3	.	60	110	62	1	32	49	.	1	6	.	324
glum spel	.	.	13	10	4	.	19	8	.	.	.	.	54
glum inde	3	.	42	57	41	.	46	33	.	1	9	.	232
glumes	.	.	2	1	.	.	1	.	.	.	.	.	4
rach brit	2	.	21	25	16	.	19	10	.	3	3	.	99
rach Hord	4	.	131	199	68	.	42	69	.	9	15	1	538
base Hord	.	.	6	21	3	.	1	.	.	.	.	.	31
chaf inde	.	.	.	.	1	.	.	.	.	.	.	.	1
awns Aven	.	.	.	.	.	.	.	.	.	.	19	.	19
flor Avef	.	.	.	.	.	.	.	.	.	.	3	.	3
culm node	3	1	.	.	1	.	2	.	.	.	1	.	8
WEEDS													
Ranu repe	.	.	1	.	.	.	.	.	.	.	1	.	2
Ranu flam	.	.	.	.	1	.	.	.	.	.	1	.	2
Raph raph	.	.	1	.	1	.	.	2	.	.	.	.	4
Bras sp.	.	.	2	3	1	.	1	1	.	.	.	.	8
Mont font	.	.	.	.	1	.	.	.	.	.	.	.	1
Stel medi	4	.	4	2	5	.	4	3	.	1	.	.	23
Sper arve	5	.	2	1	4	.	.	2	.	2	3	.	19
Atri spp.	3	.	.	.	.	.	.	2	.	.	2	.	7
Chen albu	9	3	52	159	101	7	75	23	.	1	7	.	437
Chen sp.	6	1	30	138	111	.	44	49	.	1	6	.	386
Vici Lath	.	.	.	2	.	1	3	2	.	.	.	.	8
Legumin.	6	2	1	1	.	.	.	3	.	.	2	.	15
Pote erec	2	.	1	4	3	.	1	.	.	.	.	.	11
Poly avic	1	2	1	4	1	.	3	.	.	.	.	.	12
Poly conv	.	.	.	1	1	1	2	2	.	.	.	.	7
Poly lapa	.	.	3	7	4	.	.	1	.	.	.	.	15
Poly pers	.	.	2	7	4	.	5	.	.	.	2	.	20
Poly l/p	1	.	.	5	3	.	1	1	.	.	3	.	14
Poly sp.	1	.	2	5	1	.	1	4	.	.	.	.	14
Rume acet	2	.	.	3	2	.	3	8	.	1	4	.	23
Rume spp.	.	.	10	3	6	.	4	7	.	.	.	.	30
Urti uren	1	.	.	1	.	.	.	.	.	.	.	.	2
Urti dioc	.	.	2	.	.	.	.	.	.	.	.	.	2
Vero arve	1	.	.	.	.	.	.	.	.	.	.	.	1
Ajug rept	.	.	.	.	.	.	.	.	.	1	.	.	1
Gale sp.	1	.	.	.	.	.	.	.	.	.	.	.	1
Lami a/p	1	.	.	.	.	.	.	1	.	.	.	.	2
Plan lanc	3	1	7	8	4	.	1	2	.	.	.	.	26
Aven sp.	3	2	2	.	.	.	2	.	.	2	4	.	15
Gali apar	.	.	.	.	1	.	.	.	.	.	.	.	1
Brom m/s	.	.	2	6	4	1	4	2	.	.	1	.	20
Sieg decu	.	.	2	3	2	.	2	2	.	.	.	.	11
smal gras	42	16	42	54	42	.	39	39	.	2	1	2	279
Arrh elat	.	.	.	.	.	.	1	.	.	.	.	.	1
Gramin.	3	.	.	2	.	.	.	12	.	2	3	.	22
rhiz Gram	.	.	10	.	6	.	9	2	.	.	.	.	27
Care pilu	.	.	5	9	6	.	1	1	.	.	.	.	22
Care spp.	7	.	24	36	45	1	8	12	.	.	.	.	133
OTHER													
Cory avel	93	6	1	.	2	.	2	.	.	1	1	.	106
Rosa sp.	4	.	.	.	1	.	.	.	.	.	.	.	5
Rubu sp.	.	.	1	.	.	.	1	.	.	.	.	.	2
tree buds	.	.	.	.	.	2	.	1	.	3	4	.	10
Call leaf	.	.	6	19	13	.	5	10	.	.	.	.	53
Call flow	.	.	77	139	127	.	72	112	.	.	.	.	527
Eric flow	.	.	2	3	8	.	1	2	.	.	.	.	16
Empe nigr	.	.	2	.	.	.	.	.	.	.	.	.	2
Junc sp.	.	.	2	2	6	.	1	1	.	.	1	.	13
Pter aqui	.	.	2	4	3	.	.	.	.	.	1	.	10
INDET.	10	2	8	15	13	.	7	12	2	1	5	1	76
TOTAL	235	44	746	1712	1162	163	755	668	2	48	137*	12	5684

KEY: * = only 25 per cent of this sample has been analysed.

Table 4.7 Sample contexts from Chester House.

CONTEXT	DESCRIPTION	VOLUME IN LITRES
110	lower fill of enclosure ditch	30
114	fill of context 115, pit/posthole	60
116	fill of context 139, ring groove of House 1	120
117	fill of context 119, eavesdrop gulley of House 2	270
118	fill of context 120, eavesdrop gulley of House 1	90
127	fill of context 128, posthole, House 1	25
129	fill of context 130, posthole	30
131	fill of context 132, posthole, House 1	20
133	fill of context 134, ring groove of House 2	60
135	fill of context 136, posthole	25
137	fill of context 138, posthole	15
140	fill of context 141, gulley of House 3	30
142	fill of context 143, pit	25
148	fill of context 149, palisade trench	90
TOTAL VOLUME OF SIEVED SEDIMENT IN LITRES:		890

Table 4.8 Carbonized seeds from Chester House.

CONTEXT:	110	114	116	117	118	127	129	131	133	135	137	140	142	148	TOTAL
NO. OF SAMPLES:	1	2	4	9	3	1	1	1	2	1	1	1	1	3	31
VOL. IN LITRES:	30	60	120	270	90	25	30	20	60	25	15	30	25	90	890
GRAIN															
Trit spel	.	.	.	.	.	.	.	.	.	.	.	.	.	1	1
Trit sp.	.	.	.	3	1	3	.	.	.	.	.	.	.	.	7
Hord vulg	.	.	.	16	1	3	.	4	2	3	.	1	2	2	33
Cere inde	.	1	7	21	6	7	2	3	2	2	1	2	3	3	59
CHAFF															
glum dico	.	.	.	22	1	4	.	1	.	4	3	5	7	2	49
glum spel	.	.	.	1	1	.	.	.	.	.	.	1	1	.	4
glum inde	.	.	1	8	5	3	.	.	1	4	2	2	9	.	36
rach brit	.	.	.	1	1	1	.	1	1	3	4	4	2	.	21
rach Hord	.	1	.	14	1	1	.	3	.	4	1	5	11	.	40
awns Aven	.	.	.	1	.	.	.	.	.	1	.	.	.	.	2
flor Avef	.	.	.	.	.	.	.	.	.	.	.	.	.	.	2
flor Aven	.	.	.	1	.	2	.	.	.	.	.	.	.	.	2
culm node	.	.	.	1	.	.	.	.	.	.	.	.	.	.	1
chaf inde	.	.	.	2	.	.	.	.	.	.	.	.	.	.	2
WEEDS															
Ranu Ranu	1	.	11	10	5	2	.	4	7	.	.	3	.	6	49
Crucif.	.	.	1	1	.	.	.	.	.	.	.	.	.	.	1
Sper arve	.	.	.	.	.	.	.	.	.	.	.	.	1	.	1
Atri spp.	.	.	.	.	.	.	.	.	.	.	.	.	1	.	1
Chen albu	.	.	.	1	.	.	.	.	.	.	.	1	.	.	1
Chen sp.	.	.	.	1	.	1	.	1	.	.	.	.	.	.	2
Chenop.	.	.	.	1	.	.	.	.	.	.	.	.	.	3	3
Vici Lath	1	.	2	1	1	.	.	.	.	.	.	1	.	.	8
Legumin.	.	.	2	.	.	.	.	.	.	.	.	.	.	.	1
Poly conv	.	.	.	1	.	.	.	.	.	1	.	.	.	.	1
Poly lapa	.	.	.	1	.	.	.	.	.	.	.	.	1	.	1
Poly pers	.	.	.	1	1	.	.	1	.	.	.	1	1	.	2
Poly l/p	.	.	.	.	2	.	.	1	.	1	.	.	.	.	3
Rume acet	.	.	1	.	.	1	.	.	2	.	.	.	.	.	1
Plan lanc	.	.	.	2	.	5	.	.	1	.	.	.	.	.	7
Aven sp.	.	.	.	2	1	1	.	.	.	.	.	.	1	.	4
Brom m/s	.	.	1	12	.	.	.	.	.	.	.	1	1	1	15
Sieg decu	.	.	.	8	.	.	.	.	.	1	.	2	1	1	14
smal gras	.	.	.	5	.	.	.	.	.	.	.	.	1	.	6
Gramin.	.	.	.	6	.	.	.	1	.	.	.	.	.	.	7
rhiz Gram	.	.	.	1	1	.	.	.	.	.	.	1	.	.	1
Care pilu	.	.	.	3	1	1	.	.	.	.	1	1	.	.	8
Care spp.	.	.	.	.	.	.	.	.	.	.	.	.	.	3	3
OTHER															
Cory avel	.	.	1	.	.	.	.	.	.	.	1	1	.	1	4
tree buds	.	.	.	.	1	.	.	.	.	1	1	.	.	.	1
Call leaf	.	.	.	.	1	.	.	.	.	.	.	.	.	.	1
Call flow	.	.	.	2	.	.	.	.	.	.	.	.	2	.	3
Junc sp.	.	.	.	.	.	.	.	.	.	.	.	2	.	.	2
Pter. aqui	1	.	.	.	.	.	.	.	.	.	.	.	.	.	3
INDET.	1	2	2	12	.	1	1	3	3	1	1	.	1	2	27
TOTAL	3	5	28	168	28	36	3	17	17	26	11	33	43	21	439

Table 4.9 Sample contexts from Thorpe Thewles.

CATEGORIES OF SAMPLES:

Linear samples : random samples from linear features, such as ditches and gullies (coded LS)

Point samples : random samples from point features, such as pits and postholes (coded PF)

Judgement samples : samples from features judged by the excavator to be important contexts which needed sampling (coded JS)

Masking layer samples : samples from the layers of extant stratigraphy overlying the subsoil cut features (coded ML)

PHASE	NO. OF SAMPLES PER PHASE	VOLUME IN LITRES
PHASE I	2 samples (1 LS, 1PF)	56
PHASE I/II	1 sample (1 LS)	28
PHASE II	44 samples (27 LS, 4 PF, 13 JS)	1232
PHASE II/III	10 samples (8 LS, 2 JS)	280
PHASE III	29 samples (16 LS, 3 PF, 10 JS)	812
PHASE III/IV	29 samples (1 LS, 28 ML)	812
PHASE IV	12 samples (10 LS, 2 JS)	336
TOTAL OF SIEVED SEDIMENT IN LITRES:		3556

Table 4.10a   Carbonized seeds from Thorpe Thewles.

	I		I/II	II							sub-
PHASE:											
SAMPLE NO.:	LS 268	PF 1	LS 270	LS 112	LS 120	LS 138	LS 150	LS 160	LS 233	LS 248	total
VOL. IN LITRES:	28	28	28	28	28	28	28	28	28	28	280
GRAIN											
Trit dico	.	.	.	.	.	.	.	.	.	.	.
Trit spel	3	.	.	6	13	9	4	2	.	.	37
Trit sp.	3	3	.	4	10	9	4	3	1	.	37
Hord vulg	1	5	3	7	17	15	9	2	1	7	67
Cere inde	7	3	2	9	33	17	7	3	6	3	90
CHAFF											
glum dico	.	.	.	.	.	.	.	.	.	.	.
glum spel	15	.	3	66	236	65	39	8	6	10	448
glum inde	17	1	7	54	276	43	18	11	6	25	458
glumes	3	.	1	.	20	6	.	.	.	.	30
rach brit	8	.	.	13	102	6	4	2	.	4	139
base Trit	.	.	.	.	.	.	.	.	.	.	.
rach Hord	1	35	.	.	20	2	1	.	2	.	61
base Hord	.	.	.	.	1	.	.	.	.	.	1
flor Avef	.	.	.	.	1	.	.	.	.	.	1
awns Aven	1	.	.	.	9	.	.	.	.	1	11
culm node	.	.	.	.	.	.	.	.	.	1	1
chaf inde	.	.	.	.	.	.	.	.	.	.	.
WEEDS											
Ranu acri	.	.	.	.	.	.	.	.	.	.	.
Ranu repe	.	1	.	.	.	.	.	.	.	.	1
Ranu Ranu	.	.	1	1	4	2	.	.	.	.	8
Ranu flam	.	.	.	.	.	1	.	.	.	.	1
Papa arge	.	.	.	.	.	.	.	.	.	.	.
Papa rh/d	.	.	.	.	.	.	.	.	.	.	.
Papa sp.	.	.	.	.	.	.	.	.	.	.	.
Raph raph	.	1	.	.	.	.	.	.	.	.	1
Bras sp.	.	.	.	.	1	.	.	.	.	.	1
Crucif.	.	.	.	.	1	.	.	.	.	.	1
Viol Mela	.	.	.	.	.	.	.	.	.	.	.
Mont font	5	10	.	3	55	9	1	3	3	1	90
Stel medi	.	.	.	1	1	1	.	.	.	.	3
Chen albu	7	.	.	2	10	6	3	12	.	.	40
Chen sp.	1	.	1	.	6	3	1	.	.	1	13
Atri spp.	3	.	.	1	13	1	1	.	1	.	20
Chenop.	.	.	.	.	.	.	.	.	.	.	.
Malv sylv	.	.	.	.	.	.	.	.	.	.	.
Malv sp.	.	.	.	.	.	.	.	.	.	.	.
Vici Lath	.	.	1	1	1	2	1	.	.	.	6
Legumin.	.	2	1	4	15	5	.	.	1	.	28
Apha arve	.	.	.	.	.	.	.	.	.	.	.
Pote erec	.	1	.	.	1	1	.	.	.	.	3
Hera spon	.	.	.	.	.	.	.	.	1	.	1
Poly avic	1	.	.	.	4	1	1	.	1	.	8
Poly conv	.	1	.	.	.	1	1	.	.	.	3
Poly lapa	.	.	.	.	.	2	.	.	.	.	2
Poly pers	.	.	.	.	.	1	1	.	1	1	4
Poly l/p	.	.	.	.	1	.	.	.	.	.	1
Rume acet	.	.	.	.	.	.	1	.	.	.	1
Rume spp.	3	.	.	.	31	11	5	1	.	1	52
Polygon.	2	.	.	.	1	1	.	.	.	.	4
Urti dioc	.	.	.	.	.	.	.	.	.	.	.
Urti uren	.	.	.	.	.	.	.	.	.	.	.
Sola nigr	.	.	.	.	.	.	.	.	.	.	.
Hyos nige	.	.	.	.	.	.	.	.	.	.	.
Odon vern	.	.	.	1	.	.	.	.	.	.	1
Verb sp.	.	.	.	.	.	.	1	.	.	.	1
Lami a/p	.	.	.	.	.	.	.	.	.	.	.
Ment a/a	.	.	.	.	.	.	.	.	.	.	.
Prun vulg	1	.	.	.	.	.	.	.	.	.	1
Plan lanc	.	.	.	2	3	2	.	.	.	.	7
Plan maju	.	.	.	.	.	.	.	.	1	.	1
Gali apar	.	.	.	.	3	1	.	.	.	.	4
Gali palu	.	.	.	.	.	.	.	.	.	.	.
Sher arve	.	.	.	.	.	.	.	.	.	.	.
Vale dent	.	.	.	.	.	.	.	.	.	.	.
Trip inod	2	.	.	4	13	2	1	.	.	1	23
Laps comm	.	.	.	.	.	.	.	.	.	.	.
Sonc aspe	.	.	.	.	.	.	.	.	.	.	.
Compos.	.	.	.	.	.	.	1	.	.	.	1
Aven sp.	.	.	.	2	3	.	.	.	.	.	5
Brom m/s	6	.	.	24	183	43	16	5	4	1	282
Sieg decu	27	16	12	20	168	80	22	10	39	10	404
Smal gras	3	.	1	29	37	11	10	1	1	.	93
Gramin.	5	1	3	6	.	23	8	9	7	3	65
Arrh elat	.	.	.	1	2	2	.	.	.	1	6
rhiz Gram	6	9	1	5	24	22	9	1	18	4	99
Junc sp.	.	.	.	.	.	.	.	.	.	.	.
Eleo sp.	.	.	.	.	1	.	.	.	.	.	1
Isol seta	.	.	.	.	.	.	.	.	.	.	.
Care pilu	.	.	.	.	.	2	.	.	.	.	2
Care puli	.	.	.	.	.	.	.	.	.	.	.
Care spp.	16	12	1	9	23	14	1	3	1	1	81
OTHER											
Cory avel	.	.	1	.	.	.	.	1	1	.	3
Crat mono	.	.	.	.	.	.	.	.	.	.	.
INDET.	5	4	.	3	10	9	5	2	1	.	39
TOTAL	152	105	39	278	1353	431	176	79	103	76	2792

Table 4.10b   Carbonized seeds from Thorpe Thewles.

PHASE: II	LS 304	LS 350	LS 496	LS 515	LS 523	LS 547	LS 551	LS 570	LS 573	LS 584	sub-total
VOL. IN LITRES:	28	28	28	28	28	28	28	28	28	28	280
GRAIN											
Trit dico	.	.	.	.	.	.	.	.	.	.	.
Trit spel	.	.	1	.	.	.	.	.	1	.	2
Trit sp.	.	.	.	.	.	.	.	1	.	.	1
Hord vulg	.	.	.	.	.	.	1	.	.	.	.
Cere inde	2	.	2	1	5	1	1	.	.	.	12
CHAFF											
glum dico	.	.	.	.	.	.	.	.	.	.	.
glum spel	6	.	.	.	4	.	17	5	.	1	33
glum inde	4	1	.	2	.	1	14	2	.	.	24
glumes	.	.	.	.	.	.	1	1	.	.	2
rach brit	.	.	.	1	.	.	4	2	.	.	7
base Trit	.	.	.	.	.	.	.	.	.	.	.
rach Hord	.	.	1	.	.	.	2	.	.	.	3
base Hord	.	.	.	.	.	.	.	.	.	.	.
flor Avef	.	.	.	.	.	.	.	.	.	.	.
awns Aven	.	.	.	.	.	.	.	.	.	.	.
culm node	.	.	.	.	.	.	.	.	.	.	.
chaf inde	.	.	.	.	.	1	.	.	.	.	1
WEEDS											
Ranu acri	.	.	.	.	.	.	.	.	.	.	.
Ranu repe	.	.	.	.	.	.	.	.	.	.	.
Ranu Ranu	.	.	.	.	.	.	.	.	.	.	.
Ranu flam	.	.	.	.	.	.	.	.	.	.	.
Papa arge	.	.	.	.	.	1	.	.	.	.	1
Papa rh/d	.	.	.	.	.	.	.	.	.	.	.
Papa sp.	.	.	.	.	.	.	.	.	.	.	.
Raph raph	.	.	.	.	.	.	.	.	.	.	.
Bras sp.	.	.	.	.	.	.	.	.	.	.	.
Crucif.	.	.	.	.	.	.	.	.	.	.	.
Viol Mela	.	.	.	.	.	.	.	.	.	.	.
Mont font	.	1	.	1	9	.	4	.	.	.	15
Stel medi	.	.	.	.	.	.	.	.	.	.	.
Chen albu	.	.	1	.	.	.	1	.	.	.	2
Chen sp.	.	.	.	.	.	.	.	1	.	.	1
Atri spp.	.	.	.	.	.	.	.	.	.	.	.
Chenop.	.	.	.	.	.	.	4	.	.	.	.
Malv sylv	.	.	.	.	.	.	.	.	.	.	.
Malv sp.	.	.	.	.	.	.	.	.	.	.	.
Vici Lath	.	.	.	.	.	.	.	.	.	.	.
Legumin.	.	.	.	.	5	.	4	.	.	1	10
Apha arve	.	.	.	.	.	.	.	.	.	.	.
Pote erec	.	.	.	.	.	.	.	.	.	.	.
Hera spon	.	.	.	.	.	.	.	.	.	.	.
Poly avic	.	.	.	.	.	.	.	.	.	.	.
Poly conv	.	.	.	.	.	.	.	.	.	.	.
Poly lapa	.	.	.	.	.	.	.	.	.	.	.
Poly pers	.	.	.	.	.	.	.	.	.	.	.
Poly l/p	.	.	.	.	.	.	.	.	.	.	.
Rume acet	.	1	.	.	1	.	.	.	.	.	2
Rume spp.	.	.	.	1	2	.	.	.	.	.	3
Polygon.	.	.	.	.	.	.	.	.	.	.	.
Urti dioc	.	.	.	.	.	.	.	.	.	.	.
Urti uren	.	.	.	.	.	.	.	.	.	.	.
Sola nigr	.	.	.	.	.	.	.	.	.	.	.
Hyos nige	.	.	.	.	.	.	.	.	.	.	.
Odon vern	.	.	.	.	.	.	.	.	.	.	.
Verb sp.	.	.	.	.	.	.	.	.	.	.	.
Lami a/p	.	.	.	.	.	.	.	.	.	.	.
Ment a/a	.	.	.	.	.	.	.	.	.	.	.
Prun vulg	.	.	.	.	.	.	.	.	.	.	.
Plan lanc	.	.	1	.	.	.	.	.	.	.	1
Plan majo	.	.	1	.	.	2	.	.	.	.	3
Gali apar	.	.	.	.	.	.	1	.	.	.	1
Gali palu	.	.	.	.	.	.	.	.	.	.	.
Sher arve	.	.	.	.	.	.	.	.	.	.	.
Vale dent	.	.	.	.	.	.	.	.	.	.	.
Trip inod	.	1	.	1	.	.	1	.	.	.	3
Laps comm	.	.	.	.	.	.	.	.	.	.	.
Sonc aspe	.	.	.	.	.	.	.	.	.	.	.
Compos.	.	.	.	.	.	.	.	.	.	.	.
Aven sp.	.	.	.	.	.	.	.	.	.	.	.
Brom m/s	.	.	1	.	2	2	4	2	.	.	11
Sieg decu	4	1	3	6	27	3	31	4	.	.	79
Smal gras	1	1	.	.	3	.	4	.	.	.	9
Gramin.	.	.	1	2	.	1	1	2	.	3	10
Arrh elat	.	.	.	.	.	.	.	.	.	.	.
rhiz Gram	.	.	.	1	14	.	3	1	.	.	19
Junc sp.	.	.	.	.	.	.	.	.	.	.	.
Eleo sp.	.	.	.	.	.	.	.	.	.	.	.
Isol seta	.	.	.	.	.	.	.	.	.	.	.
Care pilu	.	.	.	.	.	.	.	.	.	.	.
Care puli	.	.	.	.	.	.	.	.	.	.	.
Care spp.	.	.	.	.	4	.	2	.	.	.	6
OTHER											
Cory avel	1	.	.	.	.	.	.	.	.	.	1
Crat mono	.	.	.	.	.	.	.	.	.	.	.
INDET.	.	2	5	2	4	.	6	.	2	3	24
TOTAL	18	8	17	18	80	12	106	21	3	8	291

Table 4.10c   Carbonized seeds from Thorpe Thewles.

PHASE II	LS589	LS 614	LS 631	LS 634	LS 637	LS 652	LS 664	LS 670	LS 674	LS 677	sub-total
VOL. IN LITRES:	28	28	28	28	28	28	28	28	28	28	280
GRAIN											
Trit dico	.	.	.	.	.	.	.	.	2	.	2
Trit spel	.	.	1	.	.	.	1	.	.	.	2
Trit sp.	.	1	.	.	.	.	.	.	.	.	1
Hord vulg	.	.	1	.	.	.	.	.	.	.	1
Cere inde	1	.	4	.	2	4	1	1	2	1	16
CHAFF											
glum dico	.	.	.	.	.	.	.	.	.	.	.
glum spel	1	.	.	.	4	1	.	.	.	.	6
glum inde	2	.	.	.	2	.	.	1	3	1	9
glumes	.	.	.	.	.	.	.	.	.	.	.
rach brit	.	.	.	.	.	.	.	1	.	.	1
base Trit	.	.	.	.	.	.	.	.	.	.	.
rach Hord	.	.	.	.	1	.	.	.	1	1	3
base Hord	.	.	.	.	.	.	.	.	.	.	.
flor Avef	.	.	.	.	.	.	.	.	.	.	.
awns Aven	.	.	.	.	.	.	.	.	.	.	.
culm node	.	.	.	.	.	.	.	.	.	.	.
chaf inde	.	.	.	.	.	.	.	.	.	.	.
WEEDS											
Ranu acri	.	.	.	.	.	.	.	.	.	.	.
Ranu repe	.	.	.	.	.	.	.	.	.	.	.
Ranu Ranu	.	.	.	1	1	.	.	.	.	.	2
Ranu flam	.	.	.	.	.	.	.	.	.	.	.
Papa arge	.	.	.	.	.	.	.	.	.	.	.
Papa rh/d	.	.	.	.	.	.	.	.	.	.	.
Papa sp.	.	.	.	.	.	.	.	.	.	.	.
Raph raph	.	.	.	.	.	.	.	.	.	.	.
Bras sp.	.	.	.	.	.	.	.	.	.	.	.
Crucif.	.	.	.	.	.	.	.	.	.	.	.
Viol Mela	.	.	.	.	.	.	.	.	.	.	.
Mont font	1	.	.	.	1	.	.	.	1	.	3
Stel medi	.	.	1	.	.	.	.	.	.	1	2
Chen albu	.	.	.	.	7	.	.	.	.	.	7
Chen sp.	.	.	.	.	11	.	1	.	.	.	12
Atri spp.	.	.	.	.	.	.	.	1	.	.	1
Chenop.	.	.	.	.	.	.	.	.	.	.	.
Malv sylv	.	.	.	.	.	.	.	.	.	.	.
Malv sp.	.	.	.	.	.	.	.	.	.	.	.
Vici Lath	.	.	.	.	.	.	.	.	.	.	.
Legumin.	.	.	.	.	8	.	.	1	.	.	9
Apha arve	.	.	.	.	.	.	.	.	.	.	.
Pote erec	.	1	.	.	2	.	.	.	1	.	4
Hera spon	.	.	.	.	.	.	.	.	.	.	.
Poly avic	.	.	.	.	.	.	.	.	.	.	.
Poly conv	.	.	.	.	.	.	.	.	.	.	.
Poly lapa	.	.	.	.	.	.	.	.	.	.	.
Poly pers	.	.	.	.	.	.	.	.	.	.	.
Poly l/p	.	.	.	.	.	.	.	.	.	.	.
Rume acet	.	.	.	.	.	.	.	.	.	.	.
Rume spp.	1	.	.	.	4	.	.	.	1	.	6
Polygon.	.	.	.	.	.	.	.	.	.	.	.
Urti dioc	.	.	.	.	.	.	.	.	.	.	.
Urti uren	.	.	.	.	.	.	.	.	.	.	.
Sola nigr	.	.	.	.	.	.	.	.	.	.	.
Hyos nige	.	.	.	.	.	.	.	.	.	.	.
Odon vern	.	.	.	.	.	.	.	.	.	.	.
Verb sp.	.	.	.	.	.	.	.	.	.	.	.
Lami a/p	.	.	.	.	.	.	.	.	.	.	.
Ment a/a	.	.	.	.	.	.	.	.	.	.	.
Prun vulg	.	.	.	.	.	.	.	.	.	.	.
Plan lanc	.	.	.	.	.	.	.	.	.	.	.
Plan majo	.	.	.	.	.	.	.	.	.	.	.
Gali apar	.	.	.	.	1	.	.	.	.	1	2
Gali palu	.	.	.	.	.	.	.	.	.	.	.
Sher arve	.	.	.	.	.	.	.	.	.	.	.
Vale dent	.	.	.	.	.	.	.	.	.	.	.
Trip inod	.	.	.	.	.	.	.	.	.	.	.
Laps comm	.	.	.	.	.	.	.	.	.	.	.
Sonc aspe	.	.	.	.	.	.	.	.	.	.	.
Compos.	.	.	.	.	.	.	.	.	.	.	.
Aven sp.	.	.	.	.	.	.	.	.	.	.	.
Brom m/s	.	1	2	.	2	.	1	2	3	.	11
Sieg decu	2	2	10	.	32	.	2	.	.	1	49
Smal gras	.	2	.	.	2	.	.	.	.	.	4
Gramin.	2	.	2	1	9	3	1	.	3	.	21
Arrh elat	.	.	.	.	.	.	.	.	.	.	.
rhiz Gram	1	.	.	.	17	1	.	.	3	.	22
Junc sp.	.	.	.	.	.	.	.	.	.	.	.
Eleo sp.	.	.	.	.	.	.	.	.	.	.	.
Isol seta	.	.	.	.	.	.	.	.	.	.	.
Care pilu	.	.	.	.	.	.	.	.	.	.	.
Care puli	.	.	.	.	.	.	.	.	.	.	.
Care spp.	.	1	2	.	17	.	.	2	.	.	22
OTHER											
Cory avel	.	.	.	.	.	.	.	.	.	.	.
Crat mono	.	.	.	.	.	.	.	.	.	.	.
INDET.	.	.	1	2	6	.	1	1	1	3	15
TOTAL	11	8	24	4	129	9	8	8	21	9	231

Table 4.10d   Carbonized seeds from Thorpe Thewles.

PHASE: II	PF 29	PF 41	PF 46	PF 57	JS 2	JS 3	JS 4*	JS 7	JS 9	JS 13	sub-total
VOL. IN LITRES:	28	28	28	28	28	28	28	28	28	28	280
GRAIN											
Trit dico	.	.	.	.	1	.	.	.	.	.	1
Trit spel	.	.	.	1	5	5	.	4	.	7	22
Trit sp.	.	1	.	4	3	11	4	.	.	3	26
Hord vulg	1	1	.	7	59	12	6	59	8	7	.
Cere inde	2	7	.	18	70	33	10	72	4	14	230
CHAFF											
glum dico	.	.	.	.	.	.	.	.	.	.	.
glum spel	.	45	.	72	10	3	23	7	6	59	225
glum inde	3	41	2	117	5	2	11	.	3	56	240
glumes	.	5	.	4	.	.	.	.	.	4	13
rach brit	.	10	1	18	19	1	4	1	7	38	99
base Trit	.	.	.	.	.	.	.	.	.	2	2
rach Hord	.	.	.	3	12	.	.	.	2	15	32
base Hord	.	.	.	.	.	.	.	.	.	.	.
flor Avef	.	.	.	.	.	.	.	.	.	.	.
awns Aven	.	.	.	1	.	.	.	.	.	2	3
culm node	.	.	.	.	1	1	.	.	.	.	2
chaf inde	1	.	.	.	.	.	.	.	.	.	1
WEEDS											
Ranu acri	.	.	.	.	.	.	.	.	.	.	.
Ranu repe	.	.	.	.	.	.	.	.	.	1	1
Ranu Ranu	.	.	1	.	.	.	3	1	2	.	7
Ranu flam	.	.	.	.	.	.	.	1	.	.	1
Papa arge	.	.	.	.	.	.	.	.	.	.	.
Papa rh/d	.	.	.	.	.	.	.	.	.	.	.
Papa sp.	.	.	.	.	.	.	.	.	.	.	.
Raph raph	.	.	.	1	.	.	.	.	.	.	1
Bras sp.	.	.	.	.	.	.	.	.	.	.	.
Crucif.	.	.	.	.	1	.	.	1	.	.	2
Viol Mela	.	.	.	.	1	.	.	.	1	.	2
Mont font	8	2	4	15	325	173	321	.	12	18	878
Stel medi	.	.	.	.	3	2	.	1	.	.	6
Chen albu	2	.	1	5	.	2	5	1	4	9	29
Chen sp.	.	.	.	10	1	2	1	.	1	5	20
Atri spp.	.	.	.	2	1	2	1	.	1	4	11
Chenop.	.	.	.	.	.	.	.	.	.	.	.
Malv sylv	.	.	.	.	.	.	.	.	.	.	.
Malv sp.	.	.	.	.	.	.	.	.	.	.	.
Vici Lath	.	.	.	.	.	1	2	1	.	3	7
Legumin.	1	1	.	3	16	21	17	14	1	8	82
Apha arve	.	.	.	.	.	.	.	.	.	.	.
Pote erec	.	.	.	.	13	3	10	2	2	2	32
Hera spon	.	.	.	.	.	.	.	.	.	.	.
Poly avic	1	.	.	1	.	2	.	6	.	.	10
Poly conv	.	.	.	1	1	1	.	1	.	.	4
Poly lapa	.	.	.	1	.	.	1	9	2	1	14
Poly pers	.	.	.	3	.	.	.	.	2	.	5
Poly l/p	1	.	.	.	.	.	.	.	.	1	2
Rume acet	.	.	.	1	.	.	.	1	.	1	3
Rume spp.	.	1	.	10	5	4	.	3	9	25	57
Polygon.	.	.	.	.	1	.	.	.	.	.	1
Urti dioc	.	.	.	.	.	2	.	.	.	1	3
Urti uren	.	.	.	.	.	.	.	.	.	.	.
Sola nigr	.	.	.	.	.	.	.	.	.	.	.
Hyos nige	.	.	.	.	.	.	.	.	.	.	.
Odon vern	.	.	.	1	.	.	.	.	.	.	1
Verb sp.	.	.	.	.	.	.	.	.	.	.	.
Lami a/p	.	.	.	.	.	.	.	.	.	.	.
Ment a/a	.	.	.	.	.	.	.	.	.	.	.
Prun vulg	.	.	.	.	2	1	1	.	1	.	5
Plan lanc	.	1	.	.	4	2	5	1	1	2	16
Plan majo	.	.	.	.	.	1	.	.	.	.	1
Gali apar	.	1	.	1	.	5	.	.	.	.	7
Gali palu	.	.	.	.	.	.	.	.	.	.	.
Sher arve	.	.	.	.	.	1	1	.	.	.	2
Vale dent	.	.	.	.	.	.	.	.	.	.	.
Trip inod	.	.	.	2	5	.	.	1	2	16	26
Laps comm	.	.	.	.	.	.	.	.	.	.	.
Sonc aspe	.	.	.	.	.	.	.	.	.	.	.
Compos.	.	.	.	.	.	2	.	1	.	.	3
Aven sp.	.	1	.	.	10	.	.	2	1	1	15
Brom m/s	.	11	.	54	113	3	6	3	11	71	272
Sieg decu	28	36	2	66	260	245	167	131	47	46	1028
Smal gras	2	4	.	4	51	59	18	5	4	20	167
Gramin.	5	6	.	44	75	27	8	16	4	24	209
Arrh elat	.	.	1	1	2	6	3	.	.	.	13
rhiz Gram	.	10	.	14	79	72	40	20	19	3	257
Junc sp.	.	.	.	.	.	.	.	.	.	.	.
Eleo sp.	1	1	.	.	1	.	.	2	.	.	5
Isol seta	.	.	.	.	1	1	.	.	.	1	3
Care pilu	.	.	.	1	1	.	1	5	2	1	11
Care puli	1	.	.	2	4	2	.	1	.	1	11
Care spp.	5	2	.	22	56	42	12	29	8	22	198
OTHER											
Cory avel	.	.	.	.	1	.	.	.	.	.	1
Crat mono	.	.	.	.	.	.	.	.	.	.	.
INDET.	3	1	1	5	8	4	4	3	5	.	34
TOTAL	65	188	13	515	1226	756	685	405	172	494	4519

KEY: * = only 50 per cent of the sample has been analysed.

Table 4.10e   Carbonized seeds from Thorpe Thewles.

PHASE:	II							I+II
SAMPLE NO.:	JS 16	JS 17	JS 18	JS 19	JS 23	JS 25	JS 27	TOTAL
VOL. IN LITRES:	28	28	28	28	28	28	28	1344
GRAIN								
Trit dico	.	.	.	.	.	.	.	1
Trit spel	.	.	.	.	.	.	3	66
Trit sp.	1	1	.	.	.	1	1	69
Hord vulg	5	3	1	2	.	2	5	86
Cere inde	2	3	10	1	.	1	.	365
CHAFF								
glum dico	.	.	.	.	.	.	1	1
glum spel	3	36	22	1	3	2	25	804
glum inde	2	19	8	.	.	3	8	771
glumes	.	6	1	.	.	.	1	53
rach brit	.	10	6	.	.	1	15	278
base Trit	.	1	.	.	.	.	.	3
rach Hord	.	6	1	.	.	.	4	110
base Hord	.	.	.	.	.	.	.	1
flor Avef	.	.	.	.	.	.	.	1
awns Aven	.	2	.	.	.	.	2	18
culm node	.	.	.	.	.	.	1	4
chaf inde	.	.	.	.	.	.	.	2
WEEDS								
Ranu acri	.	.	.	.	.	.	.	.
Ranu repe	.	1	.	.	.	.	.	3
Ranu Ranu	1	1	1	.	.	.	1	21
Ranu flam	.	.	.	.	.	.	17	19
Papa arge	.	.	.	.	.	.	.	1
Papa rh/d	.	.	.	.	.	.	.	.
Papa sp.	.	.	.	.	.	.	.	.
Raph raph	.	.	.	.	.	.	.	2
Bras sp.	.	.	.	.	.	.	.	1
Crucif.	.	.	.	.	.	.	.	3
Viol Mela	.	.	.	.	.	.	.	2
Mont font	17	6	2	.	.	.	7	1018
Stel medi	.	.	.	.	.	.	.	11
Chen albu	.	3	.	.	.	1	52	133
Chen sp.	.	.	1	.	.	1	28	76
Atri spp.	.	2	1	.	.	1	.	36
Chenop.	.	.	.	.	.	.	.	.
Malv sylv	.	.	.	.	.	.	.	.
Malv sp.	.	.	.	.	.	.	.	.
Vici Lath	.	1	.	.	.	.	2	16
Legumin.	4	4	2	.	.	.	9	148
Apha arve	.	.	.	.	.	.	.	.
Pote erec	.	1	.	.	.	.	2	42
Hera spon	.	.	.	.	.	.	.	1
Poly avic	.	.	.	.	.	.	.	18
Poly conv	.	.	.	.	.	.	.	7
Poly lapa	.	.	.	.	.	.	4	20
Poly pers	.	.	.	.	.	.	4	13
Poly l/p	.	1	.	.	.	.	4	8
Rume acet	.	.	.	.	1	.	.	7
Rume spp.	.	2	1	.	.	.	5	126
Polygon.	1	.	.	.	.	.	.	6
Urti dioc	.	.	.	.	.	.	.	3
Urti uren	.	.	.	.	.	.	.	.
Sola nigr	.	.	.	.	.	.	.	.
Hyos nige	.	.	.	.	.	.	.	.
Odon vern	.	.	.	.	.	.	1	3
Verb sp.	.	.	.	.	.	.	.	1
Lami a/p	.	.	.	.	.	.	.	.
Ment a/a	.	.	.	.	.	.	.	.
Prun vulg	1	.	.	.	.	.	1	8
Plan lanc	1	.	.	.	.	.	.	25
Plan majo	.	2	.	.	.	.	.	7
Gali apar	.	.	.	.	.	.	2	16
Gali palu	.	.	.	.	.	.	1	1
Sher arve	.	.	.	.	.	.	1	3
Vale dent	.	.	.	.	.	.	.	.
Trip inod	.	1	.	.	.	.	2	55
Laps comm	.	.	.	.	.	.	.	.
Sonc aspe	.	.	.	.	.	.	.	.
Compos.	.	.	.	.	.	.	.	4
Aven sp.	.	.	.	.	.	.	.	20
Brom m/s	.	18	6	.	.	.	21	621
Sieg decu	13	5	14	2	3	4	57	1658
Smal gras	2	2	3	2	.	.	2	284
Gramin.	5	10	5	1	.	2	30	358
Arrh elat	1	.	1	.	.	.	1	22
rhiz Gram	.	3	12	1	.	.	8	421
Junc sp.	.	.	.	.	.	.	.	.
Eleo sp.	.	.	.	.	.	.	3	9
Isol seta	.	.	.	.	.	.	.	3
Care pilu	.	.	.	.	.	1	1	15
Care puli	.	.	.	.	.	.	.	11
Care spp.	2	4	4	1	.	1	213	532
OTHER								
Cory avel	.	2	.	.	.	.	.	7
Crat mono	.	.	.	.	.	.	.	.
INDET.	4	2	3	2	.	.	4	127
TOTAL	65	158	105	13	7	20	549	8585

Table 4.10f   Carbonized seeds from Thorpe Thewles.

PHASE: II/III	LS 8	LS 329	LS 340	LS 362	LS 374	LS 385	LS 397	LS 405	JS 1	JS 8*	II/III TOTAL
VOL.IN LITRES:	28	28	28	28	28	28	28	28	28	28	280
GRAIN											
Trit dico	.	.	.	.	.	.	.	.	2	17	19
Trit spel	.	.	.	.	.	.	.	.	.	13	20
Trit sp.	.	.	2	2	.	.	.	2	1	13	20
Hord vulg	1	6	1	.	.	.	1	.	18	202	229
Cere inde	4	8	7	4	.	.	1	6	8	108	146
CHAFF											
glum dico	.	.	.	.	.	.	.	.	.	1	1
glum spel	.	25	22	1	.	1	.	.	2	278	329
glum inde	.	12	28	1	.	2	2	.	13	195	253
glumes	.	3	.	.	.	.	1	.	.	4	8
rach brit	.	.	.	2	.	1	1	1	3	48	56
base Trit	.	.	.	.	.	.	.	.	.	.	.
rach Hord	1	.	.	.	.	.	.	.	14	37	52
base Hord	.	.	.	.	.	.	.	.	.	.	.
flor Avef	.	.	.	.	.	.	.	.	.	3	3
awns Aven	.	.	.	.	.	.	.	.	.	.	.
culm node	.	.	.	.	.	.	.	.	.	.	.
chaf inde	.	.	.	.	.	.	.	.	.	.	.
WEEDS											
Ranu acri	.	.	.	.	.	.	.	.	.	1	1
Ranu repe	.	.	.	.	.	.	.	.	.	1	1
Ranu Ranu	.	.	.	.	.	.	.	1	.	.	1
Ranu flam	.	.	.	.	.	.	.	.	.	.	.
Papa arge	.	.	.	.	.	.	.	.	.	.	.
Papa rh/d	.	.	.	.	.	.	.	.	.	.	.
Papaver sp.	.	.	.	.	.	.	.	.	.	1	1
Raph raph	.	.	.	.	.	.	.	.	.	1	1
Bras sp.	.	.	1	.	.	.	.	.	.	.	1
Crucif.	.	.	.	.	.	.	.	.	.	29	29
Viol Mela	.	.	.	.	.	.	.	.	.	.	.
Mont font	.	7	3	1	2	.	.	4	14	1	32
Stel medi	.	.	.	.	.	.	.	.	.	2	2
Chen albu	.	.	.	.	.	.	.	.	2	112	114
Chen sp.	.	.	.	.	.	.	.	.	4	.	4
Atri spp.	.	.	.	.	.	.	.	.	.	42	42
Chenop.	.	.	.	.	.	.	.	.	.	54	54
Malv sylv	.	.	.	.	.	.	.	.	.	.	.
Malv sp.	.	.	.	.	.	.	.	.	.	2	3
Vici Lath	.	1	.	.	.	.	.	.	.	2	3
Legumin.	.	2	1	1	.	.	.	.	3	.	7
Apha arve	.	.	.	.	.	.	.	.	.	.	.
Pote erec	.	.	.	.	.	.	.	.	.	1	1
Hera spon	.	.	.	.	.	.	.	.	.	.	.
Poly avic	.	.	.	.	.	.	.	1	3	1	5
Poly conv	.	1	.	.	.	.	.	.	.	2	3
Poly lapa	.	.	.	.	.	.	.	.	1	.	1
Poly pers	.	.	.	1	.	.	.	.	.	1	2
Poly l/p	.	.	.	.	.	.	.	.	.	1	1
Rume acet	.	.	.	.	.	.	.	.	1	.	1
Rume spp.	.	.	2	.	.	.	.	.	3	92	97
Polygon.	.	.	.	.	.	.	.	.	.	.	.
Urti dioc	.	.	.	.	.	.	.	.	.	.	.
Urti uren	.	.	.	.	.	.	.	.	.	.	.
Sloa nigr	.	.	.	.	.	.	.	.	.	.	.
Hyos nige	.	.	1	.	.	.	.	.	.	.	1
Odon vern	1	.	.	.	.	.	.	.	.	1	2
Verb sp.	.	.	.	.	.	.	.	.	.	.	.
Lami a/p	.	.	.	.	.	.	.	.	.	.	.
Ment a/a	.	.	.	.	.	.	.	.	.	.	.
Prun vulg	.	.	.	.	.	.	.	.	1	.	1
Plan lanc	.	1	.	.	.	.	.	.	.	.	1
Plan majo	.	.	.	.	.	.	.	.	.	2	2
Gali apar	.	3	1	.	.	.	.	.	.	1	5
Gali palu	.	.	.	.	.	.	.	.	.	.	.
Sher arve	.	.	.	.	.	.	.	.	.	.	.
Vale dent	.	.	.	.	.	.	.	.	.	.	.
Trip inod	.	.	1	.	1	.	.	.	2	21	25
Laps comm	.	.	.	.	.	.	.	.	.	2	2
Sonc aspe	.	.	.	.	.	.	.	.	.	5	5
Compos.	.	.	.	.	.	.	.	.	.	52	52
Aven sp.	.	.	.	.	.	.	.	.	.	.	.
Brom m/s	.	2	.	1	.	.	.	2	2	349	356
Sieg decu	8	20	16	7	.	3	6	12	40	3	115
smal gras	2	.	.	.	.	1	2	.	32	17	55
Gramin.	.	3	6	1	.	1	1	.	6	121	139
Arrh elat	.	.	.	.	.	.	.	.	.	.	.
rhiz Gram	.	6	5	.	.	.	2	5	10	.	28
Junc sp.	.	.	.	.	.	.	.	.	.	.	.
Eleo sp.	.	.	.	.	.	.	.	.	.	.	.
Isol seta	.	.	.	.	.	.	.	.	.	.	.
Care pilu	.	.	.	.	.	.	.	.	.	.	.
Care puli	.	.	.	.	.	.	.	.	.	.	.
Care spp.	.	4	6	.	1	2	.	.	3	.	16
OTHER											
Cory avel	.	3	3	.	.	.	.	.	1	.	7
Crat mono	.	.	.	.	.	.	.	.	.	.	.
INDET.	2	3	1	2	.	.	.	.	10	12	30
TOTAL	19	110	107	24	4	10	16	37	199	1836	2362

Table 4.10g   Carbonized seeds from Thorpe Thewles.

											sub-
PHASE:	III										
SAMPLE NO.:	LS 178	LS 194	LS 208	LS 229	LS 238	LS 246	LS 291	LS 422	LS 431	LS 443	total
VOL. IN LITRES:	28	28	28	28	28	28	28	28	28	28	280
GRAIN											
Trit dico	.	.	.	.	.	.	.	.	.	.	.
Trit spel	5	.	.	.	.	1	1	6	4	.	17
Trit sp.	4	2	1	.	.	1	3	9	2	.	22
Hord vulg	12	7	.	.	1	3	4	21	2	3	53
Cere inde	12	13	4	2	6	8	11	36	15	3	110
CHAFF											
glum dico	.	.	.	.	.	.	.	.	.	.	.
glum spel	33	.	1	4	22	26	56	7	16	2	167
glum inde	24	2	1	3	25	32	54	7	19	.	167
glumes	2	.	.	.	.	1	2	.	.	.	5
rach brit	3	3	.	3	10	12	14	1	5	.	51
base Trit	.	.	.	.	.	.	.	.	.	.	.
rach Hord	.	.	.	.	.	.	1	.	.	.	1
base Hord	.	.	.	.	.	.	1	.	.	.	1
flor Avef	.	.	.	.	.	.	.	.	.	.	.
awns Aven	.	.	.	2	1	.	.	.	.	.	3
culm node	.	.	.	.	.	.	.	.	.	.	.
chaf inde	.	.	.	.	.	.	.	.	.	.	.
WEEDS											
Ranu acri	.	.	.	.	.	.	.	.	.	.	.
Ranu repe	.	.	.	.	.	.	.	.	.	.	.
Ranu Ranu	.	1	1	.	.	.	1	.	2	.	5
Ranu flam	.	1	.	.	.	.	.	.	.	.	1
Papa arge	.	6	.	.	.	.	.	.	.	.	6
Papa rh/d	.	.	.	.	.	.	.	.	.	.	.
Papaver sp.	.	.	.	.	.	.	.	.	.	.	.
Raph raph	.	.	.	.	.	.	.	.	.	.	.
Bras sp.	.	.	.	.	.	.	.	.	.	.	.
Crucif.	.	.	.	.	.	.	.	.	.	.	.
Viol Mela	.	.	.	.	.	.	.	.	.	.	.
Mont font	14	87	6	1	5	2	6	.	.	.	121
Stel medi	2	1	.	.	.	.	1	.	.	.	4
Chen albu	.	2	.	.	.	1	1	2	2	.	8
Chen sp.	.	.	.	.	.	1	.	1	.	.	2
Atri spp.	.	.	.	.	.	.	2	.	.	.	2
Chenop.	.	.	.	.	.	.	.	.	.	.	.
Malv sylv	.	.	.	.	.	.	.	.	.	.	.
Malv sp.	.	.	.	.	.	.	.	.	.	.	.
Vici Lath	.	.	3	.	.	.	1	.	3	.	7
Legumin.	.	1	1	.	.	.	4	.	.	.	6
Apha arve	.	.	.	.	.	.	.	.	.	.	.
Pote erec	.	.	1	.	1	.	.	.	.	1	3
Hera spon	.	.	.	.	.	.	.	.	.	.	.
Poly avic	.	.	.	.	.	1	1	.	.	.	2
Poly conv	1	.	.	.	.	.	.	.	.	.	1
Poly lapa	.	.	.	.	.	.	.	.	.	.	.
Poly pers	.	1	.	.	.	.	1	.	.	.	2
Poly l/p	.	.	.	.	.	.	.	.	.	.	.
Rume acet	.	1	.	.	.	.	.	.	2	.	3
Rume spp.	5	5	2	.	1	2	1	1	2	1	20
Polygon.	.	1	.	.	.	.	.	.	.	.	1
Urti dioc	.	.	.	.	.	1	.	.	.	.	1
Urti uren	.	.	.	.	.	2	.	.	.	.	2
Sloa nigr	.	.	.	.	.	.	.	.	.	.	.
Hyos nige	.	12	.	.	.	.	1	.	.	.	13
Odon vern	.	.	.	.	.	.	.	1	.	.	1
Verb sp.	.	.	.	.	.	.	.	.	.	.	.
Lami a/p	.	.	.	.	.	.	3	.	.	.	3
Ment a/a	.	.	.	.	.	.	.	.	.	.	.
Prun vulg	1	.	.	.	.	.	.	.	.	.	1
Plan lanc	1	2	1	.	.	.	.	.	.	.	4
Plan majo	1	2	.	.	.	.	1	.	.	.	4
Gali apar	4	1	1	.	.	.	1	5	3	.	15
Gali palu	.	.	.	.	.	.	.	.	.	.	.
Sher arve	.	.	.	.	.	.	.	.	.	.	.
Vale dent	.	.	.	.	.	.	.	.	.	.	.
Trip inod	2	.	.	2	2	.	.	4	.	2	12
Laps comm	.	.	.	.	.	.	.	.	.	.	.
Sonc aspe	.	.	.	.	.	.	.	.	.	.	.
Compos.	.	.	.	.	.	.	1	.	.	.	1
Aven sp.	.	.	.	.	.	.	.	.	.	.	.
Brom m/s	11	10	1	2	2	2	12	5	8	2	55
Sieg decu	52	37	20	11	27	15	30	64	53	6	315
smal gras	8	3	4	1	3	.	8	2	6	3	38
Gramin.	12	6	4	4	7	4	16	11	14	1	79
Arrh elat	.	1	1	.	.	.	.	.	.	.	2
rhiz Gram	5	9	4	.	9	2	1	27	11	1	69
Junc sp.	.	.	.	.	.	.	.	.	.	.	.
Eleo sp.	1	.	.	.	1	.	.	.	1	.	3
Isol seta	.	.	.	.	.	.	.	.	.	.	.
Care pilu	.	.	1	.	.	.	.	.	.	.	1
Care puli	.	.	.	.	.	.	.	.	.	.	.
Care spp.	1	4	6	1	2	4	8	11	6	3	46
OTHER											
Cory avel	1	.	.	.	.	.	.	.	.	.	1
Crat mono	.	.	.	.	.	.	.	.	.	.	.
INDET.	4	4	2	.	7	.	5	7	3	4	36
TOTAL	221	225	66	36	132	121	253	228	179	32	1493

Table 4.10h　　Carbonized seeds from Thorpe Thewles.

	LS 450	LS 465	LS 470	LS 483	LS 499	LS 597	PF 12	PF 49	PF 78	JS 6	sub-total
PHASE: III											
VOL. IN LITRES:	28	28	28	28	28	28	28	28	28	28	280
GRAIN											
Trit dico	.	.	.	.	.	.	.	.	1	3	15
Trit spel	1	3	1	6	.	.	.	.	1	3	15
Trit sp.	.	2	.	2	.	.	.	2	.	3	9
Hord vulg	4	8	1	4	1	.	.	.	.	15	33
Cere inde	5	25	8	21	2	1	4	5	4	10	85
CHAFF											
glum dico	.	36	4	1	2	.	3	2	.	27	75
glum spel	.	36	4	1	2	1	3	.	1	25	73
glum inde	.	2	.	.	1	.	.	.	.	2	5
glumes	.	.	.	.	.	.	.	.	.	.	.
rach brit	1	5	4	.	1	.	.	.	.	13	24
base Trit	.	.	.	.	.	.	.	.	.	9	18
rach Hord	.	7	.	.	2	.	.	.	.	9	18
base Hord	.	.	.	.	.	.	.	.	.	1	2
flor Avef	.	.	1	.	.	.	.	.	.	.	2
awns Aven	.	.	.	.	.	.	2	.	.	.	2
culm node	.	.	.	.	.	.	.	.	.	.	.
chaf inde	.	.	.	.	.	.	.	.	.	3	3
WEEDS											
Ranu acri	.	.	.	.	.	.	.	1	.	.	1
Ranu repe	.	.	.	.	.	.	1	.	.	.	1
Ranu Ranu	1	2	1	.	.	.	1	.	.	.	5
Ranu flam	.	.	.	.	.	.	.	.	.	.	.
Papa arge	.	.	.	.	.	.	.	.	.	.	.
Papa rh/d	.	.	.	.	.	.	.	.	.	.	.
Papaver sp.	.	.	.	.	.	.	.	.	.	.	.
Raph raph	.	1	.	.	.	.	.	.	.	.	1
Bras sp.	.	.	.	.	.	.	.	.	.	.	.
Crucif.	.	.	.	.	.	.	.	.	.	.	.
Viol Mela	.	.	.	.	.	.	.	.	.	.	.
Mont font	1	3	1	3	.	1	3	32	12	3	59
Stel medi	.	.	.	.	.	.	.	.	.	.	.
Chen albu	.	1	.	1	.	.	.	1	1	4	8
Chen sp.	.	.	2	1	1	.	.	.	.	2	6
Atri spp.	.	3	.	.	.	.	.	1	.	.	4
Chenop.	.	.	.	.	.	.	.	.	.	.	.
Malv sylv	.	.	.	.	.	.	.	.	.	.	.
Malv sp.	.	.	.	.	.	.	.	.	.	.	.
Vici Lath	.	.	.	.	.	.	.	.	.	.	.
Legumin.	.	2	.	.	1	.	1	5	4	.	13
Apha arve	.	.	.	.	.	.	.	.	.	.	.
Pote erec	.	1	.	.	.	.	.	.	.	.	1
Hera spon	.	.	.	.	.	.	.	.	.	.	.
Poly avic	.	.	.	.	.	.	1	.	1	.	2
Poly conv	.	.	.	.	.	.	.	.	.	.	.
Poly lapa	.	.	.	.	.	.	.	.	1	.	1
Poly pers	.	1	.	2	.	.	.	.	.	.	3
Poly l/p	.	.	.	.	.	.	.	.	.	.	.
Rume acet	.	.	.	.	.	.	.	.	.	.	.
Rume spp.	.	.	1	.	.	.	.	7	.	11	19
Polygon.	.	.	.	.	.	.	.	.	.	.	.
Urti dioc	.	.	.	.	.	.	.	.	.	.	.
Urti uren	.	.	.	.	.	.	.	.	.	.	.
Sloa nigr	.	.	.	.	.	.	.	.	.	.	.
Hyos nige	.	.	.	.	.	.	.	.	.	.	.
Odon vern	.	2	.	.	.	.	.	.	.	.	2
Verb sp.	.	.	.	.	.	.	.	.	.	.	.
Lami a/p	.	.	.	.	2	.	.	.	.	.	2
Ment a/a	.	.	.	.	.	.	.	.	.	.	.
Prun vulg	.	.	.	.	.	.	.	.	.	.	.
Plan lanc	.	.	.	.	.	1	.	.	.	.	1
Plan majo	.	.	.	1	1	.	.	.	.	.	2
Gali apar	1	.	.	1	.	.	1	2	2	.	7
Gali palu	.	.	.	.	.	.	.	.	.	.	.
Sher arve	.	.	.	.	.	.	.	.	.	.	.
Vale dent	.	.	.	.	.	.	.	.	.	.	.
Trip inod	.	11	.	1	.	.	2	.	.	1	15
Laps comm	.	.	.	.	.	.	.	.	.	.	.
Sonc aspe	.	.	.	.	.	.	.	.	.	.	.
Compos.	.	.	.	2	.	.	.	.	.	1	3
Aven sp.	.	.	.	2	.	.	.	.	.	1	3
Brom m/s	2	5	1	.	1	.	5	4	2	20	40
Sieg decu	38	26	33	32	6	3	17	23	8	18	204
smal gras	2	9	3	2	4	.	5	2	4	6	37
Gramin.	2	4	3	1	1	.	3	2	2	11	29
Arrh elat	.	.	.	1	1	.	.	.	.	.	2
rhiz Gram	10	9	2	31	1	.	8	9	4	3	77
Junc sp.	.	.	.	.	.	.	.	1	.	.	1
Eleo sp.	.	2	.	.	.	.	.	.	.	.	2
Isol seta	.	.	.	.	.	.	.	.	.	.	.
Care pilu	.	.	.	.	.	.	.	.	.	.	.
Care puli	.	.	.	.	.	.	.	.	.	.	.
Care spp.	2	9	1	3	1	.	2	2	5	1	26
OTHER											
Cory avel	.	.	.	.	.	.	.	.	.	.	.
Crat mono	.	.	.	.	.	.	.	.	.	.	.
INDET.	3	4	3	5	8	.	4	3	2	5	37
TOTAL	73	219	74	122	39	7	66	104	54	197	955

Table 4.10i   Carbonized seeds from Thorpe Thewles.

	JS 10	JS 11	JS 12*	JS 14	JS 15	JS 20	JS 22	JS 26	JS 28	TOTAL
PHASE: III										III
VOL. IN LITRES:	28	28	28	28	28	28	28	28	28	1092
GRAIN										
Trit dico	.	.	.	.	.	.	.	.	.	.
Trit spel	.	3	4	9	.	5	1	.	.	54
Trit sp.	1	4	3	5	1	4	.	.	.	49
Hord vulg	11	9	12	16	1	10	.	2	1	148
Cere inde	37	27	18	50	2	13	.	3	.	345
CHAFF										
glum dico	.	.	.	.	.	.	.	.	.	.
glum spel	7	95	120	366	7	4	17	28	.	886
glum inde	5	71	98	448	5	6	11	36	1	921
glumes	.	.	7	1	1	.	.	1	.	20
rach brit	.	16	32	61	3	1	3	11	1	203
base Trit	.	.	.	.	.	.	.	.	.	.
rach Hord	.	3	8	15	.	1	.	.	.	46
base Hord	.	.	.	.	.	.	.	.	.	1
flor Avef	.	.	.	.	.	.	.	.	.	2
awns Aven	.	1	.	.	.	.	.	.	.	6
culm node	.	.	.	.	.	.	.	.	.	.
chaf inde	.	.	.	.	.	.	.	.	.	3
WEEDS										
Ranu acri	.	.	.	.	.	.	.	.	.	1
Ranu repe	.	.	.	.	.	.	.	.	.	1
Ranu Ranu	.	.	.	1	.	.	.	.	1	12
Ranu flam	.	.	.	.	.	.	.	1	.	2
Papa arge	.	.	.	.	.	.	.	.	.	6
Papa rh/d	.	.	1	.	.	.	.	.	.	1
Papaver sp.	.	.	.	.	.	.	.	.	.	.
Raph raph	.	.	.	.	.	.	.	.	.	1
Bras sp.	.	.	.	.	.	.	.	.	.	.
Crucif.	.	.	.	.	.	.	.	.	.	.
Viol Mela	.	.	.	.	.	.	.	.	.	.
Mont font	5	7	23	.	33	.	.	2	241	491
Stel medi	.	2	.	2	.	.	.	.	6	14
Chen albu	6	4	5	.	4	3	1	3	.	42
Chen sp.	2	4	8	4	2	.	.	2	4	34
Atri spp.	1	.	.	21	3	.	.	.	.	31
Chenop.	.	.	.	.	.	.	.	.	.	.
Malv sylv	.	.	.	.	.	.	.	1	.	1
Malv sp.	6	.	.	.	.	.	.	.	.	6
Vici Lath	.	.	.	.	.	.	.	.	.	7
Legumin.	1	3	2	3	14	6	.	2	32	82
Apha arve	.	.	.	.	.	.	.	.	.	.
Pote erec	.	1	.	1	.	.	.	.	2	8
Hera spon	.	.	.	.	.	.	.	.	.	.
Poly avic	1	.	.	.	1	.	.	.	3	9
Poly conv	.	.	1	5	.	.	1	.	.	8
Poly lapa	.	2	2	.	.	.	.	2	.	7
Poly pers	.	2	.	2	.	.	.	.	.	9
Poly l/p	.	2	1	4	.	1	.	.	.	8
Rume acet	1	.	1	.	.	.	.	.	.	5
Rume spp.	2	6	6	9	.	4	.	3	78	147
Polygon.	.	1	.	.	.	.	.	.	.	2
Urti dioc	.	.	.	.	.	.	.	.	.	1
Urti uren	.	.	.	.	.	.	.	.	.	2
Sloa nigr	.	.	.	.	.	.	.	.	.	.
Hyos nige	.	.	.	.	.	.	.	.	.	13
Odon vern	.	1	.	.	1	.	.	1	.	6
Verb sp.	.	.	.	.	.	.	.	.	.	.
Lami a/p	.	.	.	.	.	.	.	.	.	5
Ment a/a	.	.	.	.	.	.	.	.	.	.
Prun vulg	.	.	.	.	.	.	.	.	6	7
Plan lanc	.	.	.	.	.	1	.	.	4	10
Plan majo	.	.	.	.	.	.	.	1	1	8
Gali apar	2	1	.	1	.	.	.	.	4	30
Gali palu	.	.	.	.	1	.	.	.	.	1
Sher arve	.	.	.	.	.	.	.	.	.	.
Vale dent	.	.	.	.	.	.	.	.	.	.
Trip inod	.	4	5	45	.	.	1	1	1	84
Laps comm	.	.	.	.	.	.	.	.	.	.
Sonc aspe	.	.	.	.	.	.	.	.	.	.
Compos.	.	.	.	.	.	.	.	.	.	.
Aven sp.	.	.	.	.	.	.	.	.	.	4
Brom m/s	6	22	62	20	.	4	.	4	.	213
Sieg decu	19	64	35	266	49	51	12	38	102	1155
smal gras	1	10	12	.	.	4	1	7	11	121
Gramin.	3	17	80	29	8	2	1	5	1	254
Arrh elat	.	1	1	.	2	.	.	3	2	13
rhiz Gram	7	3	4	84	3	13	1	2	10	273
Junc sp.	.	.	.	.	.	.	.	.	.	1
Eleo sp.	.	.	.	.	.	1	.	.	.	6
Isol seta	.	.	.	.	.	.	.	.	.	.
Care pilu	.	2	1	2	.	2	.	.	.	8
Care puli	.	.	.	32	.	2	.	.	.	34
Care spp.	3	10	19	71	9	4	1	21	17	227
OTHER										
Cory avel	.	.	.	.	.	.	.	.	.	1
Crat mono	.	.	.	.	.	.	.	.	1	1
INDET.	5	7	.	.	2	2	1	.	8	98
TOTAL	132	405	571	1573	152	144	52	180	538	6195

Key: * = only 50 per cent of the sample has been analysed.

Table 4.10j Carbonized seeds from Thorpe Thewles.

PHASE:	III/IV										sub-
SAMPLE NO.:	LS 69	ML 1	ML2	ML 3	ML 4	ML 5	ML 6	ML 7	ML 8	ML 9	total
VOL. IN LITRES:	28	28	28	28	28	28	28	28	28	28	280
GRAIN											
Trit dico	.	.	.	.	.	.	.	.	.	.	.
Trit spel	3	12	4	9	1	2	1	2	.	.	34
Trit sp.	6	4	.	2	.	2	2	.	2	4	22
Hord vulg	36	9	3	7	6	2	5	1	2	12	83
Cere inde	50	24	12	33	8	12	18	2	8	14	181
CHAFF											
glum dico	1	.	.	.	.	.	.	.	.	.	1
glum spel	59	43	29	97	22	32	15	6	23	38	364
glum inde	18	12	12	56	8	26	8	1	10	15	166
glumes	2	1	.	8	4	1	.	.	.	.	16
rach brit	2	7	3	13	2	7	.	.	6	2	42
base Trit	.	.	.	.	.	.	.	.	.	.	.
rach Hord	14	.	.	2	.	.	1	.	5	2	24
base Hord	.	.	.	.	.	.	.	.	.	.	.
flor Avef	.	.	.	.	.	.	.	.	.	.	.
awns Aven	.	.	.	1	1	.	.	.	.	.	2
culm node	.	.	.	.	.	.	.	.	1	1	2
chaf inde	.	.	.	.	.	.	.	.	.	.	.
WEEDS											
Ranu acri	.	.	.	.	.	.	.	.	.	.	.
Ranu repe	.	.	.	.	.	.	.	.	.	.	1
Ranu Ranu	.	.	.	1	.	.	.	.	.	.	1
Ranu flam	.	.	.	.	.	.	.	.	1	.	1
Papa arge	.	.	.	.	.	.	.	.	.	.	.
Papa rh/p	.	.	.	.	.	.	.	.	.	.	.
Papa sp.	.	.	.	.	.	.	.	.	.	.	.
Raph raph	.	.	.	.	.	.	.	.	.	.	.
Bras sp.	.	.	.	.	.	.	.	.	.	.	.
Crucif.	.	1	.	.	.	1	.	.	.	.	2
Viol Mela	.	.	.	.	.	.	.	.	.	.	.
Mont font	34	17	8	21	6	9	7	2	5	11	120
Stel medi	1	1	.	1	.	.	.	.	.	.	3
Chen albu	1	2	1	3	1	4	7	1	5	3	28
Chen sp.	.	1	1	.	.	.	.	.	3	.	5
Atri spp.	.	.	2	.	.	1	.	1	1	1	6
Chenop.	1	.	.	.	.	.	.	.	.	1	2
Malv sylv	.	.	.	.	.	.	.	.	.	.	.
Malv sp.	.	.	.	.	.	.	.	.	.	.	.
Vici Lath	1	1	.	.	.	2	1	.	.	.	5
Legumin.	9	6	4	5	.	3	2	1	1	2	33
Apha arve	.	.	.	.	.	.	.	.	.	.	.
Pote erec	3	1	.	.	.	.	.	1	2	.	7
Hera spon	.	.	.	.	.	.	.	.	.	.	.
Poly avic	.	1	.	.	.	1	.	.	4	.	6
Poly conv	2	1	.	.	.	1	.	.	.	.	4
Poly lapa	1	.	.	1	.	.	.	.	.	1	3
Poly pers	1	.	.	.	.	.	.	.	.	.	1
Poly l/p	.	.	.	.	.	.	.	.	.	.	.
Rume acet	.	.	.	.	.	2	.	1	.	.	3
Rume spp.	8	3	3	9	.	3	.	1	.	3	30
Polygon.	.	.	.	.	.	.	.	.	.	.	.
Urti dioc	.	.	.	.	.	.	.	.	.	.	.
Urti uren	.	.	.	.	.	.	.	.	.	.	.
Sola nigr	.	.	.	.	.	.	.	.	.	.	.
Hyos nige	.	.	.	.	.	1	.	.	.	.	1
Odon vern	3	.	.	.	.	.	.	.	.	.	3
Verb sp.	.	.	.	.	.	.	.	.	.	.	.
Lami a/p	.	.	.	.	.	.	.	.	.	.	.
Ment a/a	.	.	.	.	.	.	.	.	.	.	.
Prun vulg	.	.	1	.	.	.	.	.	.	.	1
Plan lanc	1	1	.	2	.	.	.	.	2	.	6
Plan majo	.	.	.	.	.	.	.	.	.	.	.
Gali apar	.	2	.	2	.	.	.	.	1	.	5
Gali palu	.	.	.	.	.	.	.	.	.	.	.
Sher arve	.	.	.	.	.	1	.	.	.	.	1
Vale dent	.	1	.	.	.	.	.	.	.	.	1
Trip inod	.	.	.	.	.	.	1	1	1	.	3
Laps comm	.	.	.	.	.	.	.	.	.	.	.
Sonc aspe	.	.	.	.	.	.	.	.	.	.	.
Compos.	.	2	.	.	.	.	2	.	2	1	7
Aven sp.	.	.	.	.	.	.	.	.	.	.	.
Brom m/s	3	34	16	107	11	19	19	2	9	16	236
Sieg decu	72	135	38	84	28	105	58	36	50	98	704
smal gras	25	9	6	14	7	17	3	2	3	13	99
Gramin.	4	11	7	11	7	7	10	2	7	13	79
Arrh elat	.	3	.	2	2	3	2	.	.	4	16
rhiz Gram	29	46	23	10	.	12	15	21	24	33	213
Junc sp.	.	.	.	.	.	.	.	.	.	.	.
Eleo sp.	1	.	.	.	.	1	.	.	.	.	2
Isol seta	.	.	.	.	.	.	.	1	.	.	1
Care pilu	2	.	.	.	.	.	.	.	1	.	3
Care puli	.	1	1	1	.	2	.	.	.	1	6
Care spp.	26	23	18	26	4	8	12	5	5	5	132
OTHER											
Cory avel	.	2	.	.	.	.	2	1	2	1	8
Crat mono	.	.	.	.	.	.	.	.	.	.	.
INDET.	3	8	2	7	4	5	3	4	.	3	39
TOTAL	422	425	194	535	122	292	194	95	186	298	2763

Table 4.10k   Carbonized seeds from Thorpe Thewles.

PHASE:	III/IV										sub-
SAMPLE NO.:	ML 10	ML 11	ML 12	ML 13	ML 14	ML 15	ML 16	ML 17	ML 18	ML 19	total
VOL. IN LITRES:	28	28	28	28	28	28	28	28	28	28	280
GRAIN											
Trit dico	.	.	2	.	.	.	3	.	1	.	7
Trit spel	1	.	2	.	.	.	3	.	1	.	7
Trit sp.	1	.	3	.	.	.	1	1	2	1	9
Hord vulg	12	1	7	4	33	2	3	2	3	1	68
Cere inde	12	5	10	9	23	1	9	2	6	3	80
CHAFF											
glum dico	.	.	.	.	.	.	.	.	.	.	.
glum spel	11	5	14	9	12	1	18	3	13	4	90
glum inde	3	1	9	5	5	.	16	3	7	2	51
glumes	1	.	.	.	.	.	.	2	.	.	3
rach brit	2	.	3	1	2	2	4	.	1	1	16
base Trit	.	.	.	.	.	.	.	.	.	.	.
rach Hord	3	.	5	3	17	.	1	2	.	1	32
base Hord	.	.	.	.	.	.	.	.	.	.	.
flor Avef	.	.	.	.	.	.	.	.	.	.	.
awns Aven	.	.	1	.	.	.	.	1	1	.	3
culm node	.	.	.	.	.	.	.	.	.	.	.
chaf inde	.	.	.	.	.	.	.	.	.	.	.
WEEDS											
Ranu acri	.	.	.	.	.	.	.	.	.	.	.
Ranu repe	.	.	.	.	.	.	.	.	.	.	.
Ranu Ranu	1	.	.	.	.	.	.	1	.	.	2
Ranu flam	.	.	.	.	.	.	.	.	.	.	.
Papa arge	.	.	.	.	.	.	.	.	.	.	.
Papa rh/p	.	.	.	.	.	.	.	.	.	.	.
Papa sp.	.	.	.	.	.	.	.	.	.	.	.
Raph raph	1	.	.	.	.	.	.	.	.	.	1
Bras sp.	.	.	.	.	.	.	.	.	.	.	.
Crucif.	5	.	.	.	.	.	.	1	.	.	6
Viol Mela	.	.	.	.	.	.	.	.	.	.	.
Mont font	12	3	6	3	5	3	3	2	1	2	40
Stel medi	2	.	.	1	1	.	2	.	.	.	6
Chen albu	8	.	5	7	1	.	.	1	.	.	22
Chen sp.	12	.	6	.	.	.	.	.	.	.	18
Atri spp.	1	.	2	1	2	.	.	.	.	.	6
Chenop.	.	.	.	.	2	.	2	.	.	.	4
Malv sylv	.	.	.	.	.	.	.	.	.	.	.
Malv sp.	.	.	.	.	.	.	.	.	.	.	.
Vici Lath	.	.	.	.	.	.	.	.	.	.	.
Legumin.	2	.	5	2	.	.	.	.	.	.	9
Apha arve	.	.	.	.	.	.	.	.	.	.	.
Pote erec	.	.	2	.	.	.	.	3	1	.	6
Hera spon	.	.	.	.	.	.	.	.	.	.	.
Poly avic	7	1	6	.	.	.	.	.	.	.	14
Poly conv	1	.	.	1	.	.	.	.	.	.	2
Poly lapa	3	1	1	1	.	.	.	.	.	.	6
Poly pers	.	.	.	.	.	.	.	1	.	.	1
Poly l/p	3	.	1	.	.	.	.	.	.	.	4
Rume acet	.	.	.	.	.	.	.	.	.	.	.
Rume spp.	2	1	6	2	14	.	.	1	2	1	29
Polygon.	.	.	.	.	.	.	.	.	.	.	.
Urti dioc	.	.	.	.	.	.	.	.	.	.	.
Urti uren	.	.	.	.	.	.	.	.	.	.	.
Sola nigr	.	.	.	.	1	.	.	.	.	.	1
Hyos nige	.	.	.	.	.	.	.	.	.	.	.
Odon vern	.	.	6	.	.	.	.	.	1	.	7
Verb sp.	.	.	.	.	.	.	.	.	.	.	.
Lami a/p	.	.	.	.	.	.	.	.	.	.	.
Ment a/a	.	.	.	.	.	.	.	.	.	.	.
Prun vulg	.	.	.	.	1	.	.	.	.	.	1
Plan lanc	1	.	.	.	.	.	.	1	.	.	2
Plan majo	.	.	.	1	.	.	.	.	.	.	1
Gali apar	1	.	.	.	.	.	2	.	.	.	3
Gali palu	.	.	.	.	.	.	.	.	.	.	.
Sher arve	.	.	.	.	.	.	.	.	.	.	.
Vale dent	.	.	.	.	.	.	.	.	.	.	.
Trip inod	.	.	2	.	.	.	.	.	1	.	3
Laps comm	.	.	.	.	.	.	.	.	.	.	.
Sonc aspe	.	.	.	.	.	.	.	.	.	.	.
Compos.	1	.	.	.	.	.	.	1	.	.	2
Aven sp.	1	.	1	.	.	.	.	.	.	.	2
Brom m/s	8	2	9	7	9	2	16	2	8	5	68
Sieg decu	53	29	44	63	33	7	38	18	21	10	316
smal gras	17	1	8	3	3	2	6	2	3	1	46
Gramin.	6	3	11	8	10	.	15	3	4	3	63
Arrh elat	1	.	.	.	.	.	.	1	1	2	5
rhiz Gram	15	5	15	6	12	.	4	2	6	3	68
Junc sp.	.	.	.	.	.	.	.	.	.	.	.
Eleo sp.	.	.	1	.	.	.	1	.	1	.	3
Isol seta	.	.	.	.	.	.	.	.	.	.	.
Care pilu	1	.	.	.	.	.	.	3	.	.	4
Care puli	.	.	.	2	.	.	.	.	.	1	3
Care spp.	4	2	10	4	9	1	4	17	9	5	65
OTHER											
Cory avel	.	.	.	.	.	.	.	.	.	.	.
Crat mono	.	.	.	.	.	.	.	.	.	.	.
INDET.	5	2	6	5	1	1	5	3	4	4	36
TOTAL	220	62	207	148	196	22	153	79	97	50	1234

Table 4.101   Carbonized seeds from Thorpe Thewles.

PHASE: III/IV	ML 20	ML 21	ML 22	ML 23	ML 25	ML 26	ML 27	ML 28	ML 29	III/IV TOTAL
VOL. IN LITRES:	28	28	28	28	28	28	28	28	28	812
GRAIN										
Trit dico	.	.	.	.	.	.	.	.	.	.
Trit spel	.	.	1	.	1	1	1	1	3	49
Trit sp.	.	.	2	2	2	4	3	3	1	48
Hord vulg	.	.	10	6	28	11	13	8	1	228
Cere inde	4	3	10	7	30	24	14	21	4	378
CHAFF										
glum dico	.	.	.	.	.	.	.	.	.	1
glum spel	12	3	23	11	19	38	9	2	5	576
glum inde	8	4	5	10	21	18	6	1	5	295
glumes	.	.	.	2	1	5	.	.	.	27
rach brit	3	1	2	1	4	15	1	.	2	87
base Trit	.	.	.	.	.	.	.	.	.	.
rach Hord	.	.	.	1	6	2	3	2	1	71
base Hord	.	.	.	.	.	.	.	.	.	.
flor Avef	.	.	.	.	.	.	.	.	1	1
awns Aven	.	.	.	.	.	.	.	.	.	5
culm node	.	.	.	1	.	.	.	.	.	3
chaf inde	.	.	.	.	.	.	.	.	.	.
WEEDS										
Ranu acri	.	.	.	.	.	.	.	.	.	.
Ranu repe	.	.	.	.	.	.	.	.	.	.
Ranu Ranu	.	.	.	.	1	.	.	1	.	5
Ranu flam	.	.	.	1	.	1	.	.	.	3
Papa arge	.	.	.	.	.	.	.	.	.	.
Papa rh/p	.	.	.	.	.	.	.	.	.	.
Papa sp.	.	.	.	.	.	.	.	.	.	.
Raph raph	.	.	3	.	.	.	.	.	.	4
Bras sp.	.	.	.	.	.	.	.	.	.	.
Crucif.	.	.	1	.	.	.	.	.	.	9
Viol Mela	.	.	.	.	.	.	.	.	.	.
Mont font	3	.	25	11	13	4	7	21	4	248
Stel medi	.	.	.	1	.	.	.	1	.	11
Chen albu	2	1	3	1	6	6	1	4	3	77
Chen sp.	.	.	.	.	2	1	.	.	1	27
Atri spp.	.	.	.	1	.	.	.	1	.	14
Chenop.	.	.	1	.	2	.	.	1	.	10
Malv sylv	.	.	.	.	.	.	.	.	.	.
Malv sp.	.	.	.	.	.	.	.	.	.	.
Vici Lath	.	.	1	.	.	.	.	.	.	6
Legumin.	.	.	1	9	11	7	1	10	3	84
Apha arve	.	.	.	.	.	.	.	.	.	.
Pote erec	.	.	.	.	1	1	.	1	.	16
Hera spon	.	.	.	.	.	.	.	.	.	.
Poly avic	.	1	2	1	2	4	1	1	1	33
Poly conv	.	.	.	1	.	.	.	.	.	7
Poly lapa	.	.	.	.	.	2	.	.	.	11
Poly pers	.	.	.	1	1	1	.	.	.	5
Poly l/p	.	.	.	.	2	.	.	.	.	6
Rume acet	.	.	.	.	.	.	.	.	.	3
Rume spp.	2	.	.	.	3	4	1	1	1	71
Polygon.	.	.	.	.	.	.	.	.	.	.
Urti dioc	.	.	.	.	.	.	.	.	.	.
Urti uren	.	.	.	.	.	.	.	.	.	.
Sola nigr	.	.	.	.	.	.	.	.	.	1
Hyos nige	.	.	.	.	.	.	.	.	.	1
Odon vern	.	.	1	.	.	.	.	.	.	11
Verb sp.	.	.	.	.	.	.	.	.	.	.
Lami a/p	.	.	.	.	.	.	.	.	.	.
Ment a/a	.	.	.	.	.	.	.	.	.	.
Prun vulg	.	.	.	.	.	.	.	1	.	3
Plan lanc	.	.	.	.	2	.	1	.	1	12
Plan majo	.	.	.	.	.	.	.	.	.	1
Gali apar	.	1	.	5	5	2	.	.	.	21
Gali palu	.	.	.	.	.	.	.	.	.	.
Sher arve	.	.	.	.	1	.	.	.	.	2
Vale dent	.	.	.	.	.	.	.	.	.	1
Trip inod	.	.	.	.	1	.	2	.	1	10
Laps comm	.	.	.	.	.	.	.	.	.	.
Sonc aspe	.	.	.	.	.	.	.	.	.	.
Compos.	.	.	.	.	.	.	.	2	.	2
Aven sp.	.	.	.	.	.	.	.	2	.	11
Brom m/s	2	1	6	4	31	22	7	5	4	386
Sieg decu	12	16	38	68	110	183	78	93	44	1662
smal gras	1	.	2	8	16	6	5	31	5	219
Gramin.	2	.	3	9	27	30	8	15	7	243
Arrh elat	.	.	.	1	2	.	.	.	.	24
rhiz Gram	2	7	27	4	27	9	27	5	9	398
Junc sp.	.	.	.	.	.	.	.	.	.	.
Eleo sp.	.	.	.	1	.	1	1	1	.	9
Isol seta	.	.	.	.	.	.	.	.	.	1
Care pilu	.	.	1	.	.	.	.	.	.	8
Care puli	.	.	.	.	.	1	.	.	.	10
Care spp.	4	5	20	11	2	16	6	15	4	280
OTHER										
Cory avel	.	.	.	.	.	.	.	.	.	8
Crat mono	.	.	.	.	.	.	.	.	.	.
INDET.	2	3	7	5	11	4	6	9	5	127
TOTAL	59	46	195	184	391	423	202	257	116	5870

Table 4.10m   Carbonized seeds from Thorpe Thewles.

PHASE:	IV										sub-
SAMPLE NO.:	LS 1	LS 17	LS 19	LS 25	LS 28	LS 33	LS 42	LS 52	LS 58	LS 507	total
VOL. IN LITRES:	28	28	28	28	28	28	28	28	28	28	280
GRAIN											
Trit dico	.	.	.	.	.	.	.	.	.	.	.
Trit spel	5	3	.	3	.	1	2	5	.	6	25
Trit sp.	3	2	2	1	.	.	.	4	.	8	20
Hord vulg	16	2	2	2	5	1	.	3	.	4	35
Cere inde	15	2	4	3	2	4	5	20	3	23	81
CHAFF											
glum dico	.	.	.	.	.	.	.	.	.	.	.
glum spel	16	10	16	4	4	5	8	5	.	213	281
glum inde	5	4	1	.	.	4	1	1	3	200	219
glumes	.	.	2	.	.	.	.	.	.	3	5
rach brit	.	1	.	.	.	1	.	1	.	70	73
base Trit	.	.	.	.	.	.	.	.	.	.	.
rach Hord	.	.	.	.	.	.	.	1	.	4	5
base Hord	.	.	.	.	.	.	.	1	.	.	1
flor Aven	1	.	.	.	.	.	.	.	.	.	1
awns Aven	.	.	.	.	.	.	.	.	.	.	.
culm node	.	.	.	.	.	.	.	.	.	.	.
chaf inde	.	.	.	.	.	.	.	.	.	.	.
WEEDS											
Ranu acri	.	.	.	.	.	.	.	.	.	.	.
Ranu repe	.	.	.	.	.	.	.	.	.	.	.
Ranu Ranu	.	1	.	.	.	.	.	.	.	.	1
Ranu flam	.	1	.	.	.	1	.	.	.	.	2
Papa arge	.	.	.	.	.	.	.	.	.	.	.
Papa rh/d	.	.	.	.	.	.	.	.	.	.	.
Papa sp.	.	.	.	.	.	.	.	.	.	.	.
Raph raph	.	.	.	.	.	.	.	.	.	.	.
Bras sp.	.	.	.	.	.	.	.	.	.	.	.
Crucif.	.	.	.	.	.	.	.	.	.	.	.
Viol Mela	.	.	.	.	.	.	.	.	.	.	.
Mont font	2	3	2	.	5	.	.	2	.	1	15
Stel medi	.	.	.	.	.	.	.	.	.	1	1
Chen albu	1	.	1	1	.	.	.	1	.	.	4
Chen sp.	.	.	3	.	.	.	.	.	1	.	4
Atri spp.	.	.	.	1	.	.	.	.	.	.	1
Chenop.	.	.	.	.	.	.	.	.	.	4	4
Malv sylv	.	.	.	.	.	.	.	.	.	.	.
Malv sp.	.	.	.	.	1	.	.	.	.	.	1
Vici Lath	1	.	.	.	.	.	.	.	.	.	1
Legumin.	.	.	.	1	1	.	.	.	1	.	3
Apha arve	.	1	.	.	.	.	.	.	.	.	1
Pote erec	1	.	.	.	.	.	.	.	.	.	1
Hera spon	.	.	.	.	.	.	.	.	.	.	.
Poly avic	.	.	1	.	.	.	.	.	1	3	5
Poly conv	.	.	.	.	.	.	.	.	.	.	.
Poly lapa	.	.	.	.	.	.	.	.	.	4	4
Poly pers	.	1	.	.	2	.	.	.	.	.	3
Poly l/p	.	.	.	.	.	.	.	.	.	.	.
Rume acet	.	1	.	.	.	.	.	.	.	.	1
Rume spp.	4	5	4	4	1	.	.	.	.	2	20
Polygon.	.	.	.	.	.	.	.	.	.	.	.
Urti dioc	.	.	.	.	.	.	.	.	.	.	.
Urti uren	.	.	.	.	.	.	.	.	.	.	.
Sola nigr	.	.	.	.	.	.	.	.	.	.	.
Hyos nige	.	.	.	.	.	.	.	.	.	.	.
Odon vern	.	.	.	1	.	.	.	.	.	.	1
Verb sp.	.	.	.	.	.	.	.	.	.	.	.
Lami a/p	.	.	.	.	.	.	.	.	.	.	.
Ment a/a	.	1	1	.	.	.	.	.	.	.	2
Prun vulg	.	.	.	2	.	.	.	.	.	.	2
Plan lanc	2	.	.	.	.	.	.	2	.	.	4
Plan majo	1	.	.	.	.	.	.	.	.	.	1
Gali apar	1	.	.	.	.	.	.	1	.	1	3
Gali palu	.	.	.	.	.	.	.	.	.	.	.
Sher arve	.	.	.	.	.	.	.	.	.	.	.
Vale dent	.	.	.	.	.	.	.	.	.	.	.
Trip inod	.	.	1	.	.	.	.	.	.	6	7
Laps comm	.	.	.	1	.	.	.	.	.	.	1
Sonc aspe	.	.	.	.	.	.	.	.	.	.	.
Compos.	.	.	.	.	.	.	.	.	.	.	.
Aven sp.	.	.	.	.	.	.	.	.	.	1	1
Brom m/s	5	5	3	4	2	.	2	1	.	9	31
Sieg decu	32	18	32	27	20	13	13	14	3	7	179
smal gras	1	2	1	.	.	1	1	1	1	7	15
Gramin.	6	4	5	3	2	.	1	.	.	11	32
Arrh elat	.	.	1	.	.	.	.	.	.	1	2
rhiz Gram	21	19	15	14	26	3	2	18	.	.	118
Junc sp.	.	.	.	.	.	.	.	.	.	.	.
Eleo sp.	.	.	.	.	.	.	.	.	.	.	.
Isol seta	.	.	.	.	.	.	.	.	.	.	.
Care pilu	.	.	.	.	.	.	.	.	.	.	.
Care puli	2	.	.	.	.	.	.	1	.	.	3
Care spp.	25	6	5	1	5	3	1	2	1	.	49
OTHER											
Cory avel	.	.	.	.	.	.	.	.	.	.	.
Crat mono	.	.	.	.	.	.	.	.	.	.	.
INDET.	.	2	3	6	.	.	.	1	2	2	16
TOTAL	166	94	105	79	76	37	36	85	16	591	1285

Table 4.10n   Carbonized seeds from Thorpe Thewles.

PHASE:	IV		IV
CONTEXT:	JS 21	JS 24	TOTAL
VOL. IN LITRES:	28	28	336
GRAIN			
Trit dico	.	.	.
Trit spel	.	1	26
Trit sp.	.	.	20
Hord vulg	.	.	35
Cere inde	.	4	85
CHAFF			
glum dico	.	.	.
glum spel	5	1	287
glum inde	3	.	222
glumes	.	.	5
rach brit	1	.	74
base Trit	.	.	.
rach Hord	.	.	5
base Hord	.	.	1
flor Aven	.	.	1
awns Aven	.	.	.
culm node	.	.	.
chaf inde	.	.	.
WEEDS			
Ranu acri	.	.	.
Ranu repe	.	.	1
Ranu Ranu	.	.	1
Ranu flam	.	.	2
Papa arge	.	.	.
Papa rh/d	.	.	.
Papa sp.	.	.	.
Raph raph	.	.	.
Bras sp.	.	.	.
Crucif.	.	.	.
Viol Mela	.	.	.
Mont font	3	.	18
Stel medi	.	.	1
Chen albu	1	.	5
Chen sp.	3	.	7
Atri spp.	.	.	1
Chenop.	.	.	4
Malv sylv	.	.	.
Malv sp.	.	.	1
Vici Lath	.	.	1
Legumin.	2	.	5
Apha arve	.	.	1
Pote erec	.	.	1
Hera spon	.	.	.
Poly avic	.	.	5
Poly conv	.	.	.
Poly lapa	.	.	4
Poly pers	.	.	3
Poly l/p	.	.	.
Rume acet	.	.	1
Rume spp.	.	.	20
Polygon.	.	.	.
Urti dioc	.	.	.
Urti uren	.	.	.
Sola nigr	.	.	.
Hyos nige	.	.	.
Odon vern	.	.	1
Verb sp.	.	.	.
Lami a/p	.	.	.
Ment a/a	.	.	2
Prun vulg	.	.	2
Plan lanc	.	.	4
Plan majo	.	.	1
Gali apar	.	.	3
Gali palu	.	.	.
Sher arve	.	.	.
Vale dent	.	.	.
Trip inod	3	.	10
Laps comm	.	.	1
Sonc aspe	.	.	.
Compos.	.	.	.
Aven sp.	.	.	1
Brom m/s	.	1	32
Sieg decu	11	3	193
smal gras	4	.	19
Gramin.	4	.	36
Arrh elat	.	.	2
rhiz Gram	.	.	118
Junc sp.	.	.	.
Eleo sp.	.	.	.
Isol seta	.	.	.
Care pilu	.	.	.
Care puli	.	.	3
Care spp.	3	.	52
OTHER			
Cory avel	.	.	.
Crat mono	.	.	.
INDET.	.	.	16
TOTAL	43	10	1338

Table 4.10o   Carbonized seeds from Thorpe Thewles - Summary.

PHASE:	I/II	II/III	III	III/IV	IV	TOTAL
NO. OF SAMPLES:	47	10	29	29	12	127
VOL. IN LITRES:	1316	280	812	812	336	3556
GRAIN						
Trit dico	1	.	.	.	.	1
Trit spel	66	19	54	49	26	214
Trit sp.	69	20	49	48	20	206
Hord vulg	86	229	148	228	35	726
Cere inde	365	146	345	378	85	1319
CHAFF						
glum dico	1	1	.	1	.	3
glum spel	804	329	886	576	287	2882
glum inde	771	253	918	295	222	2459
glumes	53	8	23	27	5	116
rach brit	278	56	203	87	74	698
base Trit	3	.	.	.	5	8
rach Hord	110	52	46	71	1	280
base Hord	1	.	1	.	1	3
flor Aven	1	3	2	1	.	7
awns Aven	18	.	6	5	.	29
culm node	4	.	.	3	.	7
chaf inde	2	.	3	.	.	5
WEEDS						
Ranu acri	.	1	1	.	.	2
Ranu repe	3	1	1	.	.	5
Ranu Ranu	21	1	12	5	1	40
Ranu flam	19	.	2	3	2	26
Papa arge	1	.	6	.	.	7
Papa rh/d	.	.	1	.	.	1
Papa sp.	.	1	.	.	.	1
Raph raph	2	1	1	4	.	8
Bras sp.	1	1	.	.	.	2
Crucif.	3	29	.	9	.	41
Viol Mela	2	.	.	.	.	2
Mont font	1018	32	491	248	18	1807
Stel medi	11	2	14	11	1	39
Chen albu	133	114	42	77	5	371
Chen sp.	76	4	34	27	7	148
Atri spp.	36	42	31	14	1	124
Chenop.	.	54	.	10	4	68
Malv sylv	.	.	1	.	.	1
Malv sp.	.	.	6	.	1	7
Vici Lath	16	3	7	6	1	33
Legumin.	148	7	82	84	5	326
Apha arve	.	.	.	.	1	1
Pote erec	42	1	8	16	1	68
Hera spon	1	.	.	.	.	1
Poly avic	18	5	9	33	5	70
Poly conv	7	3	8	7	.	25
Poly lapa	20	1	7	11	4	43
Poly pers	13	2	9	5	3	32
Poly l/p	8	1	8	6	.	23
Rume acet	7	1	5	3	1	17
Rume spp.	126	97	147	71	20	461
Polygon.	6	.	2	.	.	8
Urti dioc	3	.	1	.	.	4
Urti uren	.	.	2	.	.	2
Sola nigr	.	.	.	1	.	1
Hyos nige	.	1	13	1	.	15
Odon vern	3	2	6	11	1	23
Verb sp.	1	.	.	.	.	1
Lami a/p	.	.	5	.	.	5
Ment a/a	.	.	.	.	2	2
Prun vulg	8	1	7	3	2	21
Plan lanc	25	1	10	12	4	52
Plan majo	7	2	8	1	1	19
Gali apar	16	5	30	21	3	75
Gali palu	1	.	1	.	.	2
Sher arve	3	.	.	2	.	5
Vale dent	.	.	.	1	.	1
Trip inod	55	25	84	10	10	184
Laps comm	.	2	.	.	1	3
Sonc aspe	.	5	.	.	.	5
Compos.	4	.	.	2	.	6
Aven sp.	20	52	4	11	1	88
Brom m/s	621	356	213	386	32	1608
Sieg decu	1658	115	1155	1662	193	4783
smal gras	284	55	121	219	19	698
Gramin.	358	139	254	243	36	1030
Arrh elat	22	.	13	24	2	61
rhiz Gram	421	28	273	398	118	1238
Junc sp.	.	.	1	.	.	1
Eleo sp.	9	.	6	9	.	24
Isol seta	3	.	.	1	.	4
Care pilu	15	.	8	8	.	31
Care puli	11	.	34	10	3	58
Care spp.	532	16	227	280	52	1107
OTHER						
Cory avel	7	7	1	8	.	23
Crat mono	.	.	1	.	.	1
INDET.	127	30	98	127	16	398
TOTAL	8585	2362	6195	5870	1338	24350

Table 4.11   Sample contexts from Stanwick.

CONTEXT	DESCRIPTION	VOLUME IN LITRES
2209	lowest fill of post pit	15
1095	fill of post pit	15
2163	deep posthole cut by pennanular gulley	10
1085	'old soil' layer	15
2065	fill of posthole or pipe in post pit	15
2201	fill of posthole or pipe in post pit	15
2064	lowest fill of post pit	15
2043	above 2064, contained much charcoal	10
1112	fill of post pit	30
1110	fill of posthole in pit (?post pit)	15
2160	gravel layer beneath industrial/domestic deposits	15
2180	fill of posthole or pipe in post pit	10
2182	fill in a post pit	10
2196	fill of west side of pennanular gulley	15
2156	burnt layer above hearth or similar (2195)	10
1027	dump layer, possibly associated with use of latest hearths	15
1084	layer from across the top of a ditch	15
2045	fill of ditch, below arching wall	15
2051	fill of ditch, below context 2045	15
2195	hearth (or similar)	15
2119	fill of several possible stakeholes	1
2192	fill of post pit	10
1064	layer across the top of a ditch	30
1078	'old soil' layer	15
1013	fill of post setting	15
1022	dump of burnt material	15
1023	spread of loam with some charcoal	15
2006	soil matrix from stone spread	15
2012	soil matrix between stones of arching wall	15
2042	soil matrix between stones of arching wall	10
TOTAL VOLUME OF SIEVED SEDIMENT IN LITRES:		431

Table 4.12a   Carbonized seeds from Stanwick.

CONTEXTS:	2209	1095	2163	1085	2065	2201	2064	2043	1112	1112	1110	2160	2180	2182	sub-total
SAMPLE NO.:	79	41	38	32	63	74	65	48	45	46	42	77	50	52	
VOL. IN LITRES:	15	15	10	15	15	15	15	10	15	15	15	15	10	10	190
GRAIN															
Trit spel	.	1	.	.	.	.	.	.	.	.	.	.	2	.	3
Trit sp.	.	3	.	.	.	.	.	.	.	.	1	.	.	2	6
Hord vulg	12	15	5	58	2	13	2	4	13	7	3	3	2	12	151
Cere inde	8	6	2	12	.	4	2	4	4	3	4	4	6	10	69
CHAFF															
glum spel	9	6	5	19	.	8	2	5	1	6	2	4	5	5	77
glum inde	9	6	3	11	.	7	3	3	4	9	3	2	7	7	74
glumes	1	2	.	2	.	2	4	.	2	2	1	.	.	1	17
rach brit	3	4	1	8	.	3	6	.	7	2	3	1	2	4	44
rach Hord	.	.	.	4	.	1	.	.	.	.	.	1	1	2	9
lemm Hord	.	.	.	.	.	.	.	.	.	.	.	.	1	.	1
flor Avef	.	.	.	.	.	.	.	.	.	.	.	.	.	.	.
flor Aven	.	.	.	.	.	.	.	.	.	.	.	.	.	.	.
lemm Aven	.	.	.	.	.	.	.	.	.	.	.	.	.	.	.
culm node	.	.	.	.	.	.	1	.	.	.	1	1	.	.	3
WEEDS															
Ranu repe	.	.	.	.	.	.	.	.	.	.	.	.	.	.	.
Ranu Ranu	.	.	.	2	.	.	.	.	.	.	.	.	.	.	2
Bras sp.	.	.	.	.	.	.	.	.	.	.	1	.	.	.	1
Raph raph	1	.	.	1	.	.	.	.	.	.	.	.	.	1	3
Stel medi	1	1	.	.	.	.	.	1	.	.	2	1	.	1	7
Stel palu	.	.	.	.	.	.	.	.	.	.	1	.	.	.	1
Mont font	2	1	1	.	2	.	.	.	.	1	.	.	2	.	8
Chen albu	.	1	2	1	.	2	1	3	1	.	.	.	1	2	15
Chen sp.	.	.	.	1	.	1	.	.	.	1	.	.	.	.	3
Atri spp.	.	.	.	.	.	.	.	.	.	.	.	.	.	.	.
Chenop.	.	.	.	.	.	.	.	.	.	.	.	.	.	.	.
Vici Lath	.	.	.	.	.	.	.	.	.	.	.	.	.	.	.
Legumin.	3	1	7	8	6	1	2	4	2	2	8	6	.	6	56
Pote erec	.	.	.	.	8	.	1	.	2	.	1	.	.	.	12
Apha arve	.	.	.	.	.	.	.	.	.	.	.	.	.	.	.
Poly avic	1	.	1	1	.	.	.	.	.	.	.	.	.	.	3
Poly conv	.	1	.	.	.	.	.	.	.	.	.	.	.	.	1
Poly lapa	.	.	.	.	.	.	.	.	.	.	.	.	.	.	.
Poly pers	.	.	.	.	.	.	.	.	.	.	.	.	.	.	.
Poly l/p	.	.	1	.	.	.	.	1	.	.	.	.	.	.	2
Poly sp.	.	.	.	.	.	.	.	.	.	.	.	.	.	.	.
Rume acet	1	.	.	1	.	.	.	1	.	.	.	.	.	.	3
Rume spp.	2	.	.	.	.	1	.	.	.	.	.	.	2	1	6
Urti uren	.	.	.	.	.	.	.	.	.	.	.	.	.	.	.
Sola nigr	.	.	.	.	.	.	.	.	.	.	.	.	.	.	.
Hyos nige	.	.	.	1	.	.	1	.	.	.	.	.	.	.	2
Vero arve	.	.	.	.	.	.	.	.	.	.	.	.	.	.	.
Rhin sp.	.	.	.	.	.	.	.	.	.	.	.	.	.	.	.
Odon vern	1	.	.	.	.	.	.	.	.	.	.	.	.	.	1
Ment a/a	.	.	.	.	.	.	.	.	.	.	.	.	.	.	.
Prun vulg	.	.	.	.	.	.	.	.	.	.	.	.	.	.	.
Gale sp.	.	.	.	.	.	.	.	.	.	.	.	.	.	.	.
Plan lanc	1	.	.	.	.	.	1	.	.	.	2	.	2	.	6
Plan majo	.	.	.	.	.	.	.	.	.	.	.	.	.	.	.
Gali apar	.	1	.	.	.	.	.	.	.	.	1	.	.	1	3
Trip inod	.	.	.	2	.	1	.	.	.	1	.	.	2	.	6
Aven sp.	1	.	.	.	.	.	.	.	.	.	.	.	.	.	1
Brom m/s	12	17	4	15	6	3	4	7	4	3	6	2	5	6	94
Sieg decu	9	19	14	24	11	7	8	26	14	5	4	9	16	27	193
smal gras	4	6	2	10	2	.	7	4	3	2	2	1	3	5	51
Gramin.	3	1	.	1	.	2	2	.	1	.	.	.	.	1	11
rhiz Gram	7	1	.	13	1	.	3	7	3	.	5	7	.	2	49
Junc squa	.	.	.	.	.	.	.	.	.	.	.	.	.	.	.
Junc sp.	.	.	.	.	.	.	.	.	.	.	.	.	.	.	.
Eleo sp.	.	2	.	.	.	.	.	.	.	.	.	.	.	1	3
Care pilu	.	1	.	2	.	1	.	2	1	1	.	.	.	3	11
Care spp.	2	.	3	6	12	1	2	5	.	1	2	.	1	3	38
OTHER															
Cory avel	1	.	2	.	.	.	.	3	.	.	.	1	1	.	8
Samb nigr	1	.	.	1	.	.	2	.	.	.	.	.	.	.	4
Rosa sp.	.	.	.	.	.	.	.	.	.	.	1	.	.	.	1
Call leaf	1	.	.	1	.	2	.	.	.	.	.	.	.	.	4
Call flow	1	1	.	2	.	.	.	.	3	.	2	.	1	1	11
Eric flow	.	1	.	.	1	.	.	.	.	.	.	.	.	.	2
Cath palu	.	.	.	.	.	.	.	.	.	.	.	.	.	.	.
INDET.	7	2	3	4	3	4	4	3	.	3	3	3	2	.	41
TOTAL	104	100	56	211	54	64	58	83	65	49	59	46	64	104	1117

Table 4.12b   Carbonized seeds from Stanwick.

CONTEXTS:	2196	2156	1027	1084	2045	2051	2195	2119	2192	1064	1064	1078	sub-total
SAMPLE NO.:	76	60	4	29	51	55	85	21	58	19	23	25	
VOL. IN LITRES:	15	10	15	15	15	15	15	1	10	15	15	15	156
GRAIN													
Trit spel	1	1	.	1	2	.	.	.	2	.	.	.	7
Trit sp.	3	2	.	1	5	2	2	.	.	.	.	.	15
Hord vulg	30	97	2	5	21	2	25	.	11	22	13	1	229
Cere inde	13	50	8	9	15	4	21	2	7	14	3	2	148
CHAFF													
glum spel	25	27	1	5	45	18	3	1	8	5	.	3	141
glum inde	20	34	2	4	43	14	2	.	5	5	1	3	133
glumes	1	7	.	1	4	2	.	.	3	.	.	.	18
rach brit	13	17	.	2	22	8	.	.	7	2	.	.	71
rach Hord	4	5	1	.	8	5	.	.	5	1	.	.	29
lemm Hord	.	.	.	.	.	.	.	.	2	.	.	.	2
flor Avef	.	.	.	.	2	.	.	.	.	.	.	.	2
flor Aven	.	.	.	.	.	.	.	.	.	.	.	.	.
lemm Aven	.	.	.	.	.	.	.	.	.	.	.	.	.
culm node	.	.	2	.	.	.	.	.	.	.	.	.	2
WEEDS													
Ranu repe	.	.	.	.	1	.	.	.	.	.	.	.	1
Ranu Ranu	1	.	.	.	.	1	1	.	.	1	.	.	4
Bras sp.	.	.	.	.	.	.	.	.	.	.	.	.	.
Raph raph	1	1	1	.	.	1	.	.	.	.	.	.	4
Stel medi	.	1	1	1	1	2	3	.	.	.	.	.	9
Stel palu	.	.	.	.	.	.	.	.	.	.	.	.	.
Mont font	.	1	2	.	7	2	.	.	.	2	.	2	16
Chen albu	4	5	.	.	8	.	4	.	.	3	1	3	28
Chen sp.	.	1	.	.	2	.	5	.	2	1	.	.	11
Atri spp.	.	1	.	.	2	.	.	.	.	.	.	.	3
Chenop.	.	.	.	.	.	.	.	.	.	.	.	.	.
Vici Lath	1	1	1	.	.	.	.	.	.	.	.	.	3
Legumin.	3	3	.	1	23	8	7	.	4	1	.	9	59
Pote erec	.	.	1	.	9	.	.	.	.	1	.	6	17
Apha arve	.	2	.	.	.	.	.	.	.	.	.	.	2
Poly avic	.	2	.	.	.	.	.	.	.	.	.	.	2
Poly conv	.	.	1	.	.	.	.	.	.	.	.	.	1
Poly lapa	.	.	.	.	1	.	.	.	.	.	.	.	1
Poly pers	.	.	.	.	2	.	.	.	.	.	.	.	2
Poly l/p	.	1	.	.	.	1	.	.	.	.	.	.	2
Poly sp.	.	7	.	.	1	.	4	.	.	.	.	.	12
Rume acet	.	12	1	.	.	6	15	.	.	.	.	.	34
Rume spp.	.	116	.	.	28	.	122	.	.	3	1	2	272
Urti uren	.	3	.	.	.	.	9	.	.	.	.	.	12
Sola nigr	.	.	.	.	.	.	1	.	.	.	.	.	1
Hyos nige	.	42	.	.	1	.	145	.	.	.	.	.	188
Vero arve	.	1	.	.	.	.	.	.	.	.	.	.	1
Rhin sp.	.	.	.	.	.	.	1	.	.	.	.	.	1
Odon vern	.	1	.	.	.	.	1	.	.	.	.	.	2
Ment a/a	.	.	.	.	1	.	.	.	.	1	.	.	2
Prun vulg	.	.	.	.	1	.	.	.	.	.	.	.	1
Gale sp.	.	.	.	.	.	.	.	.	.	.	.	.	.
Plan lanc	.	.	1	.	.	1	2	.	.	.	1	.	5
Plan majo	.	1	.	.	.	.	.	.	.	.	.	.	1
Gali apar	1	.	.	.	1	2	1	.	.	.	.	.	5
Trip inod	.	.	.	.	5	.	.	.	.	.	.	.	5
Aven sp.	.	.	.	.	.	.	.	.	.	.	.	.	.
Brom m/s	11	18	4	9	31	9	10	1	4	8	3	5	113
Sieg decu	18	12	5	14	141	36	4	.	13	16	7	45	311
smal gras	7	28	.	2	16	10	26	.	8	2	2	25	126
Gramin.	.	6	.	.	.	1	1	.	.	.	.	.	8
rhiz Gram	1	.	.	.	17	.	.	.	.	4	2	15	39
Junc squa	.	1	.	.	1	.	.	.	.	.	.	.	2
Junc sp.	.	.	.	.	.	.	.	.	.	.	.	.	.
Eleo sp.	.	.	.	.	1	.	.	.	.	.	.	.	1
Care pilu	1	.	.	1	18	.	.	.	.	.	.	11	31
Care spp.	2	.	2	5	37	7	.	.	3	1	1	9	67
OTHER													
Cory avel	.	.	.	1	12	.	.	.	.	.	.	.	13
Samb nigr	.	.	.	.	.	.	.	.	.	.	.	.	.
Rosa sp.	.	.	.	.	.	.	.	.	.	.	.	.	.
Call leaf	.	.	.	.	1	.	.	.	.	.	.	.	1
Call flow	.	.	.	.	1	.	.	.	.	.	.	.	1
Eric flow	.	.	.	.	1	.	.	.	.	.	.	.	1
Cath palu	.	.	.	.	.	.	.	.	.	.	.	.	.
INDET.	2	1	.	3	4	3	2	2	.	5	.	9	31
TOTAL	163	508	36	65	542	145	417	6	84	98	35	150	2249

Table 4.12c Carbonized seeds from Stanwick.

CONTEXTS:	1013	1022	1023	2006	2012	2042	TOTAL
SAMPLE NO.:	1	5	6	1	2	40	
VOL. IN LITRES:	15	15	15	15	15	10	431
GRAIN							
Trit spel	.	.	40	.	.	1	49
Trit sp.	1	.	38	.	2	1	63
Hord vulg	2	10	63	3	18	10	486
Cere inde	2	6	49	1	9	8	292
CHAFF							
glum spel	5	4	9	.	6	24	266
glum inde	17	3	8	1	5	14	255
glumes	2	1	1	.	1	3	43
rach brit	3	1	3	.	2	8	132
rach Hord	.	.	.	.	.	2	40
lemm Hord	.	.	.	.	.	.	3
flor Avef	.	.	.	.	.	.	2
flor Aven	.	1	.	.	.	.	1
lemm Aven	.	.	.	.	1	.	1
culm node	.	.	4	1	.	.	10
WEEDS							
Ranu repe	.	.	.	.	.	.	1
Ranu Ranu	.	.	1	.	.	.	7
Bras sp.	.	.	.	.	.	.	1
Raph raph	1	3	.	.	1	.	12
Stel medi	.	1	2	.	.	.	19
Stel palu	.	.	1	.	.	.	2
Mont font	.	5	2	.	.	3	34
Chen albu	2	1	25	.	2	2	75
Chen sp.	.	.	2	.	4	.	20
Atri spp.	1	.	.	.	1	.	5
Chenop.	2	.	.	.	.	2	4
Vici Lath	2	.	1	.	.	.	6
Legumin.	1	7	5	.	4	23	155
Pote erec	.	2	.	.	1	8	40
Apha arve	.	1	.	.	.	.	3
Poly avic	1	1	2	.	.	1	10
Poly conv	.	3	2	.	.	.	7
Poly lapa	.	.	1	1	.	.	3
Poly pers	.	.	1	.	1	.	4
Poly l/p	.	.	.	.	.	.	4
Poly sp.	.	.	.	.	.	.	12
Rume acet	.	.	1	.	.	1	39
Rume spp.	11	.	6	.	3	3	301
Urti uren	.	.	.	.	.	.	12
Sola nigr	.	.	.	.	.	.	1
Hyos nige	.	.	.	.	.	1	191
Vero arve	1	.	.	.	.	.	2
Rhin sp.	.	.	.	.	.	.	1
Odon vern	.	.	.	.	.	.	3
Ment a/a	.	.	.	.	.	.	2
Prun vulg	.	1	.	.	.	.	2
Gale sp.	.	1	1	.	.	.	2
Plan lanc	.	.	.	.	.	2	13
Plan majo	.	.	.	.	.	.	1
Gali apar	3	2	7	.	1	1	22
Trip inod	.	.	.	.	.	2	13
Aven sp.	.	.	1	.	.	.	2
Brom m/s	6	15	183	1	11	21	444
Sieg decu	15	37	13	3	48	129	749
smal gras	.	11	19	1	4	24	236
Gramin.	2	.	.	.	7	2	30
rhiz Gram	6	6	4	1	6	22	133
Junc squa	.	.	.	.	.	.	2
Junc sp.	.	.	1	.	.	.	1
Eleo sp.	1	.	.	.	.	.	5
Care pilu	1	3	.	.	3	3	52
Care spp.	7	9	6	4	12	22	165
OTHER							
Cory avel	1	.	.	.	1	3	26
Samb nigr	1	2	2	1	.	.	10
Rosa sp.	.	.	.	.	.	.	1
Call leaf	.	2	.	.	2	1	10
Call flow	1	10	3	.	2	1	29
Eric flow	1	.	.	.	1	1	6
Cath palu	1	.	.	.	.	.	1
INDET.	3	5	4	1	4	7	96
TOTAL	103	154	511	19	163	356	4672

Table 4.13   Sample contexts from Rock Castle.

```
CONTEXT DESCRIPTION                                    VOLUME IN LITRES
47      fill of ring ditch of 'early' house                   12
61      fill of ring ditch of 'early' house                   60
74      fill of ring groove of 'early' house                  16
24      fill of ring groove of 'main' house                   30
59      fill of posthole at entrance of 'main' house          30
60      fill of posthole at entrance of 'main' house          30
50      fill of pit inside surrounds of 'main' house          30
 2      fill of ditch 25                                      60
14      fill of pit/posthole                                  15
69      fill of pit/posthole                                  15

12.1    fill of ring ditch of 'main' house section 4          20
12.2    fill of ring ditch of 'main' house section 4/5        30
12.3    fill of ring ditch of 'main' house section 5/6        30
12.4    fill of ring ditch of 'main' house section 5/6        30
12.5    fill of ring ditch of 'main' house section 6/7        30
12.6    fill of ring ditch of 'main' house section 7/10       30
12.7    fill of ring ditch of 'main' house section 8          30
12.8    fill of ring ditch of 'main' house section 9/10       30
12.9    fill of ring ditch of 'main' house section 10         30
38      fill of ring ditch of 'main' house                    10
54      fill of ring ditch of 'main' house                    30

TOTAL VOLUME OF SIEVED SEDIMENT IN LITRES:                   598
```

Table 4.15   Sample contexts from Thornbrough.

```
CONTEXT DESCRIPTION                    VOLUME IN LITRES
1983

2       fill of posthole                       15
5       fill of rectangular posthole           30
10      sediment in between cobbled area       30
17      fill of gulley                         30
39      sediment in between cobbled area       25
40      fill of posthole                       10
41      fill of posthole                       30
43      fill of rectangular posthole           35
45      fill of pit below context 10           17
46      fill of posthole                       22
49      fill of posthole                       16
54a     fill of gulley, same as context 17     34
55      fill of posthole                       14

1984

44      fill of posthole                       60
46      sediment in between cobbled area       60
54b     sediment in between cobbled area       60
58      fill of posthole/trench                30
110     fill of posthole                       12
120     fill of posthole/trench                30
123     fill of posthole                        5
125     fill of posthole                       15
127     fill of posthole                       25
134     fill of posthole, below context 58     30

TOTAL VOLUME OF SIEVED SEDIMENT IN LITRES:    635
```

Table 4.17   Sample contexts from South Shields.

```
CONTEXT         DESCRIPTION                                    VOLUME IN LITRES
12236           lower deposit in between the sleeper walls
                of the forecourt granary, consisting of
                clay and clay-silt, flakes of sandstone
                and mortar, charcoal and grain. Sealed
                when the flagstone floor was coated with
                a layer of opus signinum. The grain from
                this deposit probably represents spillage
                through the cracks of the floor
                33 samples                                           465

12176           layer of debris accumulated over the
                floor of the granary
                30 samples                                           427

TOTAL VOLUME OF SIEVED SEDIMENT IN LITRES:                           892
```

Table 4.14a Carbonized seeds from Rock Castle.

CONTEXTS:	47	61a	61b	74	24	59	60	50*	2a*	2b*	14	69	sub-total
VOL. IN LITRES:	12	30	30	16	30	30	30	30	30	30	14	15	297
GRAIN													
Trit spel	2	1	.	4	6	4	14	4	8	6	3	1	53
Trit aest	.	.	.	.	.	.	.	1	.	.	.	.	1
Trit sp.	2	1	4	10	4	4	8	3	5	8	1	1	51
Hord vulg	10	4	2	19	8	12	8	19	8	1	2	.	93
Cere inde	8	5	5	30	13	7	25	19	16	10	8	2	148
CHAFF													
glum spel	13	6	13	35	21	10	52	11	23	71	1	3	259
glum dico	.	.	.	.	.	.	.	1	.	.	.	.	1
glum inde	7	6	11	48	24	12	111	44	14	58	2	9	346
glumes	.	.	.	8	5	.	13	6	4	12	.	2	50
rach brit	3	.	3	16	11	3	24	38	8	19	.	.	125
rach aest	.	.	.	.	.	.	.	125	.	.	.	.	125
awns Trit	.	.	.	.	.	.	.	.	.	.	.	.	.
rach Hord	8	.	.	12	2	6	1	12	.	1	.	.	42
flor Aven	.	.	.	.	1	.	.	.	.	.	.	.	1
awns Aven	.	.	.	1	1	.	1	2	.	.	.	.	5
culm node	.	.	.	1	2	.	.	2	.	.	.	1	6
chaf inde	.	.	.	.	.	.	7	35	1	2	.	.	45
WEEDS													
Ranu flam	1	.	.	.	.	.	.	.	1	.	.	.	2
Ranu repe	.	.	.	.	.	.	.	.	.	.	.	.	.
Ranu Ranu	1	1	2	.	.	1	.	.	1	.	.	1	7
Bras sp.	.	.	.	.	.	.	.	.	.	.	.	.	.
Raph raph	.	.	.	.	.	1	1	.	.	.	.	.	2
Viol Mela	.	.	.	.	.	.	.	.	1	.	.	.	1
Stel medi	1	.	4	2	8	2	20	3	11	3	.	1	55
Stel palu	.	.	.	.	.	.	.	.	1	.	.	.	1
Sper arve	.	.	.	.	.	.	.	.	.	.	.	1	1
Mont font	.	.	.	.	.	.	1	4	13	2	.	.	23
Chen albu	2	.	2	4	5	7	8	2	5	5	3	.	43
Chen sp.	.	.	3	3	6	13	20	9	5	3	4	21	87
Atri spp.	.	.	.	1	1	9	4	1	1	2	.	15	34
Chenop.	4	.	.	.	.	.	.	.	.	.	.	.	4
Malv sp.	.	.	.	.	.	.	.	.	.	.	.	.	.
Linu cath	.	.	.	.	2	.	.	.	.	.	.	.	2
Vici Lath	.	.	.	2	12	9	13	7	.	.	1	6	50
Legumin.	.	1	9	8	19	15	23	5	120	74	7	2	283
Pote erec	1	.	3	5	7	2	1	.	.	11	.	.	30
Apha arve	.	.	.	.	.	.	.	.	.	.	.	.	.
Anth cauc	.	.	.	.	.	.	.	.	.	.	.	.	.
Hera spon	.	.	.	.	.	.	.	.	.	.	.	.	.
Poly avic	.	1	3	1	5	4	5	4	1	.	.	.	24
Poly conv	2	.	.	.	2	1	1	1	.	.	.	1	8
Poly lapa	.	.	.	.	.	1	.	.	.	.	.	.	1
Poly pers	.	.	.	.	.	.	.	.	.	.	.	.	.
Poly l/p	.	.	.	.	.	.	.	1	.	.	.	.	1
Poly sp.	1	.	1	.	3	.	4	.	1	1	.	4	15
Rume acet	2	.	.	.	1	1	1	1	.	.	.	.	6
Rume spp.	1	.	.	.	1	9	7	10	.	.	.	.	28
Urti uren	.	.	.	.	6	.	1	.	.	.	.	.	7
Sola nigr	.	.	.	.	.	.	.	.	.	1	.	.	1
Vero arve	.	.	.	.	1	.	.	.	.	1	.	.	2
Odon vern	.	.	.	.	.	.	.	.	.	.	.	.	.
Prun vulg	.	.	.	.	1	.	.	.	1	.	.	.	2
Plan lanc	2	.	.	.	5	1	3	.	8	4	.	1	24
Gali apar	.	1	.	.	4	5	6	1	1	.	1	2	21
Gali palu	.	.	.	.	.	.	.	.	.	.	.	.	.
Trip inod	1	.	14	3	17	12	33	3	2	6	1	4	96
Hypo g/r	.	.	.	.	.	.	.	.	1	.	.	.	1
Compos.	.	.	.	.	1	.	.	.	.	.	.	.	1
Aven sp.	.	1	.	.	3	1	2	.	1	3	.	1	12
Brom m/s	6	6	3	42	8	4	5	9	20	23	1	4	131
Sieg decu	10	9	21	23	66	94	49	34	85	140	20	10	561
smal gras	22	10	18	11	34	25	39	17	38	41	11	14	280
Gramin.	.	.	.	1	8	7	7	7	2	2	.	2	36
Arrh elat	.	.	.	.	.	.	.	.	3	.	.	.	3
rhiz Gram	5	1	5	17	21	26	6	.	37	31	5	.	154
Junc squa	.	.	1	.	1	1	1	1	.	2	1	.	8
Junc sp.	1	1	.	.	.	.	.	.	2	.	.	.	4
Eleo sp.	.	.	.	.	1	.	1	.	.	.	.	.	2
Care pilu	1	.	.	1	2	1	4	.	.	2	.	.	11
Care spp.	7	2	4	7	16	17	7	5	27	11	9	.	112
OTHER													
Cory avel	.	.	.	.	.	1	2	1	2	.	.	.	6
Prun spin	.	.	.	.	.	.	.	.	.	.	.	.	.
Call leaf	4	.	.	5	.	3	2	1	9	2	2	.	28
Call flow	6	.	2	4	1	4	1	2	.	1	1	.	22
Vacc myrt	.	.	.	.	.	.	.	.	1	.	.	.	1
Viol Viol	.	.	.	.	.	.	.	.	.	1	.	.	1
Meny trif	.	.	.	.	1	.	.	.	3	.	.	.	4
Pota spp.	.	.	.	.	.	.	.	.	3	.	.	.	3
INDET.	4	5	6	6	7	9	11	11	15	7	3	3	87
TOTAL	138	62	139	331	376	344	553	462*	509*	567*	87	112	3680

KEY: * = only 50 per cent of this sample is analysed.

Table 4.14b   Carbonized seeds from Rock Castle.

CONTEXTS:	12.1	12.2	12.3	12.4	12.5	12.6	12.7	12.8	12.9	38	54	TOTAL
VOL. IN LITRES:	20	30	30	30	30	30	30	30	30	10	30	598
GRAIN												
Trit spel	.	.	.	1	2	1	10	2	2	1	2	74
Trit aest	.	.	.	.	.	.	.	.	.	.	.	1
Trit sp.	1	.	2	2	2	2	5	2	2	2	3	74
Hord vulg	3	4	1	2	8	8	9	2	6	1	7	144
Cere inde	4	3	2	4	7	14	20	7	12	3	14	238
CHAFF												
glum spel	7	6	5	27	17	14	25	2	7	9	6	384
glum dico	.	.	.	.	.	.	.	.	.	.	.	1
glum Trit	4	10	12	27	29	14	19	14	20	9	12	516
glum inde	1	5	.	2	1	2	5	1	3	.	.	70
rach brit	2	1	.	10	12	4	10	1	6	2	2	175
rach aest	.	.	.	.	.	.	.	.	.	.	.	125
awns Trit	.	.	.	.	.	.	.	.	1	.	.	1
rach Hord	.	3	1	3	4	.	.	.	3	.	2	58
flor Aven	.	.	.	.	.	.	.	.	.	.	.	1
awns Aven	2	4	.	3	3	3	1	.	1	.	1	23
culm node	.	.	.	2	2	2	1	2	10	.	.	25
chaff inde	.	.	.	.	.	.	1	.	.	.	.	46
WEEDS												
Ranu flam	.	.	.	.	.	.	.	.	1	.	1	4
Ranu repe	.	.	.	.	.	.	1	.	.	.	.	1
Ranu Ranu	1	.	1	1	1	.	1	1	.	1	.	14
Bras sp.	.	.	.	.	.	1	.	.	.	.	.	1
Raph raph	.	.	.	.	1	.	.	.	.	.	2	5
Viol Mela	.	.	.	.	.	.	.	.	.	.	.	1
Stel medi	3	4	2	3	7	26	6	7	13	3	6	135
Stel palu	.	.	.	.	.	.	.	.	.	.	.	1
Sper arve	.	.	.	.	.	.	.	.	.	.	.	1
Mont font	12	.	1	.	.	3	7	.	3	.	2	51
Chen albu	1	1	1	3	8	3	5	1	4	3	6	79
Chen sp.	3	1	3	3	19	14	8	9	14	12	8	181
Atri spp.	2	1	1	1	4	6	1	1	4	5	7	67
Chenop.	.	.	.	.	.	.	.	.	.	.	.	4
Malv sp.	.	.	.	.	.	.	.	1	.	.	.	1
Linu cath	.	.	.	.	.	.	.	.	.	.	.	2
Vici Lath	3	.	1	3	6	10	12	12	40	4	5	146
Legumin.	13	4	1	8	8	14	38	10	13	2	11	405
Pote erec	1	.	.	1	1	.	2	2	1	3	2	43
Apha arve	.	.	.	.	.	.	.	.	5	.	.	5
Anth cauc	.	.	.	.	.	.	.	1	.	.	.	1
Hera spon	.	.	.	.	.	.	1	.	.	.	.	1
Poly avic	2	.	.	.	4	2	5	2	15	4	9	67
Poly conv	.	.	.	.	.	.	1	.	.	1	1	11
Poly lapa	.	.	.	.	.	.	.	.	.	.	.	1
Poly pers	.	1	.	.	.	.	.	.	.	.	.	1
Poly l/p	.	.	.	.	.	2	.	.	2	.	.	5
Poly sp.	.	.	1	1	.	.	2	.	.	.	.	19
Rume acet	.	2	.	2	2	1	1	.	.	.	3	17
Rume spp.	.	.	2	.	1	.	2	.	1	.	3	37
Urti uren	.	.	.	.	.	.	.	.	3	.	1	11
Sola nigr	.	.	.	.	.	.	.	.	1	.	.	1
Vero arve	.	.	.	.	.	.	.	.	1	.	.	3
Odon vern	.	.	.	.	.	.	.	.	1	.	1	1
Prun vulg	.	.	.	.	.	.	.	.	1	.	.	3
Plan lanc	.	.	1	1	.	1	2	1	.	1	.	31
Gali apar	3	.	1	2	3	7	11	7	18	2	4	79
Gali palu	.	1	.	.	.	.	.	.	.	.	.	1
Trip inod	9	2	2	2	15	14	25	10	18	11	10	214
Hypo g/r	.	.	.	.	.	.	.	.	.	.	.	1
Compos.	.	.	.	.	.	.	.	.	.	.	.	1
Aven sp.	1	.	1	4	3	.	6	1	3	.	5	36
Brom m/s	2	3	3	4	6	6	7	3	8	5	7	185
Sieg decu	60	27	11	43	51	34	206	9	36	33	47	1118
smal gras	22	23	6	21	39	27	40	24	90	11	19	602
Gramin.	4	.	1	1	3	.	5	4	28	4	8	94
Arrh elat	.	.	.	.	.	.	.	.	.	.	.	3
rhiz Gram	20	4	1	3	7	4	26	2	.	9	5	235
Junc squa	3	1	.	.	1	1	.	.	.	.	1	15
Junc sp.	.	.	.	.	.	1	.	1	.	.	1	6
Eleo sp.	.	.	.	.	.	1	.	.	1	.	.	4
Care pilu	1	.	.	.	.	.	5	1	.	2	1	21
Care spp.	29	10	.	5	5	3	42	7	2	7	8	230
OTHER												
Cory avel	.	.	.	1	.	2	.	.	.	.	.	9
Prun spin	.	.	.	.	1	1	.	.	.	.	.	2
Call leaf	1	1	.	.	2	.	.	.	.	.	.	32
Call flow	2	.	.	1	5	2	1	.	.	1	.	34
Vacc myrt	.	.	.	.	.	.	.	.	.	.	.	1
Viol Viol	.	.	.	.	.	.	.	1	.	.	.	2
Meny trif	.	.	.	.	.	.	.	.	.	.	1	5
Pota spp.	.	.	.	.	.	.	.	.	.	.	.	3
INDET.	6	8	.	1	6	8	4	9	6	8	2	145
TOTAL	228	130	64	198	296	257	578	160	405	159	236	6391

Table 4.16a   Carbonized seeds from Thornbrough.

1983 CONTEXTS:	2	5	10	17	39	40	41	43	45*	46	49	54a	55	sub-total
VOL. IN LITRES:	15	30	30	30	25	10	30	35	17	22	16	34	14	308
GRAIN														
Trit spel	.	10	235	1	38	10	10	25	116	62	30	4	1	542
Trit sp.	.	.	5	2	4	3	7	14	9	1	4	.	.	49
Hord vulg	.	32	580	8	105	35	49	128	146	149	117	10	6	1365
Seca cere	.	2	20	.	2	2	.	3	8	5	5	.	.	47
Cere inde	4	36	598	7	128	23	47	129	450	125	87	11	3	1648
coleopti.	.	.	9	.	.	.	.	.	4	.	1	.	.	14
CHAFF														
glum spel	.	6	372	1	26	16	7	6	427	17	9	17	2	906
glum dico	.	.	6	.	.	.	.	.	3	.	.	.	.	9
glum inde	.	2	256	1	2	3	.	.	412	10	3	11	1	701
glumes	.	.	63	.	1	.	3	2	85	.	1	3	.	158
rach brit	.	.	150	.	2	2	.	.	181	1	.	5	.	341
base Hord	.	.	.	.	.	.	.	.	3	.	.	1	.	4
rach Hord	.	.	11	.	.	.	.	.	12	.	1	.	.	24
lemm Hord	.	.	.	.	.	.	.	.	.	.	.	.	.	.
rach Seca	.	.	15	.	.	.	.	.	23	.	.	.	.	38
awns Aven	.	.	3	.	.	.	.	.	4	.	.	.	.	7
culm node	.	.	.	2	.	1	.	1	4	.	.	10	.	18
chaf inde	.	.	15	.	.	.	.	.	11	.	.	.	.	26
WEEDS														
Ranu flam	.	.	.	.	.	.	.	.	.	.	.	1	.	1
Ranu repe	.	.	.	.	.	.	.	.	.	.	.	1	.	1
Ranu Ranu	.	.	.	.	.	.	.	1	.	.	.	.	.	1
Raph raph	.	.	.	.	.	.	2	.	.	.	.	1	.	3
Stel medi	.	.	.	.	.	.	.	.	.	.	.	.	.	.
Agro gith	.	.	1	.	.	.	.	.	1	.	.	.	.	2
Mont font	.	.	.	.	.	.	.	.	.	1	.	.	.	1
Chen albu	.	.	2	.	.	.	1	.	3	.	.	.	.	6
Chen sp.	.	.	1	.	.	.	.	.	1	.	.	.	.	2
Atri spp.	.	.	.	.	.	.	.	.	2	.	.	.	.	2
Vici Lath	.	.	.	.	.	.	.	.	.	.	.	1	.	1
Legumin.	.	.	1	.	.	.	.	.	2	1	.	.	.	4
Pote erec	.	.	.	.	.	.	.	.	1	.	.	.	.	1
Poly avic	.	.	.	.	1	.	.	.	3	.	1	1	.	6
Poly conv	.	.	1	.	.	.	.	.	2	.	.	1	.	4
Poly lapa	.	.	.	.	.	.	.	.	2	.	.	.	.	2
Poly pers	.	.	.	.	.	.	.	.	2	.	.	.	.	2
Poly l/p	.	.	.	.	.	.	.	1	.	.	.	.	.	1
Rume acet	.	.	.	.	1	.	.	.	.	.	.	1	.	2
Rume spp.	.	.	.	.	.	.	.	.	.	.	.	1	.	1
Ment a/a	.	.	.	.	.	.	.	.	2	.	.	.	.	2
Prun vulg	.	.	.	.	.	.	.	.	.	.	.	1	.	1
Plan lanc	.	.	.	.	.	.	.	2	.	.	.	1	.	3
Gali apar	.	.	1	.	.	.	.	.	2	.	.	1	1	5
Gali palu	.	.	.	.	.	.	.	.	.	.	.	.	.	.
Trip inod	.	.	1	.	.	.	.	.	1	.	.	.	.	2
Brom m/s	.	.	59	.	18	1	.	4	93	2	5	2	.	184
Aven sp.	.	.	1	.	.	.	.	1	.	.	.	.	.	2
Sieg decu	.	.	3	.	1	.	.	2	4	1	.	1	.	12
smal gras	.	.	1	1	1	.	1	.	1	2	1	14	.	22
Gramin.	.	.	27	.	.	.	.	.	23	.	.	4	.	54
Arrh elat	.	1	.	.	.	.	.	.	.	.	.	.	.	1
rhiz Gram	.	.	.	.	.	1	1	.	3	1	.	1	.	7
Junc squa	.	.	.	.	.	.	.	.	.	.	.	.	.	.
Eleo sp.	.	.	.	.	.	.	.	.	.	.	.	1	.	1
Care pilu	.	.	.	.	.	.	.	.	2	.	.	.	.	2
Care spp.	.	.	3	1	2	1	.	1	15	1	.	3	.	27
OTHER														
Linu usit	.	.	.	.	.	.	.	.	.	.	.	.	.	.
Cory avel	.	.	1	.	2	4	.	.	1	.	.	1	.	9
Rubu frut	.	.	.	.	.	.	.	.	.	.	.	1	.	1
Prun spin	.	.	.	.	.	.	.	.	.	.	.	.	.	.
Call leaf	.	.	.	.	.	.	.	.	1	.	.	.	.	1
Call flow	.	.	.	.	.	.	.	.	1	.	.	.	.	1
Eric flow	.	.	.	.	.	.	.	.	1	.	.	.	.	1
INDET.	.	.	.	2	.	.	.	.	1	.	.	4	.	7
TOTAL	4	89	2441	26	334	102	128	320	2068*	379	265	115	14	6285

KEY: * = only 50 per cent of this sample has been analysed.

Table 4.16b   Carbonized seeds from Thornbrough.

1984 CONTEXTS:	44	46	54b	58	110	120	123	125	127	134	TOTAL
VOL. IN LITRES:	60	60	60	30	12	30	5	15	25	30	635
GRAIN											
Trit spel	3	.	12	18	1	10	2	3	16	18	625
Trit sp.	1	3	2	4	2	.	2	.	.	3	66
Hord vulg	10	6	26	50	1	182	16	133	91	109	1989
Seca cere	.	.	1	1	.	.	.	.	.	1	50
Cere inde	9	8	31	77	6	143	6	51	60	69	2108
coleopti.	.	.	.	.	.	.	.	.	.	.	14
CHAFF											
glum spel	6	1	6	92	1	19	4	6	3	53	1097
glum dico	.	.	.	.	1	.	.	.	.	.	10
glum inde	3	.	1	40	2	7	.	7	2	54	816
glumes	.	.	.	11	.	1	.	2	2	4	178
rach brit	.	.	.	25	1	1	1	3	.	18	390
base Trit	.	.	.	1	.	.	.	.	.	.	5
rach Hord	.	1	.	4	.	1	.	2	.	3	35
lemm Hord	.	.	.	.	.	1	.	.	.	.	1
rach Seca	.	.	.	1	.	.	.	.	.	.	39
awns Aven	.	.	.	.	.	.	.	.	.	.	7
culm node	.	.	.	1	.	.	.	.	.	.	19
chaf inde	.	.	.	.	.	.	.	.	.	.	26
WEEDS											
Ranu flam	.	.	.	.	.	.	.	.	.	.	1
Ranu repe	.	.	.	.	.	.	.	.	.	.	1
Ranu Ranu	.	1	.	.	.	.	.	.	.	.	2
Raph raph	.	.	.	3	.	1	.	.	.	.	7
Stel medi	1	.	.	.	.	.	.	.	.	.	1
Agro gith	.	.	.	.	.	.	.	.	.	.	2
Mont font	.	.	.	.	.	.	.	.	.	.	1
Chen albu	1	.	.	5	.	1	.	.	.	3	16
Chen sp.	.	.	.	2	1	.	.	.	.	.	5
Atri spp.	1	.	.	.	.	.	.	.	.	.	3
Vici Lath	1	.	.	.	.	1	.	.	.	.	3
Legumin.	3	.	.	.	1	.	.	.	.	.	8
Pote erec	.	.	.	.	.	1	.	.	.	1	3
Poly avic	.	.	.	.	.	1	.	.	.	1	8
Poly conv	1	.	.	.	.	.	.	.	.	.	5
Poly lapa	.	.	.	.	.	1	.	.	.	.	3
Poly pers	.	.	.	.	.	.	.	.	.	.	2
Poly l/p	1	.	.	.	.	2	.	.	.	.	4
Rume acet	.	.	.	3	.	.	.	1	.	.	6
Rume spp.	.	.	.	.	.	2	.	.	.	.	3
Ment a/a	.	.	.	.	.	.	.	.	.	.	2
Prun vulg	.	.	.	.	.	.	.	.	.	.	1
Plan lanc	2	1	1	.	.	.	.	.	.	1	8
Gali apar	.	.	.	.	.	.	.	.	.	.	5
Gali palu	.	.	.	.	.	.	.	1	.	.	1
Trip inod	.	.	.	.	.	.	.	.	.	.	2
Brom m/s	1	.	1	31	.	6	1	.	.	11	235
Aven sp.	1	.	.	.	1	1	.	.	.	.	5
Sieg decu	.	2	.	2	.	4	.	.	1	1	22
smal gras	5	.	.	2	.	.	1	1	.	2	33
Gramin.	.	.	.	.	.	.	.	.	.	.	54
Arrh elat	2	.	.	4	1	.	.	.	.	.	8
rhiz Gram	1	3	.	.	1	.	.	.	.	.	12
Junc squa	.	.	.	.	.	.	.	1	.	.	1
Eleo sp.	.	.	.	1	.	.	.	.	.	.	2
Care pilu	.	.	.	.	.	.	.	.	.	.	2
Care spp.	.	2	1	2	.	.	.	.	.	.	32
OTHER											
Linu usit	1	.	.	.	.	.	.	.	.	.	1
Cory avel	2	1	.	1	.	.	.	.	1	3	17
Rubu frut	.	.	1	.	.	.	.	.	.	.	2
Prun spin	.	.	.	.	.	1	.	.	.	.	1
Call leaf	.	.	.	.	.	.	.	.	.	.	1
Call flow	.	.	.	.	.	.	.	.	.	.	1
Eric flow	.	.	.	.	.	.	.	.	.	.	1
INDET.	4	1	1	2	1	2	.	1	.	.	19
TOTAL	60	30	84	383	21	389	33	212	175	355	8027

Table 4.18a Carbonized seeds from South Shields, deposit 12236.

CONTEXT: 12236											
SAMPLE NO.:	1	2	3	4	5	6	7a	7b	8	9	10a
VOL. IN LITRES:	9.5	20	23	30	11.5	16	24	29	14.5	15	10.5
GRAIN											
Trit aest	77	186	126	348	171	64	239	38	57	105	185
Trit spel	106	256	226	316	272	89	322	64	81	145	221
Trit sp.	197	602	718	1078	912	186	347	185	205	284	757
Hord vulg	2	3	9	4	6	1	10	.	4	6	13
CHAFF											
glum spel	3	20	4	10	3	5	9	5	5	2	5
glum inde	1	11	7	5	.	1	5	6	.	2	5
rach brit						1					
rach Hord											
flor Avef								1			
flor Aven									1		
WEEDS											
Raph raph	1	1	1	1	.	2	.	1	.	1	1
Crucif.											
Mont font	1										
Agro gith		1			1		2	2		1	
Stel medi					1						
Atri spp.					1						
Chen sp.										1	
Vici hirs				2		1					
Vici Lath				1		1	2				
Legumin.	1						2		1		
Pote erec											
Coni macu											
Poly conv											
Poly lapa						1					
Rume acet											
Rume spp.				1	2		1	2	3		
Odon vern											
Plan lanc			1					1			
Gali apar	2									1	
Trip inod				1							
Cent cyan							1				1
Compos.											
Aven sp.	6	4	1	9	3	5	6	3	2	7	6
Brom m/s	11	8	9	13	10	8	9	6	6	8	13
Brom ster	1	1	4	5	1	2		1	3		2
Sieg decu			1	1	1			13			1
smal gras					1						1
Gramin.											
Eleo sp.		1								1	
Care pilu		1									
Care puli		2						2			
Care spp.	1	1	2		1		1		3		1
OTHER											
Cory avel									1		
Call leaf			1	1							
Thel sang											
INDET.	1		1		1		2	1	2	1	1
TOTAL	408	1097	1112	1796	1385	365	957	333	374	565	1214

Table 4.18b Carbonized seeds from South Shields, deposit 12236.

CONTEXT: 12236											
SAMPLE NO.:	10b	10c	10d	11	12	13	14	15	16	17	18	19
VOL. IN LITRES:	5	4	6	13	11	10	13.5	12	15	13.5	16	18
GRAIN												
Trit aest	32	51	118	105	156	149	106	325	7	119	199	433
Trit spel	30	83	193	163	200	263	136	408	4	180	265	444
Trit sp.	181	172	527	484	960	252	221	618	22	333	650	1210
Hord vulg	2	3	6	4	9	19	4	8	1	10	14	17
CHAFF												
glum spel	2	5	8	8	15	10	6	16	1	10	8	41
glum inde	2	2	2	5	11	4		17	1	4	5	18
rach brit									1	1		
rach Hord												
flor Avef												
flor Aven												
WEEDS												
Raph raph			1		1			1				1
Crucif.												
Mont font						1	1				4	3
Agro gith					1	1	1	6	1			
Stel medi												
Atri spp.												
Chen sp.	1											
Vici hirs			1									
Vici Lath			1			1		3				
Legumin.				1					1		1	1
Pote erec												
Coni macu											1	1
Poly conv											1	1
Poly lapa											2	2
Rume acet		1			2		1			1	1	1
Rume spp.												
Odon vern												
Plan lanc												
Gali apar												
Trip inod												
Cent cyan										1		
Compos.										4	2	4
Aven sp.	1	10	3	6	1	1	6	7	2	18	8	23
Brom m/s	1		13	15	22	11	8	36	4		6	2
Brom ster	2			2	3	1		1	1	10	1	2
Sieg decu			1		1	1	1					1
smal gras												
Gramin.										1		
Eleo sp.	1											
Care pilu					1						6	2
Care puli				1	2	2	3					1
Care spp.	1									4	3	
OTHER												
Cory avel												
Call leaf								1	1			
Thel sang												
INDET.										1	1	
TOTAL	254	328	874	794	1385	714	493	1447	46	699	1167	2206

Table 4.18c Carbonized seeds from South Shields, deposit 12236.

CONTEXT:	12236										
SAMPLE NO.:	21	22	23	24	25	26	27	28	29	30	TOTAL
VOL. IN LITRES:	10	12.5	15	13	10.5	12.5	13	14.5	14	10	465
GRAIN											
Trit aest	118	485	536	29	505	58	92	54	362	485	6120
Trit spel	200	723	663	13	516	89	158	55	238	987	8109
Trit sp.	541	1016	847	54	1072	169	379	109	402	1229	16919
Hord vulg	12	20	13	.	23	1	10	2	8	26	270
CHAFF											
glum spel	10	32	16	36	42	47	21	18	16	47	486
glum inde	5	15	11	22	12	23	12	12	6	44	274
rach b-it	1	2	.	2	.	3	.	2	1	.	13
rach Hord	.	.	.	1	.	.	.	.	.	.	.
flor Aref	.	2	2	1	.	.	.	.	.	1	6
flor Aven	1	.	.	1	1	.	.	.	1	.	3
WEEDS											
Raph raph	1	1	1	5	1	2	.	2	2	2	28
Crucif.	.	.	.	2	.	6	.	1	1	1	10
Mont font	.	.	.	.	.	.	.	.	.	.	1
Agro cith	3	2	1	1	5	1	3	2	1	2	46
Stel medi	1	.	.	.	.	.	.	.	.	.	4
Atri spp.	.	.	.	.	.	.	1	2	.	.	1
Chen sp.	.	.	.	.	.	.	.	.	.	.	1
Vici hirs	1	.	1	.	1	1	.	1	.	1	5
Vici lath	.	.	1	.	3	.	.	1	4	4	16
Legumin.n	.	.	.	.	.	.	.	1	.	.	7
Pote erec	.	.	.	.	.	.	.	.	.	.	2
Coni macu	.	.	.	.	.	.	.	.	1	1	1
Poly conv	.	.	.	.	.	.	.	.	.	1	1
Poly lapa	.	.	.	.	.	.	.	.	.	.	1
Rume acet	.	4	1	1	2	3	1	4	.	.	12
Rume spp.	.	.	.	15	.	11	.	8	2	3	65
Odon vern	.	.	.	.	.	.	.	.	.	.	1
Plan lanc	.	.	.	.	.	.	.	.	.	1	2
Gali apar	.	.	.	1	1	1	1	1	1	2	6
Trip inod	.	.	.	.	.	.	.	.	.	.	2
Cent cyan	.	.	.	.	.	.	.	.	.	.	2
Compcs.	2	.	.	1	11	8	1	3	11	11	142
Aven sp.	13	4	5	8	11	76	31	34	15	58	665
Brom m/s	.	33	15	84	38	.	2	1	.	.	13
Brom ster	.	.	.	1	1	3	3	5	4	3	99
Sieg decu	4	4	2	8	.	1	5	10	.	3	19
smal gras	.	.	.	1	1	1	1	1	.	2	4
Gram.n	.	.	.	.	.	.	.	.	.	.	3
Eleo sp.	.	.	.	.	.	.	.	.	2	.	7
Care pilu	.	.	.	.	.	.	1	1	7	.	51
Care puli	.	.	1	6	.	1	.	3	.	2	.
Care spp.	.	1	1	.	1	.	.	.	.	.	.
OTHER											
Cory avel	.	.	.	.	.	.	.	.	.	.	4
Call leaf	.	.	.	.	.	.	.	.	.	.	3
Thel sang	.	.	.	.	.	.	.	.	.	.	1
INDET.	.	.	.	4	1	1	.	1	1	.	19
TOTAL	908	2344	2115	295	2235	507	719	328	1069	2916	33449

Table 4.18d Carbonized seeds from South Shields, deposit 12176.

CONTEXT:	12176									
SAMPLE NO.:	1	2	3	4	5	6	7	8	9	10
VOL. IN LITRES:	15	10	10	12	14.5	16	14	16	20.5	14.5
GRAIN										
Trit aest	20	7	1	19	.	4	3	.	.	.
Trit spel	34	18	1	44	.	4	11	.	1	1
Trit sp.	324	189	32	366	39	42	49	17	28	16
Hord vulg	2	.	.	1	.	.	1	.	1	.
CHAFF										
glum spel	2	3	.	1	1	2	2	2	.	1
glum inde	.	.	1	1	1	.	.	.	.	2
rach brit	.	.	.	.	.	.	.	.	.	.
rach Hord	.	.	.	.	.	.	.	.	.	.
awns Aven	.	.	1	.	1	.	.	.	.	.
flor Aven	.	.	.	.	.	.	.	.	.	.
WEEDS										
Ranu flam	.	.	.	.	.	.	.	.	.	.
Raph raph	.	.	.	.	.	.	.	.	.	.
Crucif.	.	.	.	.	.	.	.	.	.	.
Mont font	.	.	.	.	.	.	.	.	.	.
Stel medi	.	.	.	.	.	.	.	.	.	.
Caryoph.	.	.	.	.	.	.	.	.	.	.
Atri spp.	.	.	.	.	.	.	.	.	.	.
Chen albu	.	.	.	.	.	.	.	.	.	.
Chen sp.	.	.	.	.	.	.	.	.	.	.
Vici hirs	.	.	.	.	.	.	.	.	1	.
Vici Lath	.	.	.	.	1	.	.	.	.	.
Legumin.	.	.	.	.	.	.	.	.	.	.
Coni macu	.	.	.	.	.	.	.	.	.	.
Rume acet	.	.	.	.	.	.	.	.	.	.
Rume spp.	.	.	.	1	3	2	1	3	.	.
Plan lanc	.	.	1	1	1	.	.	.	.	.
Gali apar	.	.	.	.	.	.	.	.	.	.
Compos.	.	.	.	.	.	.	.	.	.	.
Aven sp.	3	2	.	2	.	1	1	.	3	4
Brom m/s	4	3	.	4	.	1	2	5	4	4
Sieg decu	1	.	.	3	1	4	2	.	.	1
smal gras	1	.	.	.	.	.	.	.	.	.
Gramin.	.	.	.	.	.	1	.	.	1	.
rhiz Gram	.	.	.	.	.	.	.	.	.	.
Care pilu	.	.	.	.	.	.	.	.	.	.
Care puli	.	.	.	.	.	.	.	.	.	.
Care spp.	.	.	.	.	1	.	1	2	.	1
OTHER										
Cory avel	.	.	.	.	.	.	.	.	.	.
INDET.	.	.	1	1	.	.	.	1	.	.
TOTAL	393	222	37	442	48	60	71	30	39	26

Table 4.18e   Carbonized seeds from South Shields, deposit 12176.

CONTEXT:	12176									
SAMPLE NO.:	11	12	13	14	15	16	17	18	19	20
VOL. IN LITRES:	14.5	17	12	18	16	18	15	12	15	11.5
GRAIN										
Trit aest	1	2	.	.	8	1	2	.	3	1
Trit spel	4	1	2	1	9	3	2	2	3	2
Trit sp.	8	25	11	20	79	23	13	4	27	13
Hord vulg	.	1	1	5	1	.	2	.	.	.
CHAFF										
glum spel	1	1	1	5	3	.	3	.	.	2
glum inde	1	.	.	2	.	1	.	.	1	1
rach brit	.	.	.	.	.	.	.	.	.	.
rach Hord	.	.	.	.	.	1	1	.	.	.
awns Aven	.	.	.	.	.	.	.	.	.	.
flor Aven	.	.	.	.	.	.	.	.	.	.
WEEDS										
Ranu flam	.	.	.	.	.	.	.	.	.	.
Raph raph	.	.	.	.	.	.	.	.	.	.
Crucif.	.	.	.	.	.	.	.	.	.	.
Mont font	.	.	.	.	.	.	.	.	.	.
Stel medi	.	1	.	.	.	.	.	.	.	.
Caryoph.	.	.	.	.	.	.	2	.	.	.
Atri spp.	.	.	.	.	.	.	1	.	.	.
Chen albu	.	.	.	.	.	1	.	.	.	.
Chen sp.	.	.	.	.	.	.	.	1	.	.
Vici hirs	.	.	.	.	1	.	.	1	.	.
Vici Lath	1	1	.	2	1	1	1	.	1	.
Legumin.	.	.	.	.	1	1	1	.	.	.
Coni macu	.	.	.	.	.	.	.	.	.	.
Rume acet	.	.	.	.	.	.	.	.	.	.
Rume spp.	.	.	1	.	3	4	4	2	.	.
Plan lanc	1	.	.	1	.	1	1	1	1	.
Gali apar	.	3	.	.	2	1	1	.	.	.
Compos.	.	.	.	.	1	1	.	.	.	.
Aven sp.	1	.	.	5	1	.	.	.	.	.
Brom m/s	2	.	.	1	3	.	.	.	.	.
Sieg decu	5	10	4	5	5	4	5	6	8	7
smal gras	.	.	1	2	.	.	.	.	.	.
Gramin.	.	.	.	.	.	.	.	.	.	.
rhiz Gram	1	.	.	.	2	2	1	1	.	.
Care pilu	.	.	.	.	.	1	.	.	.	.
Care puli	.	.	.	.	.	1	.	.	1	.
Care spp.	5	1	.	2	2	7	2	1	1	7
OTHER										
Cory avel	.	1	.	.	.	.	.	.	.	.
INDET.	1	1	.	.	.	1	4	1	2	.
TOTAL	31	48	21	51	120	45	45	20	46	33

Table 4.18f   Carbonized seeds from South Shields, deposit 12176.

CONTEXT:	12176										
SAMPLE NO.:	21	22	23	24	25	26	27	28	29	30	TOTAL
VOL. IN LITRES:	11	12.5	13	10.5	14	12.5	15	16	14	17	427
GRAIN											
Trit aest	1	2	9	.	10	1	1	2	4	4	101
Trit spel	1	4	8	3	9	.	1	3	5	5	175
Trit sp.	4	19	55	9	107	4	1	28	6	30	1588
Hord vulg	.	2	.	.	1	2	1	1	.	1	23
CHAFF											
glum spel	2	.	2	.	1	2	.	8	4	4	53
glum inde	1	.	.	.	.	.	.	2	.	.	12
rach brit	1	.	.	.	.	.	.	.	.	.	1
rach Hord	.	.	.	.	.	.	.	.	.	.	1
awns Aven	.	.	.	.	.	.	.	.	.	.	1
flor Aven	.	.	.	.	.	.	.	.	.	.	1
WEEDS											
Ranu flam	.	.	1	.	.	.	.	.	.	.	1
Raph raph	.	.	.	.	.	.	.	2	.	.	2
Crucif.	.	.	.	1	.	.	.	.	.	.	1
Mont font	.	.	.	1	.	.	.	.	.	.	1
Stel medi	.	1	.	.	.	.	.	.	2	.	3
Caryoph.	1	.	.	1	.	.	.	.	.	.	5
Atri spp.	.	.	.	.	1	.	.	.	.	.	1
Chen albu	.	.	.	.	.	.	.	.	1	.	5
Chen sp.	.	.	.	.	.	.	.	.	.	2	1
Vici hirs	.	.	.	.	.	.	.	.	.	.	5
Vici Lath	.	.	.	.	.	.	.	.	1	.	7
Legumin.	2	.	.	.	1	.	2	1	1	2	13
Coni macu	.	.	1	.	.	.	.	.	.	4	7
Rume acet	.	.	.	.	.	.	.	.	1	1	7
Rume spp.	12	1	8	.	1	1	3	10	4	7	64
Plan lanc	3	1	.	1	.	1	.	.	.	.	4
Gali apar	3	.	.	.	.	.	.	.	1	.	19
Compos.	1	.	.	.	.	.	.	.	.	.	2
Aven sp.	.	.	1	.	.	.	.	2	.	.	20
Brom m/s	14	7	8	3	4	3	7	10	6	1	38
Sieg decu	1	.	1	.	6	.	.	7	1	12	156
smal gras	2	.	.	.	.	.	2	.	.	2	10
Gramin.	4	1	.	1	1	1	1	2	4	4	13
rhiz Gram	1	.	.	.	.	.	.	.	.	.	12
Care pilu	1	1	.	.	.	.	.	.	.	.	3
Care puli	.	1	.	.	.	.	.	.	.	.	1
Care spp.	4	1	4	1	2	.	4	6	4	8	67
OTHER											
Cory avel	1	.	.	.	.	.	.	1	.	.	2
INDET.	1	3	.	.	1	3	.	1	1	3	22
TOTAL	57	41	100	20	145	18	22	86	31	90	2438

Table 5.1   Calibrated age ranges for all radio-carbon dates.

	68.3% (1 SIGMA)	95.4% (2 SIGMA)
Hallshill		
HAR-8184	1510 cal BC-1478 cal BC	1522 cal BC-1296 cal BC
	1460 cal BC-1382 cal BC	1292 cal BC-1266 cal BC
	1344 cal BC-1320 cal BC	
HAR-8183	1302 cal BC-1286 cal BC	1386 cal BC-1342 cal BC
	1268 cal BC-1096 cal BC	1320 cal BC-1012 cal BC
OxA-1764	1252 cal BC-1246 cal BC	1308 cal BC-1280 cal BC
	1212 cal BC-1182 cal BC	1272 cal BC- 912 cal BC
	1168 cal BC-1000 cal BC	
OxA-1763	1124 cal BC-1115 cal BC	1256 cal BC-1240 cal BC
	1104 cal BC- 910 cal BC	1216 cal BC- 890 cal BC
		884 cal BC- 844 cal BC
HAR-4800	1020 cal BC- 838 cal BC	1202 cal BC-1194 cal BC
		1162 cal BC-1142 cal BC
		1136 cal BC- 806 cal BC
OxA-1765	988 cal BC- 956 cal BC	1090 cal BC-1076 cal BC
	940 cal BC- 832 cal BC	1062 cal BC- 802 cal BC
HAR-8185	920 cal BC- 810 cal BC	1016 cal BC- 794 cal BC
HAR-4789	810 cal BC- 760 cal BC	836 cal BC- 510 cal BC
	686 cal BC- 656 cal BC	492 cal BC- 488 cal BC
	636 cal BC- 592 cal BC	436 cal BC- 414 cal BC
	586 cal BC- 550 cal BC	
OxA-1766	812 cal BC- 758 cal BC	840 cal BC- 474 cal BC
	690 cal BC- 652 cal BC	446 cal BC- 412 cal BC
	642 cal BC- 542 cal BC	
HAR-4788	798 cal BC- 754 cal BC	806 cal BC- 466 cal BC
	702 cal BC- 532 cal BC	448 cal BC- 410 cal BC
Murton		
GrN-15673	3022 cal BC-3000 cal BC	3034 cal BC-2944 cal BC
	2926 cal BC-2882 cal BC	2942 cal BC-2872 cal BC
	2798 cal BC-2782 cal BC	2806 cal BC-2776 cal BC
		2720 cal BC-2702 cal BC
GrN-15672	1510 cal BC-1472 cal BC	1584 cal BC-1576 cal BC
	1466 cal BC-1414 cal BC	1528 cal BC-1372 cal BC
		1348 cal BC-1314 cal BC
HAR-6201	1308 cal BC-1280 cal BC	1408 cal BC- 990 cal BC
	1272 cal BC-1086 cal BC	950 cal BC- 946 cal BC
	1082 cal BC-1060 cal BC	
HAR-6202	358 cal BC- 290 cal BC	382 cal BC- 0 cal AD
	250 cal BC- 94 cal BC	
HAR-6200	200 cal BC- 60 cal AD	370 cal BC- 120 cal AD
OxA-1742	94 cal BC- 64 cal AD	190 cal BC- 130 cal AD
OxA-1741	46 cal BC- 116 cal AD	156 cal BC- 146 cal BC
		114 cal BC- 216 cal AD
OxA-1740	8 cal AD- 142 cal AD	88 cal BC- 62 cal BC
	164 cal AD- 200 cal AD	60 cal BC- 242 cal AD
Dod Law		
GrN-15677	394 cal BC- 360 cal BC	398 cal BC- 350 cal BC
	286 cal BC- 254 cal BC	312 cal BC- 208 cal BC
GrN-15674	386 cal BC- 354 cal BC	392 cal BC- 342 cal BC
	306 cal BC- 246 cal BC	324 cal BC- 202 cal BC
	224 cal BC- 212 cal BC	
GrN-15675	368 cal BC- 350 cal BC	382 cal BC- 198 cal BC
	314 cal BC- 274 cal BC	
	266 cal BC- 208 cal BC	
GrN-15676	172 cal BC- 102 cal BC	194 cal BC- 50 cal BC
OxA-1735	86 cal BC- 70 cal BC	164 cal BC- 138 cal BC
	54 cal BC- 88 cal AD	122 cal BC- 146 cal AD
	102 cal AD- 108 cal AD	160 cal AD- 208 cal AD
OxA-1734	46 cal BC- 116 cal AD	156 cal BC- 146 cal BC
		114 cal BC- 216 cal AD
OxA-1736	6 cal AD- 146 cal AD	102 cal BC- 254 cal AD
	158 cal AD- 212 cal AD	298 cal AD- 318 cal AD
Chester House		
GrN-15709	804 cal BC- 752 cal BC	812 cal BC- 460 cal BC
	706 cal BC- 530 cal BC	456 cal BC- 410 cal BC
GrN-15708	752 cal BC- 724 cal BC	764 cal BC- 678 cal BC
	528 cal BC- 388 cal BC	664 cal BC- 624 cal BC
		606 cal BC- 364 cal BC
		282 cal BC- 258 cal BC
GrN-15707	402 cal BC- 356 cal BC	406 cal BC- 336 cal BC
	298 cal BC- 248 cal BC	330 cal BC- 200 cal BC
OxA-1743	156 cal BC- 146 cal BC	342 cal BC- 324 cal BC
	116 cal BC- 28 cal AD	202 cal BC- 118 cal AD
	38 cal AD- 54 cal AD	

	68.3% (1 SIGMA)	95.4% (2 SIGMA)
Thorpe Thewles		
GrN-15661	980 cal BC- 964 cal BC	1096 cal BC- 782 cal BC
	932 cal BC- 810 cal BC	
GrN-15662	760 cal BC- 686 cal BC	782 cal BC- 390 cal BC
	656 cal BC- 638 cal BC	
	548 cal BC- 400 cal BC	
GrN-15663	402 cal BC- 370 cal BC	408 cal BC- 356 cal BC
		298 cal BC- 248 cal BC
OxA-1731	480 cal BC- 440 cal BC	758 cal BC- 688 cal BC
	412 cal BC- 350 cal BC	654 cal BC- 640 cal BC
	316 cal BC- 206 cal BC	546 cal BC- 186 cal BC
GrN-15658	366 cal BC- 346 cal BC	380 cal BC- 194 cal BC
	320 cal BC- 280 cal BC	
	262 cal BC- 204 cal BC	
GrN-15659	366 cal BC- 338 cal BC	390 cal BC- 166 cal BC
	328 cal BC- 278 cal BC	136 cal BC- 122 cal BC
	262 cal BC- 200 cal BC	
OxA-1732	368 cal BC- 272 cal BC	394 cal BC- 104 cal BC
	268 cal BC- 190 cal BC	
GrN-15660	354 cal BC- 308 cal BC	370 cal BC- 36 cal BC
	244 cal BC- 226 cal BC	
	212 cal BC- 98 cal BC	
OxA-1733	164 cal BC- 136 cal BC	348 cal BC- 316 cal BC
	122 cal BC- 22 cal AD	206 cal BC- 114 cal AD
OxA-1745	1228 cal AD-1300 cal AD	1172 cal AD-1330 cal AD
	1358 cal AD-1380 cal AD	1332 cal AD-1396 cal AD
Stanwick		
GrN-15664	404 cal BC- 382 cal BC	508 cal BC- 498 cal BC
		484 cal BC- 438 cal BC
		414 cal BC- 362 cal BC
		284 cal BC- 256 cal BC
GrN-15665	86 cal BC- 70 cal BC	164 cal BC- 138 cal BC
	54 cal BC- 68 cal AD	120 cal BC- 126 cal AD
GrN-15666	32 cal BC- 22 cal AD	43 cal BC- 61 cal AD
GrN-15667	44 cal BC- 28 cal AD	94 cal BC- 68 cal AD
	36 cal AD- 56 cal AD	
Rock Castle		
GrN-15668	1302 cal BC-1288 cal BC	1398 cal BC- 976 cal BC
	1268 cal BC-1030 cal BC	970 cal BC- 928 cal BC
GrN-15671	808 cal BC- 770 cal BC	820 cal BC- 760 cal BC
		684 cal BC- 658 cal BC
		634 cal BC- 596 cal BC
		578 cal BC- 552 cal BC
GrN-15669	770 cal BC- 748 cal BC	790 cal BC- 464 cal BC
	734 cal BC- 522 cal BC	452 cal BC- 410 cal BC
GrN-15670	756 cal BC- 698 cal BC	764 cal BC- 678 cal BC
	536 cal BC- 404 cal BC	666 cal BC- 624 cal BC
		606 cal BC- 398 cal BC
OxA-1738	108 cal BC- 56 cal AD	334 cal BC- 330 cal BC
		198 cal BC- 124 cal AD
OxA-1739	94 cal BC- 64 cal AD	190 cal BC- 130 cal AD
OxA-1737	86 cal BC- 70 cal AD	164 cal BC- 138 cal AD
	54 cal BC- 88 cal AD	122 cal BC- 146 cal AD
	102 cal AD- 108 cal AD	160 cal AD- 208 cal AD
OxA-2132	8 cal BC- 142 cal AD	92 cal BC- 234 cal AD
	166 cal AD- 188 cal AD	
OxA-1737}		
OxA-2132}	28 cal BC- 86 cal AD	94 cal BC- 130 cal AD
Thornbrough		
GrN-15678	796 cal BC- 762 cal BC	802 cal BC- 756 cal BC
	682 cal BC- 660 cal BC	696 cal BC- 536 cal BC
	632 cal BC- 598 cal BC	
	576 cal BC- 556 cal BC	
GrN-15679	156 cal BC- 146 cal BC	174 cal BC- 4 cal AD
	116 cal BC- 34 cal BC	
GrN-12608	232 cal AD- 270 cal AD	142 cal AD- 166 cal AD
	276 cal AD- 338 cal AD	190 cal AD- 196 cal AD
		198 cal AD- 392 cal AD
GrN-12607	340 cal AD- 424 cal AD	254 cal AD- 298 cal AD
		320 cal AD- 452 cal AD
		484 cal AD- 506 cal AD
		512 cal AD- 526 cal AD
OxA-2130	340 cal AD- 460 cal AD	246 cal AD- 562 cal AD
	474 cal AD- 532 cal AD	582 cal AD- 590 cal AD
OxA-2131	252 cal AD- 304 cal AD	142 cal AD- 164 cal AD
	314 cal AD- 418 cal AD	202 cal AD- 536 cal AD

Table 6.1 Total number of seeds for each species.

SITE:	HH86	MT83	DL85	CH85	TT81	SW85	RC87	TH84	SS84	TOTAL
NO. OF SAMPLES:	21	10	12	14	127	32	23	23	63	325
NO. OF LITRES:	252	68	308	890	3,556	431	598	635	892	7,630
GRAIN										
Trit dico	104	2	10	0	1	0	0	0	0	117
Trit spel	29	2	17	1	214	53	74	625	8284	9299
Trit aest	0	0	0	0	1	1	1	1	6221	6222
Trit sp.	123	6	41	7	206	63	74	66	18507	19093
Hord vulg	40	141	1040	33	726	486	144	1989	293	4892
Seca cere	177	74	819	59	1319	290	238	2108	0	5084
Cere inde	0	0	0	0	0	0	0	14	0	50
coleopti.	0	0	0	0	0	0	0	0	0	14
CHAFF										
glum dico	1057	37	324	49	3	266	384	10	0	1481
glum spel	35	18	54	4	2882	255	516	816	539	5279
glum inde	997	24	232	36	2459	255	516	816	286	5621
glumes	142	1	4	0	116	43	70	178	14	412
rach brit	142	15	99	21	698	132	175	390	14	1686
rach aest	6	0	0	0	3	0	125	0	0	125
base Trit	0	0	0	0	0	0	0	5	0	14
rach Hord	48	104	538	40	284	40	58	35	4	11151
base Hord	0	2	31	0	3	0	0	0	6	36
flor Avef	0	0	0	2	8	2	0	0	4	21
flor Aven	0	0	0	2	0	2	1	7	1	9
awns Aven	6	0	19	1	29	0	23	19	0	88
culm node	0	0	0	0	7	10	25	0	0	76
awns Trit	0	0	8	0	0	0	0	1	0	1
lemma Hord	0	2	0	2	0	3	0	1	0	4
chaff inde	0	0	1	0	5	0	46	26	0	82
rach Seca	0	0	0	0	0	0	0	39	0	39
WEEDS										
Ranu acri	0	0	0	0	2	0	1	0	1	2
Ranu repe	4	0	2	0	5	1	1	1	6	10
Ranu Ranu	0	2	2	49	40	7	14	2	5	117
Ranu flam	0	0	0	0	26	0	1	1	4	36
Papa arge	0	0	0	0	7	0	0	0	0	7
Papa rh/d	0	0	0	0	1	0	0	0	0	1
Papa sp.	0	0	0	0	1	0	0	0	0	1
Raph raph	1	1	4	0	8	12	5	7	30	67
Bras sp.	0	0	8	0	3	1	1	0	1	13
Crucif.	0	0	0	1	41	0	0	0	1	53
Viol Mela	0	0	0	0	1807	34	51	5	4	1896
Mont font	4	48	0	0	39	19	135	1	4	273
Stel palu	0	0	0	0	0	2	0	2	0	3
Stel medi	4	0	23	0	0	0	1	0	46	48
Agro gith	1	0	0	0	58	0	0	0	0	23
Sper arve	1	1	19	1	70	10	1	8	5	23
Caryoph.	0	0	0	0	25	0	0	0	1	2
OTHER										
Cheno albu	28	61	437	1	371	75	79	16	1	1069
Cheno sp.	13	32	386	2	148	20	181	5	6	782
Atri spp.	1	4	7	1	124	4	67	5	5	229
Chenop.	0	0	0	2	68	4	4	0	0	78
Malv sylv	0	0	0	0	1	0	0	0	0	1
Malv sp.	0	0	0	0	7	0	0	0	0	8
Linu cath	0	0	0	0	0	0	2	0	0	2
Vici hirs	0	0	4	0	0	0	0	3	0	5
Vici Lath	2	2	8	3	33	6	146	8	24	225
Legumin.	6	7	15	8	326	155	405	3	20	950
Apha arve	0	0	0	0	1	3	1	0	2	9
Pote erec	0	4	11	0	68	40	43	3	3	171
Hera spon	0	0	0	0	1	0	1	0	0	2
Coni macu	0	0	0	0	0	0	0	0	8	8
Anth cauc	1	0	0	0	0	0	1	0	0	1
Poly avic	0	9	12	0	70	10	67	8	0	176
Poly conv	1	0	7	1	25	7	11	5	1	60
Poly lapa	12	12	15	1	43	4	1	3	1	91
Poly pers	12	11	20	1	32	4	5	2	0	83
Poly l/p	7	2	14	2	23	1	1	4	0	61
Poly sp.	0	1	14	0	0	12	19	0	0	46
Rume acet	1	3	23	3	17	39	17	6	13	122
INDET.										
TOTAL										

SITE:	HH86	MT83	DL85	CH85	TT81	SW85	RC87	TH84	SS84	TOTAL
Rume spp.	4	6	30	0	461	301	37	3	129	971
Polygon.	0	0	0	0	8	0	0	0	0	8
Urti dioc	0	0	0	0	4	0	0	0	0	4
Urti uren	0	15	4	0	2	12	11	0	0	44
Sola nigr	0	0	0	0	1	0	1	0	0	3
Hyos nige	0	0	0	0	15	191	0	0	0	206
Odon vern	0	0	1	0	23	3	3	1	1	28
Vero arve	0	0	0	0	1	2	0	0	0	6
Vero scut	12	0	0	0	0	0	0	0	0	12
Rhin sp.	0	0	0	0	0	1	0	0	0	1
Verb sp.	3	0	0	0	0	0	0	0	0	4
Ajug rept	3	0	1	0	1	0	0	0	0	4
Lami a/p	1	1	2	0	5	0	0	0	0	7
Stac arve	1	0	0	0	0	0	0	2	0	2
Ment a/a	1	1	1	0	2	2	0	1	0	6
Gale sp.	0	0	0	0	0	0	0	0	0	4
Prun vulg	3	1	26	0	21	13	3	8	6	28
Plan lanc	3	6	1	1	52	1	31	8	6	146
Plan majo	3	2	0	0	19	22	0	5	25	23
Gali apar	0	0	1	0	75	2	79	1	0	207
Gali palu	0	0	0	0	0	0	0	0	0	4
Sher arve	0	0	0	0	5	1	0	2	0	5
Vale dent	0	0	0	0	1	0	0	0	1	1
Trip inod	6	1	0	0	184	13	214	2	0	415
Laps comm	0	0	0	0	5	0	0	0	0	3
Sonc aspe	0	0	0	0	3	0	1	1	0	5
Hypo g/r	0	0	0	0	0	0	0	0	2	2
Cent cyan	0	0	0	7	6	0	1	0	4	11
Compos.	3	1	2	7	88	0	36	5	162	319
Aven sp.	8	3	0	4	1608	444	185	235	703	3210
Brom m/s	0	0	0	0	0	0	0	0	13	13
Brom ster	1	14	11	15	4783	749	1118	22	255	6968
Sieg decu	46	94	279	14	698	236	602	33	29	2031
smal gras	14	26	22	6	1030	30	94	54	17	1293
Gramin.	0	0	0	7	61	3	0	8	0	73
Arrh elat	1	0	27	7	1238	133	235	12	12	1665
rhiz Gram	0	0	13	2	1	1	6	0	0	24
Junc sp.	0	0	0	0	0	2	15	1	0	18
Junc squa	0	0	0	0	24	5	4	2	0	4
Eleo sp.	0	0	0	1	31	52	21	0	10	146
38Isol seta	0	13	22	0	58	0	0	2	2	73
Care pilu	25	95	133	8	1107	165	230	32	118	1913
Care puli										3
Care spp.										
OTHER										
Linu usit	2	0	0	0	0	0	0	0	0	3
Cory avel	44	2	106	4	23	26	9	17	6	237
Crat mono	0	0	0	0	1	0	2	1	0	13
Prun spin	0	0	0	0	0	0	0	0	0	10
Samb nigr	0	0	9	1	0	10	0	1	0	10
tree buds	0	0	4	0	0	0	0	0	0	2
Rosa sp.	3	0	0	0	0	0	2	2	0	5
Rubu frut	0	2	2	3	33	0	0	0	1	225
Rubu sp.	2	7	15	8	326	10	32	1	3	950
Thel sang	0	8	53	0	1	29	34	1	3	9
Call leaf	1	4	543	0	68	6	0	1	0	171
Call flow	0	0	0	0	0	0	1	0	0	2
Eric flow	1	0	2	0	0	0	0	1	0	8
Vacc myrt	0	0	0	0	0	0	0	0	0	1
Empe nigr	44	9	12	0	70	10	67	8	0	3
Pter aqui	3	12	7	1	25	7	11	5	1	57
Lyco euro	0	11	20	1	43	4	1	3	1	91
Viol Viol	0	2	14	1	32	4	5	2	0	83
Calt palu	0	1	12	2	0	1	0	4	0	61
Meny trif	0	0	14	0	0	0	2	0	0	5
Pota spp.	1	3	23	3	17	0	5	0	0	3
INDET.	37	50	78	27	398	96	145	19	41	891
TOTAL	3,124	976	5,685	439	24,350	4,672	6,391	8,027	35,887	89,550

Table 6.2 Relative proportions of each species per category of data.

SITE:	HH86	MT83	DL85	CH85	TT81	SW85	RC87	TH84	SS84
NO. OF SAMPLES:	21	10	12	14	127	32	23	23	63
NO. OF LITRES:	252	68	308	890	3,556	431	598	635	892
GRAIN									
Trit dico	21.99	.89	.52	.00	.04	.00	.00	.00	.00
Trit spel	6.13	.89	.88	1.00	8.68	5.94	13.94	12.88	24.87
Trit aest	.00	.00	.00	.00	.00	.00	.19	1.36	18.68
Trit sp.	26.00	2.67	2.13	7.00	8.35	7.06	13.19	1.03	55.57
Hord vulg	8.46	62.67	53.97	33.00	29.44	54.48	27.12	40.99	.88
Seca cere	37.42	32.89	42.50	59.00	53.49	32.51	44.82	43.45	.00
Cere inde	.00	.00	.00	.00	.00	.00	.74	.29	.00
coleopti.	.00	.00	.00	.00	.00	.00	.00	.00	.00
GRAIN TOTAL	100.00	100.00	100.00	100.00	100.00	100.00	100.00	100.00	100.00
CHAFF									
glum dico	46.00	18.23	24.68	30.82	44.36	35.33	26.95	41.82	63.11
glum spel	1.52	8.87	4.11	2.52	37.85	33.86	36.21	31.11	33.49
glum inde	43.39	11.82	17.67	22.64	10.74	17.53	4.91	6.79	1.64
glumes	6.18	.49	7.54	13.21	1.79	.00	12.28	14.87	.00
rach brit	.00	7.39	.00	.00	.05	.00	4.07	.19	.47
rach aest	.00	.00	.97	25.16	4.37	5.31	.00	1.33	.00
base Trit	.26	.00	2.36	.00	.05	.27	.00	.00	.70
rach Hord	2.09	51.23	40.97	.00	.12	.27	.12	.00	.70
base Hord	.00	.99	.23	1.26	.05	.00	.00	.00	.47
flor Hord	.00	.00	.00	1.26	.45	.27	.07	.00	.12
flor Aven	.00	.00	1.45	.63	.11	1.33	1.61	.72	.00
awns Aven	.30	.00	.61	.00	.00	.00	1.75	.72	.00
culm node	.26	1.45	.00	.00	.00	.40	.07	.04	.00
awns Trit	.00	.00	.00	.00	.00	.00	.00	.00	.00
lemma Hord	.00	.00	.00	.00	.00	.00	3.23	.99	.00
chaff inde	.00	.99	.08	1.26	.08	.00	.00	.00	.00
rach Seca	.00	.00	.00	.00	.00	.00	.00	.00	.00
TOTAL CHAFF	100.00	100.00	100.00	100.00	100.00	100.00	100.00	100.00	100.00
WEEDS									
Ranu acri	.00	.00	.00	.00	.01	.00	.00	.00	.00
Ranu repe	1.84	.00	.12	.00	.03	.04	.02	.20	.00
Ranu Ranu	.00	.21	.00	34.75	.27	.25	.33	.39	.06
Ranu flam	.00	.41	.12	.71	.17	.00	.10	.00	.06
Papa arge	.00	.00	.00	1.42	.05	.00	.00	.00	.00
Papa rh/d	.00	.00	.00	.00	.01	.00	.00	.00	.00
Papa sp.	.00	.00	.00	.00	.05	.00	.05	.00	.00
Raph raph	.46	.21	.24	.00	.05	.42	.12	1.38	1.79
Bras sp.	.65	.00	.49	.71	.01	.04	.02	.00	.00
Crucif.	.00	.00	.00	.00	.27	.00	.02	.00	.66
Viol Mela	.00	.00	.00	.71	.27	.04	.02	.00	.12
Mont font	.00	.00	.06	.00	.00	.00	.00	.00	.00
Stel medi	1.84	9.86	1.41	.00	12.00	1.19	1.21	.20	.24
Stel palu	.00	.00	.00	.00	.26	.67	3.21	.20	.00
Agro gith	.00	.00	.00	.00	.00	.07	.02	.00	2.74
Sper arve	.46	.21	1.16	.71	.47	.00	.02	.39	.30
Caryoph.	.00	.00	.00	.00	.00	.00	.00	.00	.30
Cheno albu	12.90	12.53	26.70	.71	2.48	2.63	1.88	3.14	.06
Cheno sp.	5.92	6.57	23.58	1.42	.99	.70	4.31	.98	.36
Atri spp.	5.99	.82	.43	3.71	.83	.18	1.59	.59	.30
Chenop.	.00	.00	.00	1.42	.45	.14	.10	.00	.00
Malv sylv	.00	.00	.00	.00	.01	.00	.00	.00	.48
Malv sp.	.00	.00	.00	.00	.05	.00	.05	.00	.00
Linu cath	.00	.00	.00	.00	.00	.00	.00	.00	.00
Vici hirs	.00	1.85	.73	.71	.47	.35	1.59	1.57	.30
Vici Lath	.46	.41	.43	.71	.17	.25	.26	.98	1.43
Legumin.	.00	1.44	.49	2.13	.29	.11	.02	.59	1.19
Apha arve	2.76	.00	.92	5.67	2.18	5.44	3.47	1.43	.30
Pote erec	.00	.82	.67	.00	.45	1.40	9.64	1.57	.12
Hera spon	.00	.00	.00	.00	.01	.00	1.02	.59	.00
Coni macu	1.84	.00	.00	.00	.26	.67	.02	.00	.48
Anth cauc	.00	.00	.00	.00	.00	.00	.02	.00	.00
Poly avic	.00	1.85	.73	.00	.47	.35	1.59	1.57	.06
Poly conv	.46	.41	.43	.71	.17	.25	.26	.98	.06
Poly lapa	5.53	2.46	.92	.71	.29	.11	.02	.59	.06
Poly pers	5.53	2.26	1.22	.71	.21	.14	.02	.39	.06
Poly l/p	3.23	.41	.86	1.42	.15	.14	.12	.79	.00
Poly sp.	.00	.21	.86	.00	.00	.42	.45	.00	.00
Rume acet	.46	.62	1.41	2.13	.11	1.37	.40	1.18	.78
Rume spp.	1.84	1.23	1.83	.00	3.08	10.57	.88	.59	7.69
Polygon.	.00	.00	.00	.00	.05	.00	.00	.00	.00
Urti dioc	.00	.00	.24	.00	.03	.42	.26	.00	.00
Urti uren	.00	3.08	.00	.00	.01	.00	.02	.00	.00
Sola nigr	.00	.00	.00	.00	.01	.04	.00	.00	.00
Hyos nige	.00	.00	.06	.00	.15	6.71	.02	.00	.06
Odon vern	.00	.00	.00	.00	.10	.11	.07	.00	.00
Vero arve	5.53	.00	.06	.00	.07	.07	.02	.00	.00
Vero scut	.00	.00	.00	.00	.00	.04	.00	.00	.00
Rhin sp.	.00	.00	.00	.00	.01	.00	.00	.00	.00
Verb sp.	1.38	.00	.06	.00	.01	.00	.07	.00	.00
Ajug rept	.00	.00	.00	.00	.03	.00	.74	.00	.00
Lami a/p	.46	.21	.12	.00	.03	.00	1.88	1.57	1.49
Stac arve	.00	.21	.00	.00	.01	.00	.02	.98	.00
Ment a/a	.00	.00	.00	.00	.00	.00	.00	.20	.00
Gale sp.	.00	.00	.06	.00	.14	.07	5.09	.39	.00
Prun vulg	.00	.00	.00	.00	.35	.46	.00	.39	.00
Plan lanc	1.38	1.23	1.59	.71	.13	.04	.07	.00	.36
Plan majo	1.38	.00	.00	.00	.50	.77	.00	.20	.00
Gali apar	.00	.00	.06	1.26	.12	.27	.74	.98	.98
Gali palu	.00	.00	.00	1.26	.45	.27	.02	.00	.47
Sher arve	.30	.00	.00	.63	.11	1.33	1.61	.27	.12
Vale dent	.26	.21	.00	.00	.00	.00	1.75	.72	.00
Trip inod	.00	.00	.00	.00	.00	.40	.07	.72	.00
Laps comm	.00	.00	.08	1.26	.08	.00	.00	.04	.00
Sonc aspe	.00	.99	.00	.00	.00	.00	3.23	.99	.00
Hypo g/r	.00	.21	.00	.00	.00	.00	.00	.00	.00
Cent cyan	.00	.00	.00	.00	.00	.00	.00	.00	.12
Compos.	.00	.00	.92	.71	.04	.07	.02	.98	.24
Aven sp.	1.38	.21	1.22	4.96	.59	15.59	.86	46.17	9.66
Brom m/s	3.69	.62	.67	2.84	10.75	.07	4.40	.00	41.92
Brom ster	.46	2.87	.00	10.64	31.96	26.30	26.61	4.32	15.21
Sieg decu	21.20	19.30	17.04	9.93	4.66	8.29	14.33	6.48	1.73
smal gras	6.45	5.34	1.34	4.26	6.88	1.05	2.24	10.61	1.01
Gramin.	.00	.00	.06	4.00	.41	4.67	.07	1.57	.72
Arrh elat	.46	.00	.06	.00	.01	.04	.59	2.36	.00
rhiz Gram	.46	1.79	.79	1.42	.05	.07	.14	.00	.00
Junc sp.	.00	.00	.00	.00	.01	.18	.36	.00	.18
Junc squa	.66	.00	.00	.00	.27	.00	.10	.39	.00
Eleo sp.	.00	.00	.00	.71	.45	1.83	.50	.39	.60
Isol seta	.12	.00	1.34	.00	.26	.67	.00	.20	.12
Care pilu	.24	1.44	.00	.71	.00	.07	1.21	.39	.24
Care puli	.00	2.67	.08	.00	.21	.00	3.21	.00	.00
Care spp.	.46	19.51	8.12	5.67	7.40	5.79	5.47	6.29	7.04
TOTAL WEEDS	100.00	100.00	100.00	100.00	100.00	100.00	100.00	100.00	100.00
OTHER									
Linu usit	12.90	.00	.00	.00	.00	.00	.00	4.17	.00
Cory avel	2.02	18.18	14.54	33.33	95.83	31.33	10.23	70.83	60.00
Crat mono	44.44	4.17	.00	.00	4.17	.00	2.27	4.17	.00
Prun spin	.00	.00	.67	.00	.00	.00	.00	4.17	.00
Samb nigr	.00	2.87	17.04	10.64	.00	12.05	.00	.00	.00
tree buds	.00	19.30	1.34	9.93	.00	.00	.00	.00	.00
Rosa sp.	3.03	5.34	.06	4.26	.00	1.20	.00	8.33	.00
Rubu frut	.00	.00	1.65	4.00	.00	.00	1.14	.00	.00
Rubu sp.	3.03	.00	.55	4.96	.00	.00	.00	.00	.00
Thel sang	3.03	.00	.27	1.42	.00	.00	.00	.00	10.00
Call leaf	2.02	72.73	7.27	8.33	.00	12.05	36.36	4.17	30.00
Call flow	.00	.00	74.49	25.00	.00	34.94	38.64	4.17	.00
Eric flow	1.01	.00	.00	.00	.00	7.23	.00	.00	.00
Vacc myrt	.00	.00	.00	.00	.00	.00	1.14	.00	.00
Empe nigr	44.44	9.09	.27	.00	.00	.00	.00	1.57	.00
Pter aqui	.46	.00	1.37	.00	.00	1.20	.00	.98	.06
Lyco euro	5.53	.00	.00	25.00	.00	.00	2.27	.59	.06
Viol Viol	5.53	.00	.00	.00	.00	.00	.02	.59	.06
Calt palu	3.23	.00	.00	.00	.00	.00	.12	.39	.00
Meny trif	.00	.00	.00	.00	.00	1.20	5.68	.00	.00
Pota spp.	.00	.00	.00	1.42	.00	.00	3.41	.00	.00
TOTAL OTHER	100.00	100.00	100.00	100.00	100.00	100.00	100.00	100.00	100.00

Table 6.3 Number of seeds per 1 litre of sieved sediment.

SITE:	HH86	MT83	DL85	CH85	TT81	SW85	RC87	TH84	SS84
NO. OF SAMPLES:	21	10	12	14	127	32	23	23	63
NO. OF LITRES:	252	68	308	890	3,556	431	598	635	892
GRAIN									
Trit dico	.41	.03	.03	.00	.00	.00	.00	.00	.00
Trit spel	.12	.03	.06	.00	.06	.12	.12	.98	9.29
Trit aest	.00	.00	.00	.01	.00	.00	.00	.00	6.97
Trit sp.	.49	.09	.13	.01	.06	.15	.12	.10	20.75
Hord vulg	.16	2.07	3.38	.04	.20	1.13	.24	3.13	.33
Seca cere	.00	1.00	.00	.07	.37	.00	.00	.08	.00
Cere inde	.70	1.09	2.66	.07	.37	.67	.40	3.32	.00
coleopti.	.00	.00	.00	.00	.00	.00	.00	.02	.00
CHAFF									
glum dico	4.19	.54	1.05	.06	.00	.00	.00	.02	.00
glum spel	.14	.26	.18	.04	.81	.62	.64	1.73	.60
glum inde	3.96	.35	.75	.04	.69	.59	.86	1.29	.32
glumes	.00	.01	.01	.02	.03	.10	.12	.28	.02
rach brit	.56	.22	.32	.02	.20	.31	.29	.61	.00
rach aest	.00	.00	.00	.00	.00	.00	.21	.01	.00
base Trit	.02	.00	.00	.04	.08	.09	.10	.06	.00
rach Hord	.19	1.53	1.75	.00	.00	.00	.00	.00	.00
base Hord	.00	.03	.10	.00	.01	.00	.04	.01	.01
flor Avef	.00	.00	.01	.00	.00	.00	.04	.04	.00
flor Aven	.03	.00	.00	.00	.00	.00	.00	.03	.00
awns Aven	.02	.00	.06	.00	.01	.02	.00	.00	.00
culm node	.00	.00	.03	.00	.00	.00	.00	.00	.00
awns Trit	.00	.00	.00	.00	.00	.01	.08	.04	.00
lemma Hord	.00	.03	.00	.00	.00	.00	.00	.06	.00
chaff inde	.00	.00	.00	.00	.00	.00	.00	.00	.00
rach Seca	.00	.00	.00	.00	.00	.00	.00	.00	.00
WEEDS									
Ranu acri	.00	.00	.00	.00	.00	.00	.00	.00	.00
Ranu repe	.00	.00	.01	.00	.06	.00	.00	.00	.00
Ranu Ranu	.02	.01	.01	.06	.01	.02	.02	.00	.00
Ranu flam	.00	.00	.00	.00	.01	.00	.01	.00	.00
Papa arge	.00	.00	.00	.00	.01	.00	.01	.00	.00
Papa rh/d	.00	.00	.00	.00	.01	.00	.00	.00	.00
Papa sp.	.00	.00	.01	.00	.08	.00	.00	.00	.03
Raph raph	.00	.01	.03	.01	.00	.03	.00	.00	.00
Bras sp.	.00	.00	.00	.03	.00	.00	.00	.00	.00
Crucif.	.00	.00	.00	.00	.00	.00	.00	.00	.01
Viol Mela	.00	.00	.00	.00	.00	.00	.00	.00	.00
Mont font	.00	.00	.00	.00	.00	.08	.09	.00	.00
Stel palu	.02	.71	.07	.07	.51	.00	.00	.00	.00
Agro gith	.00	.00	.00	.00	.00	.00	.00	.00	.05
Sper arve	.00	.00	.00	.06	.00	.00	.00	.00	.05
Caryoph.	.00	.01	.06	.00	.00	.00	.00	.00	.01
Cheno albu	.11	.90	1.42	.00	.10	.17	.13	.03	.01
Cheno sp.	.01	.47	1.25	.00	.04	.05	.30	.00	.01
Atri spp.	.05	.06	.02	.00	.03	.01	.11	.00	.01
Chenop.	.00	.00	.00	.00	.02	.01	.01	.00	.00
Malv sylv	.00	.00	.00	.00	.00	.00	.00	.00	.00
Malv sp.	.00	.00	.00	.00	.00	.02	.00	.00	.00
Linu cath	.00	.00	.00	.00	.00	.02	.00	.00	.00
Vici hirs	.00	.00	.03	.00	.00	.01	.00	.00	.00
Vici Lath	.00	.03	.03	.03	.01	.01	.00	.01	.01
Legumin.	.02	.10	.05	.05	.09	.36	.24	.01	.03
Apha arve	.00	.00	.04	.01	.00	.01	.68	.00	.02
Pote erec	.00	.06	.04	.00	.02	.09	.01	.00	.00
Hera spon	.00	.00	.00	.00	.00	.00	.07	.00	.00
Coni macu	.00	.00	.00	.00	.00	.00	.00	.00	.01
Anth cauc	.00	.00	.04	.00	.02	.02	.11	.01	.00
Poly avic	.00	.13	.02	.04	.01	.02	.02	.01	.00
Poly conv	.05	.03	.05	.02	.01	.01	.00	.01	.00
Poly lapa	.03	.18	.06	.05	.01	.01	.01	.01	.00
Poly pers	.05	.16	.05	.06	.00	.01	.01	.00	.00
Poly l/p	.00	.03	.05	.05	.00	.03	.01	.01	.00
Poly sp.	.00	.01	.07	.07	.00	.09	.03	.03	.01
Rume acet	.00	.04	.10	.00	.13	.70	.03	.00	.14
Rume spp.	.02	.09	.00	.10	.00	.00	.06	.00	.00
Polygon.	.00	.00	.00	.00	.00	.00	.00	.00	.00
Urti dioc	.00	.00	.00	.00	.00	.00	.00	.00	.00

SITE:	SS84	TH84	RC87	SW85	TT81	CH85	DL85	MT83	HH86
Urti uren	.00	.00	.02	.03	.00	.00	.01	.22	.00
Sola nigr	.00	.00	.00	.00	.00	.00	.00	.00	.00
Hyos nige	.00	.00	.00	.44	.00	.00	.00	.00	.00
Odon vern	.00	.00	.01	.01	.01	.00	.00	.00	.00
Vero arve	.00	.00	.01	.00	.00	.00	.00	.00	.05
Vero scut	.00	.00	.00	.00	.00	.00	.00	.00	.00
Rhin sp.	.00	.00	.00	.00	.00	.00	.00	.00	.00
Verb sp.	.00	.01	.00	.00	.00	.00	.01	.01	.01
Ajug rept	.00	.00	.00	.00	.00	.00	.00	.00	.00
Lami a/p	.00	.00	.00	.00	.00	.00	.00	.01	.00
Stac arve	.00	.00	.00	.00	.01	.00	.00	.00	.00
Ment a/a	.01	.01	.05	.00	.01	.00	.00	.01	.00
Gale sp.	.00	.00	.00	.00	.01	.00	.00	.00	.00
Prun vulg	.00	.00	.00	.03	.01	.00	.08	.09	.01
Plan lanc	.03	.01	.13	.05	.02	.00	.00	.00	.01
Plan majo	.01	.00	.00	.00	.00	.00	.00	.00	.00
Gali apar	.00	.00	.00	.00	.00	.00	.00	.00	.00
Gali palu	.03	.00	.36	.03	.00	.00	.00	.00	.00
Sher arve	.00	.00	.00	.00	.00	.00	.00	.00	.00
Vale dent	.00	.00	.00	.00	.05	.00	.00	.01	.00
Trip inod	.00	.00	.00	.03	.00	.00	.00	.00	.00
Laps comm	.00	.00	.00	.00	.00	.00	.00	.00	.00
Sonc aspe	.00	.00	.00	.00	.00	.00	.00	.00	.00
Hypo g/r	.00	.00	.00	.00	.00	.00	.00	.00	.00
Cent cyan	.00	.00	.00	.00	.00	.00	.00	.00	.00
Compos.	.00	.00	.00	.00	.02	.01	.00	.03	.00
Aven sp.	.18	.01	.06	.00	.02	.01	.05	.01	.01
GRAIN									
Brom m/s	.79	.37	.31	1.03	.45	.00	.06	.04	.03
Brom ster	.01	.01	1.87	.00	1.35	.00	.04	.00	.00
Sieg decu	.29	.03	1.01	1.74	1.20	.02	.91	.21	.18
smal gras	.03	.05	.16	.55	.29	.01	.07	1.38	.06
Grami.	.02	.09	.39	.07	.35	.12	.09	.38	.00
Arrh elat	.01	.02	.01	.00	.00	.00	.04	.00	.00
rhiz Gram	.00	.01	.03	.31	.00	.00	.00	.00	.00
Junc sp.	.00	.00	.01	.00	.01	.00	.00	.00	.00
Junc squa	.00	.00	.00	.00	.00	.00	.00	.00	.00
Eleo sp.	.00	.00	.01	.01	.01	.00	.00	.00	.00
Isol seta	.01	.00	.04	.01	.01	.00	.07	.00	.00
Care pulu	.00	.00	.04	.12	.02	.00	.00	.10	.00
Care puli	.00	.00	.00	.00	.00	.00	.00	.00	.00
Care spp.	.13	.05	.38	.38	.31	.01	.43	1.40	.10
OTHER									
Linu usit	.00	.00	.00	.00	.00	.00	.00	.00	.01
Cory avel	.01	.03	.02	.06	.01	.00	.34	.03	.17
Crat mono	.00	.00	.00	.00	.00	.00	.00	.00	.00
Prun spin	.00	.00	.00	.00	.00	.00	.00	.00	.00
Samb nigr	.00	.00	.00	.02	.00	.00	.00	.00	.00
tree buds	.00	.00	.00	.00	.00	.00	.03	.00	.00
Rosa sp.	.00	.00	.00	.00	.00	.00	.00	.00	.00
Rubu frut	.00	.00	.00	.00	.00	.00	.00	.00	.01
Rubu sp.	.00	.00	.00	.00	.00	.00	.00	.00	.00
Thel sang	.00	.00	.00	.00	.00	.00	.00	.00	.01
Call leaf	.00	.00	.06	.00	.00	.00	.17	.12	.12
Call flow	.00	.00	.05	.07	.00	.00	1.76	.00	.00
Eric flow	.00	.00	.04	.01	.00	.00	.00	.00	.00
Vacc myrt	.00	.00	.00	.00	.00	.00	.00	.00	.00
Empe nigr	.00	.00	.00	.00	.00	.00	.01	.01	.01
Pter aqui	.00	.00	.00	.00	.00	.00	.03	.00	.17
Lyco euro	.00	.00	.00	.00	.00	.00	.00	.00	.00
Viol Viol	.00	.00	.00	.00	.00	.00	.00	.00	.00
Calt palu	.00	.00	.01	.00	.00	.00	.00	.00	.00
Meny trif	.00	.00	.00	.00	.00	.00	.00	.00	.00
Pota spp.	.00	.00	.00	.00	.00	.00	.00	.00	.00
INDET.	.05	.03	.24	.22	.11	.03	.25	.74	.15
TOTAL	40.23	12.64	10.69	10.84	6.85	.49	18.45	14.35	12.40

Table 6.4   Occurrence in arable fields (after Grose 1957, Clapham *et al.* 1962, and M. Jones 1988a).

```
** The majority of its modern find-spots is in arable fields
*  The minority of its modern find-spots is in arable fields
-  None of its modern find-spots is in arable fields

CEREALS
Triticum dicoccum (Schrank.) Schübl.                            **
Triticum spelta L.                                             **
Triticum aestivo-compactum Schiem.                             **
Hordeum vulgare L.                                             **
Secale cereale L.                                              **

WEEDS
Ranunculus acris L.                                            *
Ranunculus repens L.                                           *
Ranunculus Subgenus Ranunculus                                *
Ranunculus flammula L.                                        -
Papaver argemone L.                                           **
Papaver rhoeas/dubium L.                                      **
Papaver sp.                                                   **
Raphanus raphanistrum L.                                      **
Brassica sp.                                                   *
Viola Subgenus Melanium                                        *
Montia fontana, ssp. chondrosperma (Fenzl) Walters            *
Stellaria media (L.) Vill.                                    **
Stellaria palustris Retz.                                     -
Spergula arvensis L.                                          **
Agrostemma githago L.                                         **
Chenopodium album L.                                          **
Chenopodium sp.                                               **
Atriplex spp.                                                  *
Chenopodiaceae indet.                                          *
Malva sylvestris L.                                            *
Malva sp.                                                      *
Linum catharticum L.                                           *
Vicia hirsuta (L.) S. F. Gray                                  *
Vicia/Lathyrus                                                 *
Leguminosae indet. (small)                                     *
Aphanes arvensis agg.                                          *
Potentilla cf. erecta (L.) Räusch                             *
Heracleum spondylium L.                                        *
Conium maculatum L.                                            *
Anthriscus caucalis Bieb.                                      *
Polygonum aviculare agg.                                      **
Polygonum convolvulus L.                                      **
Polygonum lapathifolium L.                                    **
Polygonum persicaria L.                                       **
Polygonum lapathifolium/persicaria                            **
Rumex acetosella agg.                                          *
Rumex spp.                                                     *
Urtica dioica L.                                               *
Urtica urens L.                                                *
Solanum nigrum L.                                              *
Hyoscyamus niger L.                                            *
Odontites verna (Bell.) Dum.                                   *
Veronica arvensis L.                                          **
Veronica cf. scutellata L.                                    -
Rhinanthus sp.                                                 *
Verbascum sp.                                                  *
Ajuga reptans L.                                               *
Lamium album/purpureum L.                                      *
Stachys arvensis (L.) L.                                       *
Mentha arvensis/aquatica L.                                    *
Galeopsis sp.                                                  *
Prunella vulgaris L.                                           *
Plantago lanceolata L.                                         *
Plantago major L.                                             **
Galium aparine L.                                             **
Galium palustre L.                                            -
Sherardia arvensis L.                                         **
Valerianella dentata (L.) Poll.                              **
Tripleurospermum inodorum(L.) Schultz Bip.                   **
Lapsana communis L.                                            *
Sonchus asper (L.) Hill                                       **
Hypocheris glabra/radicata L.                                  *
Centaurea cf. cyanus L.                                       **
Compositae indet.                                              *
Avena sp.                                                     **
Bromus mollis/secalinus                                        *
Bromus sterilis L.                                             *
Sieglingia decumbens (L.)Bernh.                               -
small grasses (including Poa annua)                          **
Gramineae indet.                                             **
Arrhenatherum elatius,ssp. bulbosum (Willd.) Spenn.          **
Juncus spp.                                                   -
Eleocharis sp.                                                -
Isolepis setacea (L.) R. Br.                                  -
Carex pilulifera L.                                           -
Carex pulicaris L.                                            -
Carex spp.                                                    -

OTHER
Linum cf. ussitatissimum                                      **
Corylus avellana L.                                           -
Crataegus cf. monogyna Jacq.                                  -
Prunus spinosa L.                                             -
Sambucus nigra L.                                             -
tree buds indet.                                              -
Rosa sp.                                                      -
Rubus fruticosus agg.                                         -
Rubus sp.                                                     -
Thelycrania sanguinea (L.) Fourr.                             -
Calluna vulgaris (L.) Hull                                    -
Erica sp.                                                     -
Vaccinium myrtillus L.                                        -
Empetrum nigrum L.                                            -
Pteridium aquilinium (L.) Kuhn                                =
Lycopus europaeus L.                                          -
Viola Subgenus Viola                                          -
Caltha palustris L.                                           -
Menyanthes trifoliata L.                                      -
Potamogeton spp.                                              -
```

Table 7.1   Crop processing sequence.

```
A  - FREE THRESHING CEREALS

Harvesting          - to remove the crop from the field
Threshing           - to release the grain from the straw and chaff
Raking              - to remove the large straw fragments
Winnowing           - to remove the light chaff and straw fragments
                      and the light weed seeds
Coarse Sieving      - to remove weed heads, large weeds, unthreshed
                      ears and straw nodes
Fine Sieving        - to remove the small weed seed
GRAIN STORE         - seeds of similar size as the grains need to be
                      removed by hand (this stage may, in fact, take
                      place before fine sieving)


B  - GLUME WHEATS

Harvesting          - to remove the crop from the field
Threshing           - to break the ear into spikelets
Raking              - to remove the large straw fragments
1st Winnowing       - to remove the light chaff and straw fragments and
                      the light weed seeds
1st Coarse Sieving  - to remove weed heads, large weeds, untreshed ears
                      and straw nodes
(SPIKELET STORE)    - from this point domestic processing is often
                      done on a day to day basis
Parching            - to render the glumes brittle
Pounding            - to release the grains from the glumes
2nd Winnowing       - to remove the light chaff and light weed seeds
2nd Coarse Sieving  - to remove remaining weed heads, large weeds,
                      straw nodes etc.
Fine Sieving        - to remove the glume bases and small weed seeds
GRAIN STORE         - weed seeds of similar size as the grains need to
                      be removed by hand (this stage may, in fact, be
                      omitted if the sequence 'parching to fine sieving'
                      is done piecemeal, just before use)
```

Table 7.2   Simplified crop processing sequence.

A - SIMPLIFIED CROP PROCESSING SEQUENCE

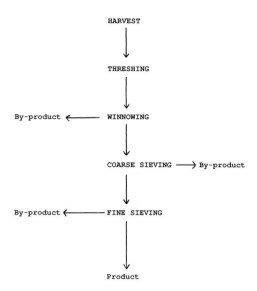

B - PROCESSING SEQUENCE INDICATING EFFECTS ON WEED SEED CATEGORIES

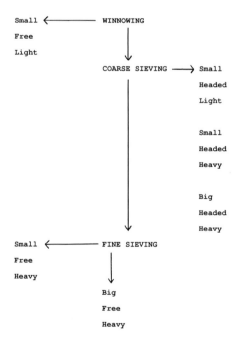

Table 7.3   Classification of samples to crop processing stages, using ratios (Method 1).

Ratio 1: no. of glume bases to no. of glume wheat grains.
Ratio 2: no. of barley rachis internodes to no. of barley grains (unless otherwise stated).
Ratio 3: no. of weed seeds to no. of cereal grains.

SITE AND CONTEXT	RATIO 1	RATIO 2	RATIO 3
Hallshill			
18	139: 3=46.3	9: 2= 4.5	9: 5= 1.8
26	104: 3=34.7	1: 0= -	6: 3= -
27	74: 0=74	2: 0= -	15: 0=15
30	77: 0=77	3: 4= -	12: 4= 3
8	5: 79= 0.1	0: 9= -	4: 88= 0.1
23A	189: 10=18.9	6: 6= 1	18: 16= 1.1
23B	90: 14= 6.4	0: 9= -	21: 23= 0.9
23C	131: 12=10.9	2: 0= -	11: 23= 0.9
23D	241: 11=21.9	3: 8= 0.4	42: 19= 2.2
23E	141: 14=10.1	4: 4= -	39: 18= 2.2
23F	190: 5=38	6: 2= -	20: 7= 2.9
25L	253:78= 3.2	4: 7= 0.6	14: 85= 0.2
25U	338:171= 2	2: 8= 0.3	15:178= 0.1
Murton			
623	10: 1=10	39: 46= 0.8	45: 47= 1
624	13: 1=13	28: 69= 0.4	320: 70= 4.6
625	53: 15= 3.5	12: 52= 0.2	118: 67= 1.8
630	0: 0= -	6: 17= 0.4	27: 17= 1.6
Dod Law			
45(8)	6: 2= -	4: 9= 0.4	112: 11=10.2
40(10)	115: 11=10.5	137:151= 0.9	206:162= 1.3
40(11)	177: 13=13.6	220:630= 0.3	179:643= 0.7
38(7)	107: 27= 3.9	71:406= 0.2	367:433= 0.9
25(3)	1: 13= 0.1	0:136= 0	111:149= 0.1
25(5)	97: 23= 4.2	43:267= 0.2	212:290= 0.7
30(6)	90: 24= 3.8	69:154= 0.5	199:178= 1.1
51(13)	15: 0=15	15: 29= 0.5	45: 29= 1.6
Chester House			
117	31: 8= 3.9	14: 33= 0.4	68: 41= 1.7
Thorpe Thewles			
LS268	35: 12= 2.9	1: 2= -	87: 14= 6.2
PF1	1: 4= -	35: 7= 5	49: 11= 4.5
LS112	120: 15= 8	0: 11= 0	114: 26= 4.4
LS120	512: 42=12.2	21: 31= 0.7	591: 73= 8.1
LS138	108: 27= 4	2: 23= 0.1	237: 50= 4.7
LS150	57: 11= 5.2	1: 13= 0.1	81: 24= 3.4
LS160	19: 7= 2.7	0: 3= -	46: 10= 4.6
LS233	12: 4= 3	2: 4= -	62: 8= 7.8
LS248	35: 0=35	0: 10= 0	21: 10= 2.1
LS523	4: 0= -	0: 2= -	57: 5=11.4
LS551	31: 0=31	2: 2= -	63: 2=31.5
LS637	6: 0= -	0: 0= -	103: 2=51.5
PF29	3: 0= -	1: 3= -	58: 3=19.3
PF41	86: 4=21.5	0: 4= -	68: 9= 7.6
PF57	189: 13=14.5	3: 17= 0.2	256: 30= 8.5
JS2	15: 18= 0.8	12:120= 0.1	961:138= 7
JS3	5: 35= 0.2	0: 26= 0	616: 61=10.1
JS4	34: 8= 4.3	0:127= 0	587: 20=29.4
JS7	7: 8= 0.9	2: 12= 0.2	242:135= 1.8
JS9	9: 0= -	15: 13= 1.2	123: 32=10.3
JS13	115: 18= 6.4	10: 13= 1	284: 31= 9.2
JS16	5: 1= -	6: 11= 1.1	52: 8= 6.5
JS17	55: 2=27.5	1: 1= 1	66: 7= 9.4
JS18	30: 0=30	4: 5= -	44: 11= 4
JS27	34: 4= 8.5	4: 5= -	475: 9=52.8
LS178	57: 14= 4.1	0: 19= 0	120: 33= 3.6
LS194	2: 5= -	0: 17= 0	189: 22= 8.6
LS208	2: 5= -	0: 0= -	55: 5=11
LS238	47: 0=47	0: 7= -	58: 7= 8.3
LS246	58: 5=11.6	0: 8= -	35: 13= 2.7
LS291	110: 9=12.2	2: 9= 0.2	105: 19= 5.5
LS422	14: 30= 0.5	0: 42= 0	114: 72= 1.6
LS431	35: 17= 2.1	0: 6= -	105: 23= 4.6
LS450	0: 2= -	0: 8= -	52: 10= 5.2
LS465	72: 15= 4.8	7: 23= 0.3	86: 38= 2.3
LS470	8: 5= 1.6	0: 5= -	346: 10=34.6
LS483	22: 22= 0.1	0: 11= 0	56: 33= 1.7
PF12	6: 0= -	0: 0= -	46: 4=11.5
PF49	1: 5= -	0: 0= -	86: 7=12.3
PF78	52: 9= 5.8	9: 22= 0.4	44: 5= 8.8
JS6	166: 19= 8.7	0: 45= 0	83: 31= 2.7
JS10	218: 14=15.6	8: 23= 0.3	64: 49= 1.3
JS11	814: 38=21.4	15: 42= 0.4	173: 43= 4
JS12	12: 2= 6	0: 2= -	265: 37= 7.2
JS14	10: 15= 0.7	1: 17= 0.1	518: 80= 6.5
JS15	28: 1=28	0: 0= -	129: 4=32.3
JS20	64: 0=64	0: 5= -	87: 32= 2.7
JS22	1: 0= -	1: 0= -	19: 1=19
JS26	78: 22= 3.5	14: 73= 0.2	97: 5=19.4
JS28	55: 31= 1.8	0: 18= 0	524: 1=524
LS69	41: 11= 3.7	0: 8= -	202: 95= 2.1
ML1	153: 31= 4.9	2: 20= 0.1	256: 49=52
ML2	30: 2=15	0: 13= 0	108: 19= 5.7
ML3	58: 12= 4.8	0: 6= -	297: 51= 5.8
ML4	23: 10= 2.3	1: 16= 0.1	70: 15= 4.7
ML5	7: 3= 2.3	5: 6= 0.8	196: 18=10.9
ML6	33: 6= 5.5	3: 22= 0.1	127: 26= 4.9
ML7	53: 8= 6.5	3: 22= 0.1	61: 5=12.2
ML8	14: 4= 3.5	5: 13= 0.4	103: 12= 8.6
ML9	6: 0= -	3: 13= 0.2	176: 30= 5.9
ML10	23: 9= 2.6	17: 56= 0.3	159: 26= 6.1
ML11	14: 0=14	1: 7= -	45: 6= 7.5
ML12	17: 0=17	2: 3= -	138: 22= 6.3
ML13	34: 9= 3.8	1: 6= -	111: 13= 8.5
ML14	6: 2= -	0: 0= -	92: 56= 1.6
ML16	20: 6= 3.3	1: 11= 0.1	94: 16= 5.9
ML17	6: 2= -	2: 28= 0.1	61: 12= 4.8
ML18	20: 0=20	3: 24= 0.1	57: 12= 4.8
ML20	28: 5= 5.6	2: 22= 0.1	34: 4= 8
ML23	21: 4= 5.3	0: 18= 0.1	32: 8= 4
ML25	56: 12= 4.7	1: 11= 0.1	142: 23= 6.2
ML26	15: 7= 2.1	2: 55= 0.1	143: 15= 9.5
ML27	305:40= 7.6	2: 28= 0.1	279: 61= 4.6
ML28	152:31= 4.9	3: 24= 0.1	305: 40= 7.6
ML29	219:33= 6.6	2: 22= 0.1	152: 31= 4.9
LS1	93: 9=10.3	0: 26= 0	219: 33= 6.6
LS17	84: 39= 2.2	0: 3= -	93: 9=10.3
LS19	51: 9= 5.7	0: 3= -	84: 39= 2.2
LS25	14: 6= 4.3	0: 4= -	51: 9= 5.7
LS28	17: 4= 4.3	0: 3= -	63: 8= 7.9
LS52	4: 6= 0.7	2: 8= 0.3	52: 9= 5.8
LS507	413: 32=12.9	4: 9= 0.4	60: 41= 1.5

SITE AND CONTEXT	RATIO 1	RATIO 2	RATIO 3
49	12: 53= 0.2	1:182= 0.005	7: 243= 0.03
54	28: 7= 4	0: 18= 0	40: 25= 1.6
44	9: 7= 1.3	0: 16= 0	24: 23= 1
54	7: 24= 0.3	0: 46= 0	4: 72= 0.06
58	132: 45= 2.9	4:103= 0.04	57: 150= 0.4
120	26: 17= 1.5	1:318= 0.003	23: 335= 0.07
125	13: 4= 3.3	2:183= 0.01	5: 187= 0.03
127	4: 25= 0.2	0:142= 0	1: 167= 0.006
134	107: 32= 3.3	3:166= 0.02	20: 200= 0.1

Ratio 2 for rye:

		RATIO 2
10		15: 32= 0.5
23		23: 21= 1.1

South Shields – Deposit 12236

Ratio 2 is that for bread wheat.

CONTEXT	RATIO 1	RATIO 2	RATIO 3
1	4: 220= 0.02	0: 160= 0	22: 380= 0.06
2	31: 605= 0.05	0: 439= 0	19:1044= 0.02
3	11: 686= 0.02	0: 384= 0	22:1070= 0.02
4	15: 833= 0.02	0: 909= 0	34:1742= 0.02
5	3: 438= 0.01	0: 344= 0	21: 782= 0.03
6	6: 197= 0.03	0: 142= 0	18: 339= 0.05
7a	14: 520= 0.03	0: 388= 0	25: 908= 0.03
7b	11: 181= 0.06	0: 106= 0	34: 287= 0.1
8	5: 202= 0.02	0: 141= 0	20: 343= 0.06
9	4: 310= 0.01	0: 224= 0	21: 534= 0.04
10a	10: 630= 0.02	0: 533= 0	28:1163= 0.04
10b	7: 117= 0.04	0: 126= 0	7: 243= 0.03
10c	7: 190= 0.04	0: 116= 0	12: 306= 0.04
10d	10: 520= 0.02	0: 318= 0	20: 838= 0.02
11	13: 458= 0.03	0: 294= 0	25: 752= 0.03
12	25: 738= 0.03	0: 578= 0	34:1316= 0.03
13	14: 424= 0.03	0: 240= 0	17: 664= 0.03
14	6: 269= 0.02	0: 203= 0	20: 463= 0.04
15	33: 346= 0.1	0: 272= 0	54: 618= 0.09
17	14: 380= 0.04	0: 252= 0	41: 632= 0.06
18	13: 635= 0.02	0: 479= 0	29:1114= 0.03
19	59:1061= 0.06	0:1026= 0	43:2087= 0.02
21	15: 541= 0.03	0: 318= 0	21: 859= 0.02
22	47:1333= 0.04	0: 891= 0	49:2224= 0.02
23	27:1129= 0.02	0: 917= 0	27: 204= 0.1
24	58: 66= 1.9	0: 66= 0	138: 96= 1.4
25	54:1063= 0.05	0:1030= 0	64:2093= 0.03
26	70: 192= 0.4	0: 124= 0	116: 316= 0.4
27	33: 397= 0.08	0: 232= 0	47: 629= 0.07
28	30: 109= 0.3	0: 109= 0	75: 218= 0.3
29	22: 399= 0.06	0: 603= 0	34:1002= 0.03
30	91:1810= 0.05	0: 891= 0	97:2701= 0.04

South Shields – Deposit 12176

Ratio 2 is that for bread wheat.

CONTEXT	RATIO 1	RATIO 2	RATIO 3
1	2: 238= 0.01	0: 140= 0	11: 378= 0.03
2	3: 154= 0.02	0: 60= 0	5: 214= 0.02
4	1: 299= 0.01	0: 129= 0	11: 428= 0.03
6	2: 25= 0.08	0: 25= 0	7: 50= 0.1
7	7: 21= 0.3	0: 13= 0	6: 63= 0.09
14	3: 51= 0.06	0: 0= -	18: 21= 0.9
15	3: 3= -	0: 45= 0	18: 96= 0.2
21	3: 3= -	0: 3= -	43: 6= 7.2
23	2: 34= 0.06	0: 38= 0	26: 72= 0.4
25	1: 59= 0.02	0: 67= 0	17: 126= 0.1
28	10: 20= 0.5	0: 13= 0	41: 33= 1.2
30	4: 22= 0.2	0: 17= 0	46: 39= 1.2

SITE AND CONTEXT	RATIO 1	RATIO 2	RATIO 3
Stanwick			
2209	18: 0=18	0: 20= 0	51: 20= 2.6
1095	12: 5= 2.4	0: 20= 0	54: 25= 2.2
2163	8: 0= -	0: 7= -	38: 7= 5.4
1085	30: 0=30	4: 70= 0.1	80: 70= 1.1
2063	0: 0= -	0: 2= -	50: 2=25
2201	15: 0=15	1: 17= 0.1	24: 17= 1.4
2064	5: 0= -	0: 4= -	33: 4= 8.3
2043	8: 0= -	0: 8= -	57: 8= 7.1
1112(4E)	5: 0= -	0: 17= 0	28: 17= 1.6
1110	5: 2= -	0: 6= -	33: 8= 4.1
2180	12: 5= 2.4	1: 5= -	36: 10= 3.6
2182	12: 3= 4	2: 21= 0.1	58: 24= 2.4
2196	45: 6= 7.5	4: 41= 0.1	52: 47= 1.1
2156	61: 5=12.2	5: 145= 0.03	268: 150= 1.8
1084	9: 5= 1.8	0: 11= 0	9: 4= 2.3
2045	88: 11= 8	8: 32= 0.3	343: 43= 8
2051	32: 4= 8	5: 4= -	90: 8=11.3
2195	5: 3= -	5: 45= 0.1	364: 48= 7.7
2193	13: 3= 4.3	5: 17= 0.3	34: 20= 1.7
1064(15)	10: 0=10	1: 36= 0.03	45: 36= 1.3
1078	6: 0= -	0: 3= -	126: 3=42
1013	22: 2=11	0: 3= -	60: 5=12
1022	7: 0= -	0: 16= 0	108: 16= 6.8
1023	17:104= 0.2	0: 84= 0	287:188= 1.5
2012	11: 3= 3.7	0: 26= 0	107: 29= 3.7
2042	38: 3=12.7	2: 17= 0.1	257: 20=12.9
Rock Castle			
47	20: 6= 3.3	8: 16= 0.5	69: 22= 3.1
61a	12: 4= 3	0: 4= -	37: 11= 3.4
61b	24: 7= 3.4	0: 4= -	93: 11= 8.5
74	83: 27= 3.1	12: 36= 0.3	121: 458= 0.9
24	45: 17= 2.6	2: 14= 0.1	254: 31= 8.2
59	22: 11= 2	6: 15= 0.4	251: 27= 9.3
60	163: 40= 4.1	0: 3= -	277: 55= 5
50	56: 12= 4.7	12: 33= 0.4	135: 45= 3
2a	37: 23= 1.6	0: 14= 0	365: 37= 9.9
2b	129: 23= 5.6	1: 2= -	343: 25=13.7
14	3: 9= 0.3	0: 5= -	61: 14= 4.4
69	12: 4= 3	0: 0= -	93: 4=23.3
12.1	11: 2= 5.5	0: 6= -	178: 8=22.3
12.2	16: 0=16	3: 7= 0.4	88: 7=12.6
12.3	17: 3= 5.7	1: 2= -	40: 5= 8
12.4	54: 6= 9	3: 3= -	110: 9=12.2
12.5	46: 7= 6.6	4: 18= 0.3	193: 19=10.2
12.6	28: 7= 4	0: 18= 0	183: 25= 7.3
12.7	44: 28=16	0: 16= 0	445: 44=10.1
12.8	18: 8= 2.3	0: 5= -	124: 13= 9.5
12.9	27: 9= 3	3: 13= 0.2	332: 22=15.1
38	18: 5= 3.6	0: 13= 0.2	122: 17=17.4
54	18: 11= 1.6	2: 15= 0.1	179: 26= 6.9

Ratio 2 for bread wheat

		RATIO 2	
50		125: 7=17.9	

Thornbrough

	RATIO 1	RATIO 2	RATIO 3
5	8: 18= 0.4		1: 80= 0.01
10	634:413= 1.5	0: 58= 0.01	102:1438= 0.07
39	28: 78= 0.4	11.993= 0.01	24: 277= 0.09
40	19: 19= 1	0: 196= 0	2: 73= 0.03
41	7: 29= 0.2	0: 51= 0	4: 113= 0.03
43	6: 69= 0.09	0: 84= 0	12: 3= -
45	842:328= 2.6	12:380= 0.03	163: 729= 0.2
46	27: 99= 0.3	0: 235= 0	8: 342= 0.02

Table 7.4   Weed seed categories relevant to crop processing.

Big-Headed-Heavy (BHH)

Raphanus raphanistrum

Big-Free-Heavy (BFH)

Ranunculus Subgenus Ranunculus
Agrostemma githago
Vicia hirsuta
Vicia/Lathyrus
Polygonum convolvulus
Galeopsis sp.
Galium aparine
Bromus mollis/secalinus
Avena sp.
Gramineae

Small-Headed-Heavy (SHH)

-

Small-Headed-Light (SHL)

-

Small-Free-Heavy (SFH)

Ranunculus flammula	Veronica arvensis
Stellaria media	Veronica cf. officinalis
Montia fontana	Urtica urens
Chenopodium album	Hyoscyamus niger
Chenopodium sp.	Brassica sp.
Atriplex spp.	Spergula arvensis
Potentilla cf. erecta	Ajuga reptans
Polygonum aviculare	Sieglingia decumbens
Polygonum lapathifolium	Carex pilulifera
Polygonum persicaria	Carex pulicaris
Rumex acetosella	Carex spp.
Rumex spp.	Eleocharis sp.
Prunella vulgaris	small grasses ??
Plantago lanceolata	
Plantago major	

Small-Free-Light (SFL)

Tripleurospermum inodorum
Conium maculatum
Odontites verna
Bromus sterilis
small grassess ??

Table 7.5   Classification of samples to crop processing group using weed seed categories (Method 2).

KEY:

1	= winnowing by-product	2	= coarse-sieve by-product
3	= fine-sieve by-product	4	= fine-sieve product
SFH	= Small-Free-Heavy	SFL	= Small-Free-Light

Column A = classification
Column B = probability of the classification
Column C = next most probable classification

	ANALYSIS 1 small grasses = SFH			ANALYSIS 2 small grasses = SFL		
	A	B	C	A	B	C
Hallshill						
18	3	1.0000	(1)	3	1.0000	(1)
26	3	0.9999	(1)	1	0.5580	(3)
30	3	1.0000	(4)	3	0.9749	(4)
8	3	0.9883	(4)	3	0.9883	(4)
23A	3	1.0000	(4)	3	0.3817	(1)
23B	3	1.0000	(1)	3	0.9687	(1)
23C	3	1.0000	(1)	3	1.0000	(1)
23D	3	0.9999	(4)	3	0.9991	(4)
23E	3	1.0000	(4)	3	0.9997	(1)
23F	3	1.0000	(4)	3	0.9999	(1)
25L	3	1.0000	(4)	3	0.7726	(1)
25U	3	0.9983	(4)	1	0.9304	(3)
Murton						
623	3	1.0000	(4)	3	0.9993	(1)
624	3	1.0000	(4)	3	0.9998	(1)
625	3	1.0000	(4)	3	1.0000	(4)
630	3	0.9999	(4)	3	0.9922	(1)
Dod Law						
45 (8)	3	1.0000	(4)	3	0.9910	(1)
40 (10)	3	1.0000	(4)	3	0.9998	(1)
40 (11)	3	1.0000	(4)	3	1.0000	(1)
38 (7)	3	1.0000	(4)	3	1.0000	(4)
25 (3)	3	0.7040	(4)	3	0.7040	(4)
25 (5)	3	1.0000	(4)	3	0.9997	(1)
30 (6)	3	1.0000	(4)	3	0.9991	(1)
51 (13)	3	0.9996	(4)	3	0.9986	(4)
Chester House						
117	3	0.9537	(4)	4	0.5337	(3)
Thorpe Thewles						
ML1	3	0.9976	(4)	3	0.9894	(4)
ML2	3	0.9998	(4)	3	0.9988	(4)
ML3	3	0.9894	(4)	3	0.9433	(4)
ML4	3	0.9935	(4)	3	0.9138	(4)
ML5	3	0.9997	(4)	3	0.9966	(4)
ML6	3	0.9696	(4)	3	0.9029	(4)
ML7	3	0.9999	(4)	3	0.9992	(1)
ML8	3	0.9998	(4)	3	0.9992	(4)
ML9	3	0.9995	(4)	3	0.9950	(4)
ML10	3	1.0000	(4)	3	0.9998	(4)
ML11	3	0.9997	(4)	3	0.9990	(4)
ML12	3	1.0000	(4)	3	0.9987	(1)
ML13	3	0.9998	(4)	3	0.9994	(4)
ML14	3	0.9996	(4)	3	0.9996	(4)
ML16	3	0.9690	(4)	3	0.8066	(4)
ML17	3	1.0000	(4)	3	0.9999	(4)
ML18	3	0.9974	(4)	3	0.9602	(1)

	ANALYSIS 1 small grasses = SFH			ANALYSIS 2 small grasses = SFL		
	A	B	C	A	B	C
ML19	3	0.9973	(4)	3	0.9884	(4)
ML20	3	0.9998	(4)	3	0.9992	(4)
ML22	3	0.9999	(4)	3	0.9996	(4)
ML23	3	0.9996	(4)	3	0.9972	(4)
ML25	3	0.9996	(4)	3	0.9968	(4)
ML26	3	0.9996	(4)	3	0.9750	(4)
ML27	3	0.9999	(4)	3	0.9987	(4)
ML28	3	0.9999	(4)	3	0.9962	(1)
ML29	3	0.9998	(4)	3	0.9980	(4)
LS69	3	1.0000	(4)	3	0.9973	(1)
LS268	3	1.0000	(4)	3	0.9992	(1)
PF1	3	0.9996	(4)	3	0.9998	(4)
LS112	3	0.9557	(4)	1	0.4825	(4)
LS120	3	0.9990	(4)	3	0.9925	(4)
LS138	3	0.9981	(4)	3	0.9893	(4)
LS150	3	0.9946	(4)	3	0.9116	(4)
LS160	3	0.9957	(4)	3	0.9864	(4)
LS233	3	0.9989	(4)	3	0.9970	(4)
LS248	3	0.9978	(4)	3	0.9978	(4)
LS523	3	1.0000	(4)	3	0.9991	(1)
LS551	3	0.9995	(4)	3	0.9876	(1)
LS637	3	0.9999	(4)	3	0.9996	(4)
PF29	3	1.0000	(4)	3	0.9999	(4)
PF41	3	0.9737	(4)	3	0.8453	(4)
PF57	3	0.9966	(4)	3	0.9910	(4)
JS2	3	0.9992	(4)	3	0.9949	(4)
JS3	3	1.0000	(4)	3	0.9996	(1)
JS4	3	1.0000	(4)	3	0.9998	(4)
JS7	3	0.9999	(4)	3	0.9998	(4)
JS9	3	1.0000	(4)	3	0.9999	(4)
JS13	3	0.9978	(4)	3	0.9822	(4)
JS16	3	0.9997	(4)	3	0.9981	(4)
JS17	3	0.9053	(4)	3	0.6991	(4)
JS18	3	0.9932	(4)	3	0.9427	(4)
JS27	3	1.0000	(4)	1	1.0000	(4)
LS178	3	0.9960	(4)	3	0.9695	(4)
LS194	3	0.9999	(4)	3	0.9998	(4)
LS208	3	0.9982	(4)	3	0.9844	(4)
LS238	3	0.9990	(4)	3	0.9829	(1)
LS246	3	0.9997	(4)	3	0.9997	(4)
LS291	3	0.9862	(4)	3	0.8789	(4)
LS422	3	0.9720	(4)	3	0.9083	(4)
LS431	3	0.9140	(4)	3	0.6186	(4)
LS450	3	0.9852	(4)	3	0.9301	(4)
LS465	3	0.9962	(1)	1	0.6452	(3)
LS470	3	0.9990	(4)	3	0.9906	(4)
LS483	3	0.9996	(4)	3	0.9984	(1)
PF12	3	0.9998	(4)	4	0.6961	(1)
PF49	3	0.9995	(4)	4	0.9982	(3)
PF78	3	0.9997	(4)	3	0.9959	(4)

Table 7.5 (cont.)

Left block — ANALYSIS 1 (small grasses = SFH) and ANALYSIS 2 (small grasses = SFL)

	ANALYSIS 1 (SFH)			ANALYSIS 2 (SFL)		
	A	B	C	A	B	C
JS6	3	0.9740	(4)	3	0.8112	(4)
JS10	3	0.9998	(4)	3	0.9995	(4)
JS11	3	0.9997	(4)	3	0.9977	(4)
JS12	3	0.9555	(4)	3	0.8026	(4)
JS14	3	0.9992	(4)	3	0.9992	(4)
JS15	3	1.0000	(4)	3	1.0000	(4)
JS20	3	1.0000	(4)	3	0.9998	(4)
JS22	3	0.9953	(4)	3	0.8438	(1)
JS26	3	1.0000	(1)	3	0.9978	(1)
JS28	3	1.0000	(4)	3	1.0000	(1)
LS1	3	0.9995	(4)	3	0.9974	(4)
LS17	3	0.9998	(4)	3	0.9993	(4)
LS19	3	0.9982	(4)	3	0.9982	(4)
LS25	3	0.9997	(4)	3	0.9997	(4)
LS28	3	1.0000	(4)	3	0.9997	(4)
LS52	3	1.0000	(4)	3	0.9997	(4)
LS507	3	0.9353	(4)	4	0.4764	(3)
Stanwick						
2209	3	0.9967	(4)	3	0.9404	(4)
1095	3	0.9645	(4)	3	0.6536	(4)
2163	3	1.0000	(4)	3	0.9990	(1)
1085	3	0.9975	(4)	3	0.8606	(1)
2063	3	0.9956	(4)	3	0.9956	(4)
2201	3	0.9996	(4)	3	0.9468	(1)
2064	3	1.0000	(4)	3	0.9992	(1)
2043	3	0.9953	(4)	3	0.9698	(4)
1112 (45)	3	0.9979	(4)	3	0.9745	(4)
1110	3	0.9997	(1)	3	0.9580	(1)
2180	3	0.9997	(4)	3	0.9990	(4)
2182	3	0.9588	(4)	3	0.5151	(4)
2196	3	0.9986	(4)	3	0.9869	(1)
2156	3	1.0000	(4)	3	0.9999	(1)
1084	3	0.9994	(4)	3	0.9866	(4)
2045	3	0.9985	(4)	3	0.9992	(1)
2051	3	0.9997	(4)	3	0.9683	(4)
2195	3	0.9998	(4)	3	0.6308	(1)
2193	3	1.0000	(4)	3	0.9985	(4)
1064 (19)	3	1.0000	(4)	3	0.9899	(1)
1078	3	0.9978	(4)	3	0.9978	(4)
1013	3	0.9992	(4)	3	0.9835	(4)
1022	3	0.7921	(4)	4	0.6685	(3)
1023	3	0.9997	(4)	3	0.9893	(4)
2012	3	0.9997	(4)	3	0.9983	(4)
2042	3	1.0000	(4)	3	0.9986	(1)
Rock Castle						
47	3	1.0000	(4)	3	0.9649	(1)
61A	3	0.9294	(4)	4	0.7464	(1)
61B	3	0.9992	(1)	1	0.8991	(3)
74	3	0.9977	(4)	3	0.9688	(4)
24	3	0.9999	(4)	3	0.9971	(4)
59	3	0.9999	(4)	3	0.9955	(4)
60	3	0.9998	(4)	3	0.9763	(1)
50	3	0.9996	(4)	3	0.9992	(1)
2A	3	1.0000	(4)	3	0.9979	(1)
2B	3	0.9998	(4)	3	0.9865	(1)
14	3	0.9998	(4)	3	0.7851	(1)
69	3	0.9466	(4)	4	0.5527	(1)
12.1	3	1.0000	(4)	3	0.9872	(4)
12.2	3	0.9998	(1)	3	0.6490	(1)
12.3	3	0.9942	(4)	3	0.8799	(4)
12.4	3	0.9987	(4)	3	0.9404	(4)
12.5	3	0.9994	(4)	3	0.8390	(1)
12.6	3	0.9999	(4)	3	0.9817	(1)
12.7	3	0.9993	(4)	3	0.9889	(4)
12.8	3	0.9835	(4)	3	0.5253	(4)
12.9	3	0.9949	(4)	3	0.7045	(4)
38	3	0.9995	(4)	3	0.9912	(4)
54	3	0.9997	(4)	3	0.9952	(4)

Right block — ANALYSIS 1 / ANALYSIS 2 (small grasses = SFH) and (small grasses = SFL)

	SFH			SFL		
	A	B	C	A	B	C
Thornbrough						
10	4	0.9994	(3)	4	0.9997	(3)
39	3	0.7662	(4)	3	0.5939	(3)
40	3	0.5873	(4)	3	0.5873	(4)
41	1	0.8010	(3)	1	0.9999	(2)
43	3	0.6179	(4)	3	0.6179	(3)
45	4	0.9161	(3)	4	0.9531	(3)
46	3	0.9920	(4)	1	0.8222	(3)
49	4	0.6066	(3)	4	0.8831	(1)
54	3	0.8874	(4)	4	0.8926	(1)
54	3	0.9735	(4)	4	0.7657	(3)
44	3	0.9576	(4)	4	0.8172	(4)
58	3	0.9817	(4)	3	0.9817	(4)
120	3	0.9995	(1)	1	0.9971	(3)
125	3	0.9955	(1)	3	0.9955	(1)
127	3	0.9904	(4)	3	0.8904	(4)
134	3			3		
5	—			—		
South Shields (12236)						
1	4	0.9996	(3)	4	0.9996	(3)
2	4	0.9990	(3)	4	0.9990	(3)
3	3	0.7969	(4)	4	0.5761	(4)
4	4	0.9985	(1)	4	0.9996	(1)
5	4	0.9503	(3)	4	0.9911	(3)
6	4	0.9996	(3)	4	0.9996	(3)
7A	4	0.9984	(3)	4	0.9994	(3)
7B	3	0.8722	(3)	3	0.8722	(4)
8	3	0.8087	(4)	3	0.8087	(4)
9	4	1.0000	(3)	4	1.0000	(3)
10A	4	0.9998	(3)	4	0.9998	(3)
10B	4	0.8936	(3)	4	0.8936	(3)
10C	4	0.8643	(3)	4	0.9657	(1)
10D	4	0.9657	(1)	4	0.9997	(1)
11	4	0.9997	(3)	4	0.9941	(3)
12	4	0.9941	(3)	4	0.9948	(3)
13	4	0.9683	(3)	4	0.9992	(3)
14	4	0.9992	(3)	4	0.9998	(3)
15	4	0.9998	(3)	3	0.4635	(1)
17	3	0.9784	(4)	4	0.8429	(3)
18	4	0.8429	(3)	4	0.9294	(3)
19	4	0.7369	(3)	4	0.9983	(3)
21	4	0.9908	(3)	4	1.0000	(3)
22	4	1.0000	(3)	4	0.9893	(3)
23	4	0.9893	(3)	4	0.9990	(3)
24	4	0.9990	(3)	4	0.9493	(3)
25	4	0.9070	(3)	4	1.0000	(3)
26	4	0.9830	(3)	4	0.9916	(3)
27	4	0.9923	(3)	4	0.9972	(3)
28	4	0.6329	(3)	4	0.8122	(3)
29	3	0.6136	(3)	3	0.6136	(4)
30	4	0.9998	(3)	4	0.9999	(3)
South Shields (12176)						
1	4	0.9975	(3)	4	0.9997	(3)
2	4	0.9998	(3)	4	0.9998	(3)
4	4	0.8393	(3)	4	0.8393	(3)
6	3	0.7191	(4)	3	0.7191	(4)
7	4	0.9984	(4)	4	0.9881	(4)
14	4	0.8664	(3)	4	0.9881	(3)
15	4	0.6623	(3)	4	0.6623	(3)
21	3	0.9995	(4)	3	0.9980	(4)
23	3	0.9985	(4)	3	0.9783	(1)
28	3	0.8299	(4)	3	0.9813	(4)
30	3	0.9984	(1)	3	0.9337	(1)

Table 9.1   Ellenberg's indicator values (Ellenberg 1979).

LIGHT

L1 full shadow plant
L2 between 1 and 3
L3 shadow plant
L4 between 3 and 5
L5 half shadow plant
L6 between 5 and 7
L7 half light plant
L8 between 7 and 9
L9 full light plant
Lx indifferent

CONTINENTALITY

K1 euoceanic
K2 oceanic
K3 between 2 and 4
K4 suboceanic
K5 intermediate
K6 subcontinental
K7 between 6 and 8
K8 continental
K9 eucontinental
Kx indifferent

ACIDITY

R1 in very acid soils
R2 between 1 and 3
R3 mostly in acid soils
R4 between 3 and 5
R5 in weakly acid soils
R6 between 5 and 7
R7 mostly in neutral soils
R8 between 7 and 9
R9 neutral or basic soils
Rx indifferent

TEMPERATURE

T1 only in cold climate
T2 between 1 and 3
T3 mostly in cold climate
T4 between 3 and 5
T5 intermediate
T6 between 5 and 7
T7 mostly in warm climate
T8 between 7 and 9
T9 in very warm climate
Tx indifferent

MOISTURE

F1 in extremely dry soils
F2 between 1 and 3
F3 in dry soils
F4 between 3 and 5
F5 in fresh soils
F6 between 5 and 7
F7 in moist soils
F8 between 7 and 9
F9 in wet soils
F10 freq. inundated soils
F11 water plant
F12 underwater plant
Fx indifferent

NITROGEN

N1 very poor in nitrogen
N2 between 1 and 3
N3 mostly in poor soils
N4 between 3 and 5
N5 in intermediate soils
N6 between 5 and 7
N7 rich in mineral nitrogen
N8 nitrogen indicator
N9 very rich in nitrogen
Nx indifferent

Table 9.2   Runhaar's codes for ecological groups (Runhaar *et al.* 1987)

STRUCTURE OF THE VEGETATION AND STAGE OF SUCCESSION

G - grassland
H - woodland and shrub
P - pioneer vegetation
R - tall herb vegetation
V - semi aquatic helophytic vegetation
W - water vegetation

MOISTURE REGIME (first figure)

1 - aquatic
2 - wet
4 - moist
6 - dry

NUTRIENT AVAILABILITY AND ACIDITY (second figure)

1 - low nutrient availability, acid
2 - low nutrient availability, moderately acid to neutral
3 - low nutrient availability, basic
4 - low nutrient availability
7 - moderate nutrient availability
8 - high nutrient availability
9 - moderate to high nutrient availability

Table 9.3   Fitter's habitat information (Fitter 1978).

WETNESS - DRYNESS

(1)   Standing water above the surface for all or most
      of the year.
(2)   Wet soils which are saturated with water for
      most of the year.
(3)   Damp soils which may be occasionally wet.
(4)   'Normal' moist soils, such as a typical field
      soil.
(5)   Dry soils which crumble to the touch and are
      usually found on high ground or above very
      porous rock.

ACIDITY

(1)   Very acid soils with no chalk or limestone and
      usually found in sandy or peaty places.
(2)   Lightly acid soils which are often again found
      on sands and peats, but also on milder soils
      which have become acid because of the plants
      growing on them, such as pines, gorse and
      sometimes beech.
(3)   Neutral soils typical of lowland meadows and
      river-plains. These soils tend to be farmed.
(4)   Slightly calcareous soils formed over chalks and
      limestones but without bits of rock in the soil.
(5)   Very calcareous soils which are usually very
      thin, formed on chalk and limestone and with
      pieces of the rock visible in the soil or lying
      on the surface. Limestone cliffs and pavements
      fit in here too, and saltmarsh soils are
      included for convenience.

FERTILITY

(1)   Very fertile soils often fertilized, with
      vigorous, tall vegetation or trees with dense
      undergrowth.
(2)   Fertile soils, usually in lowland sites or on
      alluvial deposits.
(3)   Intermediate fertility, typical of well-
      developed but unfertilized soils.
(4)   Poor soils, usually with a complete plant cover,
      but of short plants, or trees with little
      undergrowth.
(5)   Very poor soils, often with large patches of
      bare ground.

SHADE

(1)   Very dense shade, as in some beechwoods and
      conifer plantations.
(2)   Most woodlands fall into this category with full
      shade cast particularly in summer.
(3)   Open woods with the trees well-spaced, so that
      the sun still reaches the ground at times, e.g.
      natural pinewoods.
(4)   Hedges, open scrub, and woodland edges, where
      the light is still bright but the full sun may
      be shielded off.
(5)   Open habitats with no trees or tall shrubs, e.g.
      grassland, lakes, heaths.

Table 10.1   Summary results of PCA-1-4.

```
Principal Components Analysis - Prehistoric Assemblages
ANALYSIS   VARIABLES                          TRANSFORMATION
PCA-1      grain/chaff/weeds (weeds>10%)      square root
PCA-2      grain/chaff/weeds (weeds> 5%)      square root
PCA-3      grain/chaff/weeds (weeds>10%)      octave scale
PCA-4      grain/chaff/weeds (weeds> 5%)      octave scale


EIGENVALUE      % OF VAR.   FIVE HIGHEST LOADINGS ON FIRST AXIS
PCA-1:

Axis 1  5.53     14.2       Sieg decu  -0.88   glum dico  0.87
Axis 2  3.67      9.4       glum spel  -0.81   Trit dico  0.67
Axis 3  2.83      7.3       Mony font  -0.54   Chen albu  0.67
   CUM. % VAR    30.9       Brom m/s   -0.39   Poly l/p   0.62
                            Arrh elat  -0.31   smal gras  0.50

PCA-2:

Axis 1  5.89     13.7       Sieg decu  -0.88   glum dico  0.87
Axis 2  3.74      8.7       glum spel  -0.81   Chen albu  0.66
Axis 3  2.91      6.8       Mont font  -0.54   Trit dico  0.65
   CUM. % VAR    29.2       Brom m/s   -0.40   Poly l/p   0.60
                            Arrh elat  -0.31   Sper arve  0.51

PCA-3:

Axis 1  5.22     13.4       Sieg decu  -0.90   glum dico  0.89
Axis 2  3.78      9.7       glum spel  -0.72   Trit dico  0.72
Axis 3  2.49      6.4       Mont font  -0.63   Poly l/p   0.58
   CUM. % VAR    29.5       Brom m/s   -0.47   Chen albu  0.53
                            Arrh elat  -0.32   Atri spp.  0.40

PCA-4:

Axis 1  5.65     13.2       Sieg decu  -0.89   glum dico  0.89
Axis 2  3.83      8.9       glum spel  -0.72   Trit dico  0.69
Axis 3  2.57      6.0       Mont font  -0.62   Sper arve  0.55
   CUM. % VAR    28.0       Brom m/s   -0.47   Poly l/p   0.55
                            Arrh elat  -0.32   Chen albu  0.51
```

Table 10.2   Summary results of CA-1-4.

```
Cluster Analysis - Prehistoric Assemblages
ANALYSIS   VARIABLES                          TRANSFORMATION
CA-1       grain/chaff/weeds (weeds>10%)      square root
CA-2       grain/chaff/weeds (weeds> 5%)      square root
CA-3       grain/chaff/weeds (weeds>10%)      octave scale
CA-4       grain/chaff/weeds (weeds> 5%)      octave scale

CA-1:
Group A and B form separate clusters, which only join at fusion
 coefficient 25. Sample Thorpe Thewles PF1 'wrongly' classified.
CA-2:
Group A and B form separate clusters, which only join at fusion
 coefficient 25. Sample Thorpe Thewles PF1 'wrongly' classified.
CA-3:
Group A and B form separate clusters, which only join at fusion
 coefficient 25. Sample Murton 630 'wrongly' classified.
CA-4:
Group A and B form separate clusters, which only join at fusion
 coefficient 25. Sample Murton 630 'wrongly' classified.
```

Table 10.3   Summary results of CA-5-8.

```
Cluster Analysis - Prehistoric Assemblages
ANALYSIS   VARIABLES         TRANSFORMATION
CA-5       weeds (>10%)      square root
CA-6       weeds (> 5%)      square root
CA-7       weeds (>10%)      octave scale
CA-8       weeds (> 5%)      octave scale

CA-5:
Group A and B form separate clusters, which only join at fusion
 coefficient 25. Samples Murton 630 and Chester House 117 'wrongly'
 classified.
CA-6:
Group A and B form separate clusters, which only join at fusion
 coefficient 25. Samples Murton 630 and Chester House 117 'wrongly'
 classified.
CA-7:
Group A and B form separate clusters, which only join at fusion
 coefficient 25. Samples Murton 630, Chester House 117, and
 Rock Castle 47 'wrongly' classified.
CA-8:
Group A and B form separate clusters, which only join at fusion
 coefficient 25. Samples Murton 630 and Chester House 117 'wrongly' classified.
```

Table 10.4   Summary results of DA-1-4.

```
Discriminant Analysis - Prehistoric Assemblages
ANALYSIS         VARIABLES                               TRANSFORMATION
DA-1             grain/chaff/weeds (weeds>10%)           square root
DA-2             grain/chaff/weeds (weeds> 5%)           square root
DA-3             grain/chaff/weeds (weeds>10%)           octave scale
DA-4             grain/chaff/weeds (weeds> 5%)           octave scale

                         FIVE HIGHEST DISCRIMINANT SCORES
DA-1:

eigenvalue          25.97      glum dico  -0.55      Sieg decu  +0.30
Wilk's Lambda        0.04      Trit dico  -0.18      glum spel  +0.24
% correct clas.    100.00      Poly l/p   -0.16      Mont font  +0.09
                               Chen albu  -0.14      Brom m/s   +0.08
                               rach Hord  -0.08      Gali apar  +0.06

DA-2:

eigenvalue          29.29      glum dico  -0.52      Sieg decu  +0.28
Wilk's Lambda        0.08      Trit dico  -0.17      glum spel  +0.23
% correct clas.    100.00      Poly l/p   -0.15      Mont font  +0.09
                               Chen albu  -0.13      Brom m/s   +0.08
                               Sper arve  -0.09      Gali apar  +0.06

DA-3:

eigenvalue          26.21      glum dico  -0.64      Sieg decu  +0.39
Wilk's Lambda        0.04      Trit dico  -0.31      glum spel  +0.21
% correct clas.    100.00      Poly l/p   -0.14      Mont font  +0.12
                               Chen albu  -0.08      Brom m/s   +0.09
                               rach Hord  -0.07      Gali apar  +0.06

DA-4:

eigenvalue          28.47      glum dico  -0.61      Sieg decu  +0.37
Wilk's Lambda        0.03      Trit dico  -0.20      glum spel  +0.20
% correct clas.    100.00      Poly l/p   -0.13      Mont font  +0.11
                               Sper arve  -0.10      Brom m/s   +0.09
                               Chen albu  -0.08      Gali apar  +0.06
```

Table 10.5   Summary results of DA-5-8.

```
Discriminant Analysis - Prehistoric Assemblages

ANALYSIS         VARIABLES              TRANSFORMATION
DA-5             weeds (>10%)           square root
DA-6             weeds (> 5%)           square root
DA-7             weeds (>10%)           octave scale
DA-8             weeds (> 5%)           octave scale

                         FIVE HIGHEST DISCRIMINANT SCORES
DA-5:

eigenvalue           9.43      Poly l/p   -0.28      Sieg decu  +0.50
Wilk' Lambda         0.10      Chen albu  -0.24      Mont font  +0.16
% correct clas.    100.00      smal gras  -0.13      Brom m/s   +0.15
                               Atri spp.  -0.11      Gali apar  +0.11
                               Rume acet  -0.08      Trip inod  +0.10

DA-6:

eigenvalue          10.07      Poly l/p   -0.26      Sieg decu  +0.49
Wilk's Lambda        0.09      Chen albu  -0.23      Mont font  +0.15
% correct clas.    100.00      Sper arve  -0.15      Brom m/s   +0.14
                               smal gras  -0.12      Gali apar  +0.10
                               Atri spp.  -0.10      Trip inod  +0.10

DA-7:

eigenvalue           8.96      Poly l/p   -0.24      Sieg decu  +0.68
Wilk's Lambda        0.10      Chen albu  -0.15      Mont font  +0.20
% correct clas.     99.00      Atri spp.  -0.09      Brom m/s   +0.17
                               Rume acet  -0.09      Gali apar  +0.11
                               smal gras  -0.07      Trip inod  +0.10

DA-8:

eigenvalue           9.78      Poly l/p   -0.23      Sieg decu  +0.65
Wilk's Lambda        0.09      Sper arve  -0.16      Mont font  +0.20
% correct clas.    100.00      Chen albu  -0.13      Brom m/s   +0.17
                               Bras sp.   -0.10      Gali apar  +0.11
                               Atri spp.  -0.08      Trip inod  +0.10
```

Table 10.6 Ellenberg's indicator values for climatic factors (Ellenberg 1979; see also Table 9.1).

LIGHT

L1, L2, L3, L4, L5, L9 = no species

L6 = Ranunculus repens, Raphanus raphanistrum, Stellaria media, Atriplex spp., Potentilla erecta, Polygonum lapathifolium, Polygonum persicaria, Odontites verna, Plantago lanceolata, Avena fatua, Bromus mollis/secalinus, Carex pilulifera, (Spergula arvensis)

L7 = Ranunculus flammula, Montia fontana, Polygonum aviculare, Polygonum convolvulus, Rumex spp., Prunella vulgaris, Galium aparine, Tripleurospermum inodorum, small grasses (including Poa annua), (Urtica urens)

L8 = Rumex acetosella, Sieglingia decumbens, Arrhenatherum elatius, Eleocharis sp., Carex pulicaris, Juncus sp., (Plantage major)

Lx = Chenopodium album

TEMPERATURE

T1, T2, T3, T7, T8, T9 = no species

T4 = Carex pilulifera

T5 = Raphanus raphanistrum, Atriplex spp., Polygonum persicaria, Rumex acetosella, Rumex spp., Galium aparine, Arrhenatherum elatius, Carex pulicaris

T6 = Montia fontana, Polygonum lapathifolium, (Urtica urens)

Tx = Ranunculus repens, Ranunculus flammula, Stellaria media, Chenopodium album, Potentilla erecta, Polygonum aviculare, Polygonum convolvulus, Odontites verna, Prunella vulgaris, Plantago lanceolata, Tripleurospermum inodorum, Avena fatua, Bromus mollis/secalinus, Sieglingia decumbens, small grasses (including Poa annua), Eleocharis sp., (Spergula arvensis, Plantago major)

CONTINENTALITY

K1, K7, K8, K9 = no species

K2 = Montia fontana, Sieglingia decumbens, Carex pilulifera, Carex pulicaris

K3 = Ranunculus flammula, Raphanus raphanistrum, Potentilla erecta, Polygonum persicaria, Rumex acetosella, Rumex spp., Odontites verna, Prunella vulgaris, Plantago lanceolata, Galium aparine, Tripleurospermum inodorum, Bromus mollis/secalinus, Arrhenatherum elatius, (Spergula arvensis)

K4 = Polygonum lapathifolium

K5 = small grasses (including Poa annua)

K6 = Avena fatua

Kx = Ranunculus repens, Stellaria media, Chenopodium album, Atriplex spp., Polygonum aviculare, Polygonum convolvulus, Eleocharis palustris, (Urtica urens, Plantago major)

Species in brackets occur in <10%, but >5% of the samples.

Table 10.7 Ellenberg's indicator values for edaphic factors (Ellenberg 1979; see also Table 9.1).

MOISTURE

F1, F2, F11, F12 = no species

F3 = Polygonum persicaria

F4 = Stellaria media, Chenopodium album

F5 = Atriplex spp., Rumex acetosella, Odontites verna, Arrhenatherum elatius, Carex pilulifera, (Spergula arvensis, Urtica urens, Plantago major)

F6 = Rumex spp., Avena fatua, small grasses (including Poa annua)

F7 = Ranunculus repens, Polygonum lapathifolium, Juncus sp.

F8 = Montia fontana

F9 = Ranunculus flammula, Carex pulicaris

F10 = Eleocharis sp.

Fx = Raphanus raphanistrum, Potentilla erecta, Polygonum aviculare, Polygonum convolvulus, Prunella vulgaris, Plantago lanceolata, Galium aparine, Tripleurospermum inodorum, Bromus mollis/secalinus, Sieglingia decumbens

ACIDITY

R1, R5, R8, R9 = no species

R2 = Rumex acetosella, (Spergula arvensis)

R3 = Ranunculus flammula, Montia fontana, Sieglingia decumbens, Carex pilulifera

R4 = Raphanus raphanistrum, Prunella vulgaris

R6 = Galium aparine, Tripleurospermum inodorum

R7 = Stellaria media, Atriplex spp., Avena fatua, Arrhenatherum elatius

Rx = Ranunculus repens, Chenopodium album, Potentilla erecta, Polygonum aviculare, Polygonum convolvulus, Polygonum lapathifolium, Polygonum persicaria, Rumex spp., Odontites verna, Plantago lanceolata, Bromus mollis/secalinus, small grasses (including Poa annua), Eleocharis sp., Carex pulicaris, (Urtica urens, Plantago major)

NITROGEN

N1, N3, N9 = no species

N2 = Ranunculus flammula, Potentilla erecta, Rumex acetosella, Sieglingia decumbens

N4 = Montia fontana

N5 = Raphanus raphanistrum, Rumex spp., Carex pilulifera

N6 = Tripleurospermum inodorum, (Spergula arvensis, Plantago major)

N7 = Chenopodium album, Atriplex spp., Polygonum persicaria, Arrhenatherum elatius

N8 = Stellaria media, Polygonum lapathifolium, Galium aparine, small grasses (including Poa annua), (Urtica urens)

Nx = Ranunculus repens, Polygonum aviculare, Polygonum convolvulus, Odontites verna, Prunella vulgaris, Plantago lanceolata, Avena fatua, Bromus mollis/secalinus

Species in brackets occur in <10%, but >5% of the samples.

213

Table 10.8 Ellenberg's indicator values for edaphic factors, grouped into broader categories (see also Tables 9.1 and 10.7).

```
MOISTURE
F3+4   =  Polygonum persicaria, Stellaria media,
          Chenopodium album

F5+6   =  Atriplex spp., Rumex acetosella, Odontites
          verna, Arrhenatherum elatius, Carex pilulifera,
          Rumex spp., Avena fatua, small grasses
          (including Poa annua), (Spergula arvensis,
          Urtica urens, Plantago major),

F7+8   =  Ranunculus repens, Polygonum lapathifolium,
          Juncus sp., Montia fontana

F9+10  =  Ranunculus flammula, Carex pulicaris, Eleocharis
          sp.

Fx     =  Raphanus raphanistrum, Potentilla erecta,
          Polygonum aviculare, Polygonum convolvulus,
          Prunella vulgaris, Plantago lanceolata,
          Galium aparine, Tripleurospermum inodorum,
          Bromus mollis/secalinus, Sieglingia decumbens

ACIDITY
R2+3   =  Rumex acetosella, Ranunculus flammula, Montia
          fontana, Sieglingia decumbens, Carex
          pilulifera, (Spergula arvensis)

R4+5+6 =  Raphanus raphanistrum, Prunella vulgaris,
          Galium aparine, Tripleurospermum inodorum

R7+8   =  Stellaria media, Atriplex spp., Avena fatua,
          Arrhenatherum elatius

Rx     =  Ranunculus repens, Chenopodium album, Potentilla
          erecta, Polygonum aviculare, Polygonum
          convolvulus, Polygonum lapathifolium, Polygonum
          persicaria, Rumex spp., Odontites verna,
          Plantago lanceolata, Bromus mollis/secalinus,
          small grasses (including Poa annua), Eleocharis
          sp., Carex pulicaris, (Urtica urens, Plantago
          major)

NITROGEN
N2+3   =  Ranunculus flammula, Potentilla erecta, Rumex
          acetosella, Sieglingia decumbens

N4+5+6 =  Montia fontana, Raphanus raphanistrum, Rumex
          spp., Carex pilulifera, Tripleurospermum
          inodorum, (Spergula arvensis, Plantago major)

N7+8   =  Chenopodium album, Atriplex, Polygonum
          persicaria, Arrhenatherum elatius, Stellaria
          media, Polygonum lapathifolium, Galium aparine,
          small grasses (including Poa annua), (Urtica
          urens)

Nx     =  Ranunculus repens, Polygonum aviculare,
          Polygonum convolvulus, Odontites verna, Prunella
          vulgaris, Plantago lanceolata, Avena fatua,
          Bromus mollis/secalinus

Species in brackets occur in <10%, but >5% of the samples.
```

Table 10.9 Edaphic categories according to Runhaar (Runhaar *et al.* 1987; see also Table 9.2).

```
WET/MODERATE-HIGH NUTRIENT AVAILABILITY    (12+17+18+20+27+28)
      Eleocharis sp.

WET/LOW-MODERATE NUTRIENT AVAILABILITY     (22+23+27)
      Ranunculus flammula

WET/LOW NUTRIENT AV., MODERATELY ACID TO NEUTRAL    (22)
      Carex pulicaris

WET-MOIST/MODERATELY HIGH NUTRIENT AV.     (27+28+47+48)
      Ranunculus repens

WET-MOIST/MODERATE NUTRIENT AVAILABILITY   (17+27+47)
      Montia fontana

WET-MOIST/LOW NUTRIENT AV., ACID-WEAKLY ACID    (21+22+41+42)
      Potentilla erecta

MOIST/HIGH NUTRIENT AVAILABILITY                (48)
      Atriplex spp., Polygonum aviculare, Polygonum
      lapathifolium, Polygonum persicaria, Rumex spp.,
      Tripleurospermum inodorum, (Plantago major)

MOIST/MODERATE-HIGH NUTRIENT AVAILABILITY       (47+48)
      Arrhenatherum elatius

MOIST/MODERATE NUTRIENT AVAILABILITY            (47)
      Prunella vulgaris

MOIST/MODERATE-LOW NUTRIENT AVAILABILITY        (43)
      Odontites verna

MOIST-DRY/HIGH NUTRIENT AVAILABILITY            (48+68)
      Stellaria media, Chenopodium album, small
      grasses (including Poa annua), (Urtica urens)

MOIST-DRY/MODERATE-HIGH NUTRIENT AVAILABILITY   (48+69)
      Galium aparine

MOIST-DRY/MODERATE NUTRIENT AVAILABILITY        (47+67)
      Raphanus raphanistrum, Plantago lanceolata,
      Avena fatua

MOIST-DRY/LOW NUTRIENT AVAILABILITY             (41+42+61+62)
      Sieglingia decumbens, Carex pilulifera

MOIST-DRY/NUTRIENT INDIFFERENT                  (47+48+63+67+69)
      Polygonum convolvulus

DRY/MODERATE NUTRIENT AVAILABILITY              (61+62+67)
      Rumex acetosella

DRY/MODERATE-LOW NUTRIENT AVAILABILITY          (67)
      Bromus mollis/secalinus, (Spergula arvensis)

Species in brackets occur in <10%, but >5% of the samples.
```

Table 10.10   Weed species in Ellenberg's phytosociological Classes (Ellenberg 1979).

```
PHRAGMITETEA
Eleocharis sp.
SCHEUCHZERIO-CARICETEA
Ranunculus flammula, Carex pulicaris
ISOETO NANOJUNCETEA
Montia fontana
BIDENTETEA
Polygonum lapathifolium
CHENOPODIETEA
Stellaria media, Chenopodium album, Atriplex spp., Polygonum persicaria, Tripleurospermum inodorum, (Spergula arvensis, Urtica
urens)
SECALIETEA
Raphanus raphanistrum, Polygonum convolvulus, Avena fatua, Bromus mollis/secalinus
ARTEMISIETEA
Galium aparine
PLANTAGINETEA
Ranunculus repens, small grasses (including Poa annua), (Plantago major)
NARDO-CALLUNETEA
Potentilla erecta, Rumex acetosella, Sieglingia decumbens, Carex pilulifera
MOLINIO-ARRHENATHERETEA
Odontites verna, Prunella vulgaris, Plantago lanceolata, Arrhenatherum elatius
Species in brackets occur in <10%, but >5% of the samples.
```

Table 10.11   Preferred time of germination for the weed species according
to the Geigi Weed Tables (Häfliger and Brun-Hool 1968-1977).

	ANNUALS			PERENNIALS
	SPRING	*BOTH*	*AUTUMN*	
Ranunculus repens				*
Ranunculus flammula				*
Raphanus raphanistrum		*		
Montia fontana (no info)				
Stellaria media		*		
Chenopodium album	*			
Atriplex spp.	*			
Potentilla erecta				*
Polygonum aviculare	*			
Polygonum convolvulus	*			
Polygonum lapathifolium	*			
Polygonum persicaria	*			
Rumex acetosella				*
Rumex spp.				*
Odontites verna		*		
Prunella vulgaris				*
Plantago lanceolata				*
Galium aparine		*		
Tripleurospermum inodorum		*		
Avena fatua	*			
Bromus mollis/secalinus		*		
Sieglingia decumbens				*
Poa annua		*		
Arrhenatherum elatius				*
Juncus sp. (no info)				
Eleocharis sp.				*
Carex pilulifera				*
Carex pulicaris				*
Carex spp.				*

215

Table 10.12   Maximum flowering height of the weed species (after Clapham *et al.* 1962).

```
H1 = (1-30 cm)
     Potentilla erecta, Odontites verna, Prunella
     vulgaris, Poa annua, Carex pilulifera, Carex
     pulicaris.

H2 = (31-40 cm)
     Stellaria media, Sieglingia decumbens

H3 = (41-50 cm)
     Ranunculus flammula, Montia fontana, Plantago
     lanceolata

H4 = (51-60)
     Ranunculus repens, Raphanus raphanistrum,
     Tripleurospermum inodorum, Bromus mollis/secalinus,
     Eleocharis sp.

     (N.B. there are no species within the range 61-70 cm)

H5 = (71-80 cm)
     Polygonum persicaria

H6 = (81-90 cm)
     Avena fatua

H7 = (91-100 cm)
     Chenopodium album, Atriplex spp., Polygonum
     lapathifolium, Prunella vulgaris

H8 = (101-120 cm)
     Galium aparine, Arrhenatherum elatius

H9 = (121-200 cm)
     Polygonum aviculare, Polygonum convolvulus
```

Table 10.13   Summary results of DA-15-20.

```
Discriminant Analysis - Roman Assemblages
(using the Prehistoric assemblages as the control groups)

ANALYSIS          VARIABLES                        TRANSFORMATION

DA-15             F3.4, F5.6, F7.8, F9.10, Fx      square root
                  R2.3, R4.5.6, R7.8, Rx
                  N2.3, N4.5.6, N7.8, Nx
                  (weeds>10%) - Thornbrough
                  RESULTS: 5 samples classified
                           as Group A -29.4%
                           12 samples classified
                           as Group B -70.6%

DA-16             F3.4, F5.6, F7.8, F9.10, Fx      square root
                  R2.3, R4.5.6, R7.8, Rx
                  N2.3, N4.5.6, N7.8, Nx
                  Annuals, Perennials
                  (weeds> 5%) - Thornbrough
                  RESULTS: 4 samples classified
                           as Group A -23.5%
                           13 samples classified
                           as Group B -76.5%

DA-17             as DA-15, but for South Shields,
                  deposit 12236
                  RESULTS: 3 samples classified
                           as Group A - 9.4%
                           29 samples classified
                           as Group B -90.6%

DA-18             as DA-16, but for South Shields,
                  deposit 12236
                  RESULTS: 2 samples classified
                           as Group A - 6.3%
                           30 samples classified
                           as Group B -93.8%

DA-19             as DA-15, but for South Shields,
                  deposit 12176
                  RESULTS: 2 samples classified
                           as Group A -16.7%
                           10 samples classified
                           as Group B -83.3%

DA-20             as DA-16, but for South Shields,
                  deposit 12176
                  RESULTS: 1 sample  classified as
                           Group A - 8.3%
                           11 samples classified as
                           Group B -91.7%
```

216

Table 11.1   Soil associations within a 1 km radius of each site.

MURTON

60% Dunkeswick -   Typical stagnogley soil
                                          slowly permeable seasonally waterlogged
                                          fine loamy and fine loamy over clayey
                                          soils
40% Nercwys -   Stagnogleyic brown earth
                                          deep fine loamy soils with slowly
                                          permeable subsoils and slight seasonal
                                          waterlogging

DOD LAW

35% Alun -   Typical brown alluvial soil
                                          deep stoneless permeable coarse loamy
                                          soils
20% Dunkeswick -   Typical stagnogley soil
                                          slowly permeable seasonally waterlogged
                                          fine loamy and fine loamy over clayey
                                          soils
15% Anglezarke -   Humo-ferric podzol
                                          well drained very acid coarse loamy
                                          soils over sandstone with a bleached
                                          surface horizon
15% Newport 1 -   Typical brown sand
                                          deep well drained sandy and coarse
                                          loamy soils
15% Wick 1 -   Typical brown earth
                                          deep well drained coarse loamy and
                                          sandy soils

CHESTER HOUSE

70% Brickfield 3   Cambic   stagnogley soil
                                          slowly permeable seasonally waterlogged
                                          fine loamy, fine loamy over clayey and
                                          clayey soils
10% Wick 1 -   Typical brown earth
                                          deep well drained coarse loamy and
                                          sandy soils
20% unclassified   built-up area

THORPE THEWLES

60% Crewe -   Pelo-stagnogley soil
                                          slowly permeable seasonally waterlogged
                                          reddish clayey and fine loamy over
                                          clayey soils
40% Salop -   Typical stagnogley soil
                                          slowly permeable seasonally waterlogged
                                          reddish fine loamy over clayey, fine
                                          loamy and clayey soils

STANWICK

60% Dunkeswick -   Typical stagnogley soil
                                          slowly permeable seasonally waterlogged
                                          fine loamy and fine loamy over clayey
                                          soils
20% Wick 1 -   Typical brown earth
                                          deep well drained coarse loamy and
                                          sandy soils
10% Waltham -   Typical brown earth
                                          well drained fine loamy soils over
                                          limestone, locally deep
10% Dale -   Pelo-stagnogley soil
                                          slowly permeable seasonally waterlogged
                                          clayey, fine loamy over clayey and fine
                                          silty soils

ROCK CASTLE

30% Brickfield 2   Cambic stagnogley soil
                                          slowly permeable seasonally waterlogged
                                          fine loamy soils
30% Wick 1   Typical brown earth
                                          deep well drained coarse loamy and
                                          sandy soils
30% East Keswick 1   Typical brown earth
                                          deep well drained fine loamy soils
10% Wharfe -   Typical brown alluvial soil
                                          deep stoneless permeable fine loamy
                                          soils

Bibliography

Applebaum, S. 1954 The agriculture of the British early Iron Age, as exemplified at Figheldean Down, Wiltshire. *Proceedings of the Prehistoric Society* 20, 103–114.

Applebaum, S. 1972 Roman Britain. In H. P. R. Finberg (ed.) *The Agrarian History of England and Wales* I, part 2, AD 43 – 1042. Cambridge, Cambridge University Press.

Arnolds, E. and Maarel, E. van der 1979 De ecologische groepen in de Standaardlijst van der Nederlandse flora 1975. *Gorteria* 9, 303–312.

Bailiff, I. K. 1987 The thermoluminescence dating of the Iron Age pottery. In D. H. Heslop: *The Excavation of an Iron Age Settlement at Thorpe Thewles, Cleveland, 1980–1982*. London, Council for British Archaeology, Research Report No. 65, 71–72.

Bakels, C. C. 1981 De bewoningsgeschiedenis van de Maaskant I: plantenresten uit de Bronstijd en Romeinse tijd gevonden te Oss, IJsselstraat, prov. Noord-Brabant. *Analecta Praehistorica Leidensia* 13, 115–131.

Bakels, C. C. and Ham, R. W. J. M. van der 1981 Verkoold afval uit een Midden-Bronstijd en een Midden-IJzertijd nederzetting op de Hooidonksche Akkers, gem. Son en Breugel, prov. Noord-Brabant. *Analecta Praehistorica Leidensia* 13, 81–91.

Bannink, J. F., Leys, H. N. and Zonneveld, I. S. 1974 *Akkeronkruid Vegetatie als Indicator van het Milieu, in het bijzonder de Bodemgesteldheid*. Wageningen, Bodemkundige Studies 11.

Bartley, D. D., Chambers, C. and Hart-Jones, B. 1976 The vegetational history of parts of south and east Durham. *New Phytologist* 77, 437–468.

Behre, K. E. 1975 Wikingerzeitlicher Ackerbau in der Seemarsch bei Elisenhof, Schleswig-Holstein (Deutsche Nordseeküste). *Folia Quaternaria* 46, 49–62.

Behre, K. E. 1981a Pflanzenreste der Zeit um 1400 n. Chr. aus dem Lüneburger St.-Michaelis Kloster. *Nachrichten aus Niedersachsens Urgeschichte* 50, 321–327.

Behre, K. E. 1981b The interpretation of anthropogenic indicators in pollen diagrams. *Pollen et Spores* 23, part 2, 225–245.

Behre, K. E. 1986a Analysis of botanical macro-remains. In O. van de Plassche (ed.) *Sea-level Research: A Manual for the Collection and Evaluation of Data*. Norwich 1986, 413–433.

Behre, K. E. 1986b Ackerbau, Vegetation und Umwelt im Bereich früh- und hochmittelalterlicher Siedlungen im Flussmarschgebiet der unteren Ems. *Probleme der Küstenforschung im südlichen Nordseegebiet* 16, 99–125.

Beijerinck, W. 1947 *Zadenatlas der Nederlandsche Flora*. Wageningen, (Facsimile edition 1976, Amsterdam, Backhuys en Meesters).

Bennett, J. 1990 The Setting, Development and Function of the Hadrianic Frontier in Britain. PhD Thesis, University of Newcastle upon Tyne.

Berggren, G. 1969 *Atlas of Seeds, Part 2, Cyperaceae*, Stockholm, Swedish Natural Science Research Council.

Berggren, G. 1981 *Atlas of Seeds, Part 3, Salicaceae-Cruciferae*. Stockholm, Swedish Museum of Natural History.

Bidwell, P. 1989 South Shields. In C. Daniels (ed.) *The Eleventh Pilgrimage of Hadrian's Wall*, Newcastle upon Tyne, 83–89.

Bidwell, P. forthcoming The Roman Fort at South Shields. The Excavations of the Headquarters and the South West Gate.

Boardman, S. and Jones, G. 1990 Experiments on the effects of charring on cereal plant components. *Journal of Archaeological Science* 17, 1–11.

Bottema, S. 1984 The composition of modern charred seed assemblages. In W. van Zeist and W. A. Casparie (eds.) *Plants and Ancient Man*. Rotterdam, Balkema, 207–212.

Boyd, W. E. 1988 Cereals in Scottish antiquity. *Circaea* 5, No. 2, 101–110.

Bowman, A. K. and Thomas, J. D. 1983 *Vindolanda: The Latin Writing-Tablets*. London, Society for the Promotion of Roman Studies, Britannia Monograph Series No. 4.

Bradley, R. 1978 *The Prehistoric Settlement of Britain*. London, Routledge and Kegan Paul.

Bradley, R. 1984 *The Social Foundations of Prehistoric Britain*. London, Longman.

Braun-Blanquet, J. 1964 *Pflanzensoziologie*. Wien, Springer (third edition).

Breeze, D. J. 1980 Roman Scotland during the reign of Antonius Pius. W. S. Hanson and L. J. F. Keppie, *Roman Frontier Studies 1979*. Oxford, British Archaeological Reports, International Series 71 i, 45–60.

Breeze, D. J. 1982 *The Northern Frontiers of Roman Britain*. London, Batsford.

Breeze, D. J. 1984 Demand and supply on the northern frontier. R. Miket and C. Burgess (eds.) *Between and Beyond the Walls. Essays on the Prehistory and History of North Britain in Honour of George Jobey*. Edinburgh, John Donald Publishers Ltd., 264–286.

Breeze, D. J. and Dobson, B. 1985 Roman military deployment in north England. *Britannia* 16, 1–19.

Brenchley, W. E. and Warington, K. 1930 The weed seed population of arable soil. I: Numerical estimation of viable seeds and observations on their natural dormancy. *Journal of Ecology* 18, 235–272.

Buckland, P. C. 1978 Cereal production, storage and population: a caveat. In S. Limbrey and J. G. Evans (eds.) *The Effect of Man on the Landscape: the Lowland Zone*. London, Council for British Archaeology, Research Report No. 21, 43–45.

Butler, S. in press The regional palaeoenvironment. In O. Owen (ed.) *Southern Scottish Hillforts*. Edinburgh, Scottish Development Department.

Carruthers, W. 1986 The late Bronze Age midden at Potterne. *Circaea* 4, No. 1, 16–17.

Caseldine, A. E. 1989 The archaeobotanical evidence from a Roman corndrier at Dan-y-Graig, Glamorgan. Unpublished typescript.

Casey, J. 1982 Civilians and soldiers – friends, Romans, countrymen? In P. Clack and S. Haselgrove (eds.) *Rural Settlement in the Roman North*. Durham, Council for British Archaeology, Group 3, 123–132.

Cavers, P. B. and Harper, J. L. 1964 *Rumex crispus* L. and *Rumex obtusifolius* L. *Journal of Ecology* 52, 737–766.

Chambers, C. 1978 A radiocarbon dated pollen diagram from Valley Bog, on the Moor House National Nature Reserve. *New Phytologist* 80, 273–280.

Chambers, F. M. 1989 The evidence for early rye cultivation in north west Europe. In A. Milles, D. Williams, and N. Gardner (eds.) *The Beginnings of Agriculture*. Oxford, British Archaeological Reports, International Series 496, 165–175.

Champion, T., Gamble, C., Shennan, S. and Whittle, A. 1984 *Prehistoric Europe*. London, Academic Press.

Chapman, J. C. and Mytum, H. C. 1983 *Settlement in North Britain 1000 BC – AD 1000*. Oxford, British Archaeological Reports, British Series 118.

Chapman, S. B. 1964 The ecology of Coom Rigg Moss, Northumberland. *Journal of Ecology* 52, 299–315.

Clack, P. 1984 Excavations at Thornbrough Scar. *Archaeological Reports for 1983*, Durham, Universities of Durham and Newcastle upon Tyne, 43–44.

Clack, P. and Haselgrove, S. 1982 *Rural Settlement in the Roman North*. Durham, Council for British Archaeology, Group 3.

Clapham A. R., Tutin, T. G. and Warburg, E. F. 1962 *Flora of the British Isles*. Cambridge, Cambridge University Press (second edition).

Coggins, D. 1986 *Upper Teesdale. The Archaeology of a North Pennine Valley*. Oxford, British Archaeological Reports, British Series 150.

Coppock, J. T. 1971 *An Agricultural Geography of Great Britain*. London, G. Bell and Sons.

Corbet, G. B. 1971 Provisional distribution maps of British mammals. *Mammal Review* 1, No. 4/5, 95–142.

Cunliffe, B. W. 1983 The Iron Age of northern Britain: a view from the South. J. C. Chapman and H. C. Mytum (eds.) *Settlement in North Britain 1000 BC – AD 1000*. Oxford, British Archaeological Reports, British Series 118, 83–102.

Cunliffe, B. W. 1984 *Danebury – An Iron Age Hillfort in Hampshire*. Volume 1. The excavations 1964–1978: the site. London, Council for British Archaeology, Research Report No. 52.

Dagnelie, P. 1973 L'analyse factorielle. In R. H. Whittaker (ed.) *Ordination and Classification of Plant Communities*. Handbook of Vegetation Science, part V. The Hague, Junk Publishers, 223–248.

Davies, G. and Turner, J. 1979 Pollen diagrams from Northumberland. *New Phytologist* 82, 783–804.

Davies, M. S. and Hillman, G. C. 1988 Effects of soil flooding on growth and grain yield of populations of tetraploid and hexaploid species of wheat. *Annals of Botany* 62, 597–604.

Davies, R. W. 1971 The Roman military diet. *Britannia* 2, 122–142.

Dennell, R. W. 1972 The interpretation of plant remains: Bulgaria. In E. S. Higgs (ed.) *Papers in Economic History*. Cambridge, Cambridge University Press, 149–159.

Dennell, R. W. 1974 Prehistoric crop processing activities. *Journal of Archaeological Science* 1, 275–284.

Dennell, R. W. 1976 The economic importance of plant resources represented on archaeological sites. *Journal of Archaeological Science* 3, 229–247.

Dickson, C. A. 1970 The study of plant macrofossils in British Quaternary deposits. In D. Walker and R. G. West (eds.) *Studies in the Vegetational History of the British Isles*. Cambridge, Cambridge University Press, 233–254.

Dickson, C 1989 The Roman army diet in Britain and Germany. *Archaeobotanik. Dissertationes Botanicae* 133, 135–154.

Digby, P. G. N. and Kempton, R. A. 1987 *Multivariate Analysis of Ecological Communities*. London, Chapman and Hall.

Dobroruka, L. J. 1988 *A Field Guide in Colour to Mammals*. London, Octopus Books.

Donaldson, A. M. 1982 Botanical Report. In G. Jobey, The settlement at Doubstead and Romano-British Settlement on the coastal plain between Tyne and Forth. *Archaeologia Aeliana* 5th series, 10, 1–24.

Donaldson, A. M. 1983 Pollen Analysis. In D. Coggins, K. J. Fairless and C. E. Batey, Simy Folds: an early Medieval settlement site in Upper Teesdale, Co. Durham. *Medieval Archaeology* 27, 16–18.

Donaldson, A. M. and Turner, J. 1977 A pollen diagram from Hallowell Moss, near Durham city, U. K. *Journal of Biogeography* 4, 25–33.

Drewett, P. L. 1982 Later Bronze Age downland economy and excavations at Black Patch, East Sussex. *Proceedings of the Prehistoric Society* 48, 321–400.

Edwards, K. J. 1982 Man, space and the woodland edge – speculations on the detection and interpretation of human impact in pollen profiles. In S. Limbrey and M. Bell (eds.) *Archaeological Aspects of Woodland Ecology*. Oxford, British Archaeological Reports, International Series 146, 5–22.

Edwards, K. J. 1989 The cereal pollen record and early agriculture. In A. Milles, D. Williams and N. Gardner (eds.) *The Beginnings of Agriculture*. Oxford, British Archaeological Reports, International Series 496, 113–135.

Ellenberg, H. 1950 *Landwirtschaftliche Pflanzensoziologie* I: Unkrautgemeinschaften als Zeiger für Klima und Boden. Stuttgart/Ludwigsburg, Ulmer.

Ellenberg, H. 1979 Zeigerwerte der Gefässpflanzen Mitteleuropas. *Scripta Geobotanica* 9, Göttingen (second edition), 1–122.

Evans, J. G. 1975 *The Environment of Early Man in the British Isles*. London, Paul Elek.

Faechem, R. W. 1973 Ancient agriculture in the highland of Britain. *Proceedings of the Prehistoric Society* 39, 332–353.

Field, N. H., Matthews, C. L. and Smith, I. F. 1964 New neolithic sites in Dorset and Bedfordshire, with a note on the distribution of neolithic storage pits. *Proceedings of the Prehistoric Society* 30, 352–381.

Fitter, A. 1978 *An Atlas of the Wild Flowers of Britain and Northern Europe*. London, Collins.

Fowler, P. J. 1983 *The Farming of Prehistoric Britain*. Cambridge, Cambridge University Press.

Fox, C. 1932 *The Personality of Britain*. Its influence on habitat and invader in prehistoric and early historic times. Cardiff, National Museum of Wales.

Frere, S. 1978 *Britannia. A History of Roman Britain*. London, Routledge and Kegan Paul (second edition).

Froud-Williams, R. J. and Chancellor, R. J. 1982 A survey of grass weeds in cereals in central southern England. *Weed Research* 22, 163–171.

Gates, T. 1982a Farming on the frontier: Romano-British fields in Northumberland. In P. Clack and S. Haselgrove (eds.) *Rural Settlement in the Roman North*. Durham, Council for British Archaeology, Group 3, 21–42.

Gates, T. 1982b Excavations at Hallshill Farm, East Woodburn, Northumberland. *Archaeological Reports for 1981*, Durham, Universities of Durham and Newcastle upon Tyne, 7–9.

Gates, T. 1983 Unenclosed settlements in Northumberland. In J. C. Chapman and H. C. Mytum (eds.) *Settlement in North Britain 1000 BC – AD 1000*.

Oxford, British Archaeological Reports, British Series 118, 103–148.

Gates, T. forthcoming The excavations at Hallshill, Northumberland. *Archaeologia Aeliana.*

Gauch, H. G. 1984 *Multivariate Analysis in Community Ecology.* Cambridge, Cambridge University Press.

Gidney, L. J. 1989 The animal bone. In B. E. Vyner and R. Daniels, Further investigation of the Iron Age and Romano-British settlement at Catcote, Hartlepool, Cleveland, 1987. *Durham Archaeological Journal* 5, 11–34.

Gill, N. T. and Vear, K. C. 1980 *Agricultural Botany.* Vol. 2 Monocotyledonous Crops. London, Duckworth.

Gillespie, R. 1984 *Radiocarbon User's Handbook.* Oxford, Oxford University Committee for Archaeology, Monograph No. 3.

Godwin, H. 1975 *History of the British Flora.* Cambridge, Cambridge University Press (second edition).

Green, F. J. 1981 Iron Age, Roman and Saxon crops: the archaeological evidence from Wessex. In M. Jones and G. Dimbleby (eds.) *The Environment of Man: the Iron Age to the Anglo-Saxon Period.* Oxford, British Archaeological Reports, British Series 87, 129–153.

Gregg, S. A. 1988 *Foragers and Farmers.* Population Interaction and Agricultural Expansion in Prehistoric Europe. Chicago/London, Chicago University Press.

Greig, J. 1988a The interpretation of some Roman well fills from the Midlands of England. In H. Küster (ed.) *Der Prähistorische Mensch und Seine Umwelt.* Forschungen und Berichte zur Vor- und Frühgeschichte in Baden-Württemberg, Band 31, Stuttgart, Theiss, 367-378.

Greig, J. 1988b Some evidence of the development of grassland plant communities. In M. Jones (ed.) *Archaeology and the Flora of the British Isles.* Oxford, Oxford University Committee for Archaeology, Monograph No. 14, 39–54.

Greig, J. 1990 The prehistoric riverside settlement at Runnymede, Surrey: the botanical story (Final Report). *Ancient Monuments Laboratory Report* No. 40/90.

Grieg-Smith, P. 1948 *Urtica* L. genus. *Journal of Ecology* 36, 339–355.

Groenman-van Waateringe, W. 1980 Urbanization and the north-west frontier of the Roman Empire. W. S. Hanson and L. J. F. Keppie (eds.) *Roman Frontier Studies 1979.* Oxford, British Archaeological Reports, International Series 71 iii, 1037–1044.

Groenman-van Waateringe, W. 1988 New trends in palynoarchaeology in Northwest Europe or the frantic search for local pollen data. In R. E. Webb (ed.) *Recent developments in Environmental Analysis in Old and New World Archaeology.* Oxford, British Archaeological Reports, International Series 416, 1–19.

Groenman-van Waateringe, W. 1989 Food for soldiers, food for thought. In J. C. Barrett, A. P. Fitzpatrick and L. Macinnes (eds.) *Barbarians and Romans in North-West Europe.* Oxford, British Archaeological Reports, International Series 471, 96–107.

Grose, J.D. 1957. *Flora of Wiltshire.* Devizes.

Haaster, H. van forthcoming Weeds, a comparative study of recent vegetation relevés and palaeobotanical information. Proceedings of the Symposium of the International Workgroup of Palaeoethnobotany, Nitra, 1989.

Häfliger, E. and Brun-Hool, J. 1968–1977 *Ciba-Geigy Weed Tables.* A synoptic presentation of the flora accompanying agricultural crops. Basle, Documenta

Ciba-Geigy

Harding, D. W. 1979 Air survey in the Tyne-Tees region, 1969–79. In Higham, N. J. (ed.) *The Changing Past.* Manchester, Manchester University Department of Extra Mural Studies, 21–30.

Harding, D. W. 1982 *Later Prehistoric Settlement in South-East Scotland.* Edinburgh, University of Edinburgh, Occasional Paper No. 8.

Harper, J. L. 1957 *Ranunculus acris* L., *R. repens* L. and *R. bulbosus* L. *Journal of Ecology* 45, 289–342.

Harper, J. L. 1977 *The Population Biology of Plants.* London, Academic Press.

Hartley, B. and Fitts, L. 1988 *The Brigantes.* Gloucester, Alan Sutton, Peoples of Roman Britain Series.

Haselgrove, C. 1982 Indigenous settlement patterns in the Tyne-Tees Lowlands. In P. Clack and S. Haselgrove (eds.) *Rural Settlement in the Roman North.* Durham, Council for British Archaeology, Group 3, 57–104.

Haselgrove, C. 1984 The later pre-Roman Iron Age between the Humber and the Tyne. In P. R. Wilson, R. F. J. Jones and D. M. Evans (eds.) *Settlement and Society in the Roman North.* Bradford, School of Archaeological Sciences, 9–25.

Haselgrove, C. 1990 Stanwick. *Current Archaeology* 119, 380–385.

Haselgrove, C. and Turnbull, P. 1983 *Stanwick: excavation and fieldwork; interim report 1981–83.* Durham, University of Durham, Department of Archaeology, Occasional Paper 4.

Haselgrove, C. and Turnbull, P. 1984 *Stanwick: excavation and fieldwork; second interim report.* Durham, University of Durham, Department of Archaeology, Occasional Paper 5.

Haselgrove, C., Fitts, L., Turnbull, P. and Willis, S. 1988 Excavations in Tofts Field, Stanwick, North Yorkshire, 1988. Unpublished typescript.

Haselgrove, C., Fitts, L., Turnbull, P. and Willis, S. 1989 Excavations in The Tofts, Stanwick, North Yorkshire, 1989. Unpublished typescript.

Hedges, R. E. M. and Gowlett, J. A. J. 1986 Radiocarbon dating by accelerator mass spectrometry. *Nature* 254, No. 1, 100–107.

Helbaek, H. 1952 Early crops in southern Britain. *Proceedings of the Prehistoric Society* 18, 194–233.

Helbaek, H. 1964 The Isca grain. *New Phytologist* 63, 158–164.

Helbaek, H. 1977 The Fyrkat grain, a geographical and chronological study of rye. In O. Olsen and H. Schmidt (eds.) *Fyrkat, en jysk vikingeborg.* I. Borgen og bebyggelsen. Nordiske Fortidsminder Serie B, Bd. 3, 1–44.

Heslop, D. H. 1987 *The Excavation of an Iron Age Settlement at Thorpe Thewles, Cleveland, 1980–1982.* London, Council of British Archaeology, Research Report No. 65.

Higham, N. 1986 *The Northern Counties to AD 1000.* London, Longman.

Hill, P. 1987 Traprain Law: The Votadini and the Romans. *Scottish Archaeological Review* 4, part 2, 85–91.

Hillman, G. 1978 Remains of crops and other plants from Roman Camarthen (Church Street). *Cambrian Archaeological Association, Monographs and Collections* Vol. 1, 107–112.

Hillman, G. 1981a Reconstructing crop husbandry practices from charred remains of crops. In R. Mercer (ed.) *Farming Practice in British Prehistory.* Edinburgh, Edinburgh University Press, 123–162.

Hillman, G. 1981b Possible evidence of grain-roasting at Iron-Age Pembrey. Appendix to G. Williams, Survey and excavation on Pembrey Mountain. *The Camarthenshire Antiquary*, 25–32.

Hillman, G. 1982 Crop husbandry at the Medieval farmstead, Cefn Graeanog: reconstructions from charred plant remains. Appendix 4 to R. S. Kelly, The excavation of a Medieval farmstead at Cefn Graeanog, Clynnog, Gwynnedd. *The Bulletin of the Board of Celtic Studies* 29, part 4, 901–907.

Hillman, G. 1984 Interpretation of archaeological plant remains: the application of ethnographic models from Turkey. In W. van Zeist and W. A. Casparie (eds.) *Plants and Ancient Man*. Rotterdam, Balkema, 1–41.

Hillman, G. forthcoming a Crop husbandry in the late Iron Age at Breiddin Hillfort: the charred cereal remains from a second century BC hut. In C. R. Musson, W. J. Britnell and A. G. Smith (eds.) The Breiddin Hillfort, Powys: excavations 1969–76. *Cambrian Archaeological Association, Monographs and Collections.*

Hillman, G. forthcoming b Crop husbandry at Cefn Graeanog. In R. B. White (ed.) Cefn Graeanog II. An Iron Age, Roman and post-Roman Farmstead in Arvon, Gwynedd. *Cambrian Archaeological Association, Monographs and Collections.*

Hingley, R. 1989 *Rural Settlement in Roman Britain*. London, Seaby.

Hinton, M.P. 1991 (for 1990) Weed associates of recently grown *Avena strigosa* Schreber from Shetland, Scotland. *Circaea* 8, Number 1, 49-54.

Hodgson, G. W. I. 1967 Summary of report on animal remains. In G. Jobey: Excavation at Tynemouth Priory and Castle. *Archaeologia Aeliana* 4th series, 45, 33–104.

Hodgson, G. W. I. 1968 A comparative account of the animal remains from Corstopitum and the Iron Age site of Catcote, near Hartlepool, County Durham. *Archaeologia Aeliana* 4th series, 46, 127–162.

Hodgson, G. W. I. 1970 Report on faunal remains. In G. Jobey: An Iron Age settlement and homestead at Burradon, Northumberland. *Archaeologia Aeliana* 4th series, 48, 51–95.

Hodgson, G. W. I. 1973 The faunal remains. In G. Jobey: A native settlement at Hartburn and the Devil's Causeway, Northumberland. *Archaeologia Aeliana* 5th series, 1, 11–53.

Holbrook, N. 1988 The settlement at Chester House, Northumberland. *Archaeologia Aeliana* 5th series, 16, 47–59.

Holzner, W. 1978 Weed species and weed communities. *Vegetatio* 38, 13–20.

Hubbard, C. E. 1984 *Grasses. A guide to their Structure, Identification, Uses and Distribution in the British Isles*. Revised by J. C. E. Hubbard. Middlesex, Penguin Books.

Huntley, J. P. 1989 The plant remains. In B. E. Vyner and R. Daniels: Further investigation of the Iron Age and Romano-British settlement site at Catcote, Hartlepool, Cleveland, 1987. *Durham Archaeological Journal* 5, 11–34.

Hutchinson, C. S. and Seymour, G. B. 1982 *Poa annua* L. *Journal of Ecology* 70, 887–901.

Inman, R., Brown, D. R., Goddard, R. E. and Spratt, D. A. 1985 Roxby Iron Age settlement and the Iron Age in North-East Yorkshire. *Proceedings of the Prehistoric Society* 51, 181–213.

Jacomet, S. 1980 Botanische Makroreste aus der neolithischen Seeufersiedlungen des Areals 'Pressehaus Ringier' in Zürich (Schweiz). Stratigraphische und vegetationskundliche Auswertung. *Vierteljahrschrift der Naturforschenden Gesellschaft in Zürich* 125/2, 73–163.

Jacomet, S. 1987a *Praehistorische Getreidefunde*. Basel, Botanisches Institut der Universität.

Jacomet, S. 1987b Ackerbau, Sammelwirtschaft und Umwelt der Egolzwiler und Cortaillod Siedlungen. *Berichte der Zürcher Denkmalpflege*, Monographien 3, 144–166 and 234–241.

Jacomet, S., Felice, N. and Füzesi, B. 1988 Verkohlte Samen und Früchte aus der hochmittelalterlichen Grottenburg Riedfluh bei Eptingen, Kanton Baselland (Nordwest-Schweiz). *Schweizer Beiträge zur Kulturgeschichte und Archäologie des Mittelalters* 15, 169–243.

Jaquat, C. 1986 Römerzeitliche Pflanzenfunde aus Oberwinterthur (Kanton Zürich, Schweiz). *Berichte Zürcher Denkmalpflege*, Monographien 2, 240–264.

Jarvis, R. A., Bendelow, V. C., Bradley, R. I., Carroll, D. M., Furness, R. R., Kilgour, I. N. L. and King, S. J. 1984 *Soils and their Use in Northern England*. Harpenden, Soil Survey of England and Wales, Bulletin No. 10.

Jobey, G. 1966 Homesteads and settlements of the frontier area. C. Thomas (ed.) *Rural Settlement in Roman Britain*. London, Council for British Archaeology, Research Report No. 7, 1–14.

Jobey, I. & G. 1987 Prehistoric, Romano-British and later remains on Murton High Crags, Northumberland. *Archaeologia Aeliana* 5th series, 15, 151–198.

Jones, G. E. M. 1983a *The use of ethnographic and ecological models in the interpretation of archaeological plant remains: case studies from Greece*. PhD Thesis, University of Cambridge.

Jones, G. E. M. 1983b The ethnoarchaeology of crop processing: seeds of a middle-range methodology. *Archaeological Review from Cambridge* 2, part 2, 17–26.

Jones, G. E. M. 1984 Interpretation of archaeological plant remains: Ethnographic models from Greece. In W. van Zeist and W. A. Casparie (eds.) *Plants and Ancient Man*. Rotterdam, Balkema, 43–61.

Jones, G. E. M. 1987 A statistical approach to the archaeological identification of crop processing. *Journal of Archaeological Science* 14, 311–323.

Jones, G. E. M. 1989 The application of present-day cereal processing studies to charred archaeobotanical remains. *Circaea* 6, No. 2, 91–96.

Jones, G. E. M. forthcoming Weed phytosociology and crop husbandry. Identifying a contrast between ancient and modern practice. *Review for Palaeobotany and Palynology.*

Jones, M. K. 1981 The development of crop husbandry. M. Jones and G. Dimbleby (eds.) *The Environment of Man: the Iron Age to the Anglo-Saxon Period*. Oxford, British Archaeological Reports, British Series 87, 95–127.

Jones, M. K. 1984a Regional patterns in crop production. In B. Cunliffe and D. Miles (eds.) *Aspects of the Iron Age in Central Southern Britain*. Oxford, Oxford University Committee for Archaeology, Monograph No. 2, 120–125.

Jones, M. K. 1984b *The ecological and cultural implications of carbonised seed assemblages from selected archaeological contexts in southern Britain*. DPhil Thesis, University of Oxford.

Jones, M. K. 1985 Archaeobotany beyond subsistence reconstruction. In G. W. Barker and C. Gamble (eds.) *Beyond Domestication in Prehistoric Europe*. London, Academic Press, 107–128.

Jones, M. K. 1986 The carbonised plant remains. In D. Miles (ed.) *Archaeology at Barton Court Farm, Abingdon, Oxon*. London, Council for British Archaeology, Research Report 50, Fiche 9:A1–9:B5.

Jones, M. K. 1988a The phytosociology of early arable weed communities with special reference to southern England. In H. Küster (ed.) *Der Prähistorische Mensch und Seine Umwelt*. Forschungen und Berichte zur Vor- und Frühgeschichte in Baden-Württemberg, Band 31, Stuttgart, Theiss, 43–51.

Jones, M. K. 1988b The arable field: a botanical battleground. In M. K. Jones (ed.) *Archaeology and the Flora of the British Isles*. Oxford, Oxford University Committee for Archaeology, Monograph No. 14, 86–92.

Jones, R. 1966 Effect of seed-bed preparation on the weed flora of spring barley. *Proceedings of the Eighth British Weed Control Conference* 1, 227–228.

Kay, Q. O. N. 1969 The origin and distribution of diploid and tetraploid *Tripleurospermum inodorum* (L.) Schultz Bip. *Watsonia* 7, part 3, 130–141.

Kay, Q. O. N. 1971a *Anthemis cotula* L. *Journal of Ecology* 59, 623–636.

Kay, Q. O. N. 1971b *Anthemis arvensis* L. *Journal of Ecology* 59, 637–648.

Keepax, C. 1977 Contamination of archaeological deposits by seeds of modern origin with special reference to flotation machines. *Journal of Archaeological Science*, 4, 221–229.

Kenward, H. K. 1979 The insect remains. In H. K. Kenward and D. Williams (eds.) *Biological Evidence from the Roman Warehouses in Coney Street*. The Past Environment of York 14/2. London, Council for British Archaeology, 62–78.

King, A. C. 1978 A comparative survey of bone assemblages from Roman sites in Britain. *Bulletin of the Institute of Archaeology* 15, 207–232.

King, A. C. 1984 Animal bones and dietary identity of military and civilian groups in Roman Britain, Germany and Gaul. In T. F. C. Blagg and A. C. King (eds.) *Military and Civilian in Roman Britain*. Oxford, British Archaeological Reports, British Series 136, 187–217.

King, L. J. 1966 *Weeds of the World: Biology and Control*. London, Leonard Hill.

Kirby, R. H. 1963 *Vegetable Fibres: Botany, Cultivation, and Utilization*. London, Leonard Hill.

Knörzer, K. H. 1964 Über die Bedeutung von Untersuchungen subfossiler pflanzlicher Grossreste. *Bonner Jahrbücher* 164, 202–214.

Knörzer, K. H. 1971a Urgeschichtliche Unkräuter im Rheinland. Ein Beitrag zur Entstehungsgeschichte der Segetalgesellschaften. *Vegetatio* 23, 89–111.

Knörzer, K. H. 1971b Eisenzeitliche Pflanzenfunde im Rheinland. *Bonner Jahrbücher* 171, 40–58.

Knörzer, K. H. 1973a Die pflanzlichen Grossreste. In W. Göbel, K. H. Knörzer, J. S. Schalich, R. Schütrumpf and P. Stehli, Naturwissenschaftliche Untersuchungen an einer spälhallstattzeitlichen Fundstelle bei Langweiler, Kr. Düren. *Bonner Jahrbücher* 173, 289–315.

Knörzer, K. H. 1973b Römerzeitliche Pflanzenreste aus einem Brunnen in Butzbach (Hessen). *Saalburg Jahrbuch* 30, 71–111.

Knörzer, K. H. 1976 Späthallstattzeitliche Pflanzenfunde bei Bergheim, Erftkreis. *Rheinische Ausgrabungen* 17, 151–185.

Knörzer, K. H. 1984a Pflanzenfunde aus fünf eisenzeitlichen Siedlungen im südlichen Niederrheingebiet. *Bonner Jahrbücher* 184, 285–315.

Knörzer, K. H. 1984b Veränderungen der Unkrautvegetation auf rheinische Bauernhöfen seit der Römerzeit. *Bonner Jahrbücher* 184, 479–503.

Knörzer, K. H. 1987 Geschichte der Synanthropen Vegetation von Köln. *Kölner Jahrbuch für Vor- und Frühgeschichte* 20, 271–388.

Körber-Grohne, U. 1967 *Geobotanische Untersuchungen auf der Feddersen Wierde*. Wiesbaden, Steiner.

Körber-Grohne, U. 1979 *Nutzpflanzen und Umwelt im römischen Germanien*. Stuttgart, Kleine Schriften zur Kenntnis der römischen Besetzungsgeschichte Südwestdeutschlands, No. 21.

Körber-Grohne, U. 1987 *Nutzpflanzen in Deutschland. Kulturgeschichte und Biologie*. Stuttgart, Theiss.

Küster, H. 1985a Neolithische Pflanzenreste aus Hochdorf, Gemeinde Eberdingen (Kreis Ludwigsburg). *Forschungen und Berichte zur Vor- und Frühgeschichte in Baden-Württemberg* 19, 15–83.

Küster, H. 1985b Herkunft und Ausbreitungsgeschichte einer Secalietea-Arten. *Tuexenia* 5, 89–98.

Küster, H. 1988 Urnenfelderzeitliche Pflanzenreste aus Burkheim, Gemeinde Vogtsburg, Kreis Breisgau-Hochschwarzwald (Baden-Württemberg). In H. Küster (ed.) *Der Prähistorische Mensch und Seine Umwelt*. Forschungen und Berichte zur Vor- und Frühgeschichte in Baden-Württemberg, Band 31, Stuttgart, Theiss, 261–280.

Küster, H. 1989 Phytosociology and Archaeobotany. Lecture presented at the 10th Anniversary Conference of the Association of Environmental Archaeology, London.

Lamb, H. H. 1981 Climate from 1000 BC to 1000 AD. In M. Jones and G. Dimbleby (eds.) *The Environment of Man: the Iron Age to the Anglo-Saxon Period*. Oxford, British Archaeological Reports, British Series 87, 53–65.

Lambrick, G. and Robinson, M. 1988 The development of floodplain grassland in the Upper Thames Valley. In M. Jones (ed.) *Archaeology and the Flora of the British Isles*. Oxford, Oxford University Committee for Archaeology, Monograph No. 14, 55–75.

Lange, A. G. 1988 *Plant Remains from a Native Settlement at the Roman Frontier: De Horden near Wijk bij Duurstede. A Numerical Approach*. Rijksuniversiteit Groningen, PhD Thesis.

Lischka, J. J. 1975 Broken K revisited: a short description of factor analysis. *American Antiquity*, 40, 220–222.

Lundström-Baudais, K. 1984 Palaeo-ethnobotanical investigation of plant remains from a Neolithic lakeshore site in France: Clairvaux, Station III. In W. van Zeist and W. A. Casparie (eds.) *Plants and Ancient Man*. Rotterdam, Balkema, 293–305.

Lynch, A. and Paap, N. 1982 Untersuchungen an botanische Funden aus der Lübecker Innenstadt - Ein Vorbericht. *Lübecker Schriften zur Archäologie und Kulturgeschichte* 6, 339–360.

Mackney, D. 1974 *Soil Type and Land Capability*. Harpenden, Soil Survey of England and Wales, Technical Monograph No. 4.

Maguire, D. J. 1983 The identification of agricultural activity using pollen analysis. In M. Jones (ed.) *Integrating the Subsistence Economy*. Oxford, British

Archaeological Reports, International Series 181, 5–18.

Manby, T. G. 1980 Bronze Age settlement in Eastern Yorkshire. In J. Barrett and R. Bradley (eds.) *The British Later Bronze Age*. Oxford, British Archaeological Reports, British Series 83, 307–370.

Manning, W. H. 1975 Economic influences on land use in the military areas of the Highland Zone during the Roman period. J. G. Evans, S. Limbrey and H. Cleere (eds.) *The Effect of Man on the Landscape: the Highland Zone*. London, Council for British Archaeology, Research Report No. 11, 112–116.

Mannion, A. M. 1978 Late Quaternary deposits from Linton Loch, Southeast Scotland. I. Absolute and relative pollen analysis of limnic deposits. *Journal of Biogeography* 5, 193–206.

Mannion, A. M. 1979 A pollen-analytical investigation at Threepwood Moss. *Transactions of the Botanical Society of Edinburgh* 43, 105–114.

Maxwell, G. S. 1980 The native background to the Roman occupation of Scotland. W. S. Hanson and L. J. F. Keppie (eds.) *Roman Frontier Studies 1979*. Oxford, British Archaeological Reports, International Series 71 i, 1–13.

McNaughton, I. H. and Harper, J. L. 1964 *Papaver* L. genus. *Journal of Ecology* 52, 767–793.

Miket, R. and Burgess, C. 1984 *Between and Beyond the Walls. Essays on the Prehistory and History of North Britain in Honour of George Jobey*. Edinburgh, John Donald Publishers.

Miller, N. F. and Smart, T. L. 1984 Intentional burning of dung as fuel: a mechanism for the incorporation of charred seeds into the archaeological record. *Journal of Ethnobiology* 4, part 1, 15–28.

Miller, T. E. 1987 Systematics and evolution. In F. G. H. Lupton (ed.) *Wheat Breeding*. London, Chapman and Hall, 1–30.

Millett, M. 1983 Excavations at Cowdery's Down, Basingstoke, Hampshire, 1978–81. *Archaeological Journal* 140, 151–279.

Millett, M. 1990 *The Romanization of Britain*. Cambridge, Cambridge University Press.

Moffett, L. 1989 Economic activities at Rocester, Staffordshire, in the Roman, Saxon and Medieval periods; the evidence from the charred plant remains. *Ancient Monuments Laboratory Report* No. 15/89.

Mook, W. G. and Waterbolk, H. T. 1985 *Radiocarbon Dating*. Handbook for Archaeologists No. 3. Strasbourg, European Science Foundation.

Murphy, P. 1984 The charred cereals. In P. Crummy, *Excavations at Lion Walk, Balkerne Lane, and Middleborough, Colchester, Essex*. Colchester Archaeological Reports 3, Colchester, 110.

Murphy, P. forthcoming Plant macrofossils. In P. Crummy, The Excavations at Culver Steet, Colchester.

New, J. K. 1961 *Spergula arvensis* L. *Journal of Ecology* 49, 205–215.

Nilsson, O. and Hjelmqvist, H. 1967 Studies on the nutlet structure of South Scandinavian species of *Carex*. *Botaniska Notiser* 120, 460–485.

Norusis, M. J. 1985 *SPSSx Advanced Statistics Guide*. Chicago, SPSS Inc.

Oberdorfer, E. 1962 *Pflanzensoziologische Exkursionsflora für Süddeutschland und die angrenzenden Gebiete*. Stuttgart, Ulmer (second edition).

O'Connor, T. P. 1989 *Bones from the General Accident Site*. The Archaeology of York 15/2. London, Council for British Archaeology.

Paap, N. 1984 Botanische Analysen in Lübeck – Eine Zwischenbilanz. *Lübecker Schriften zur Archäologie und Kulturgeschichte* 8, 49–55.

Pals, J. P. 1984 Plant remains from Aartswoud, a neolithic settlement in the coastal area. In W. van Zeist and W. A. Casparie (eds.) *Plants and Ancient Man*. Rotterdam, Balkema, 313–321.

Pals, J. P. 1987 Reconstruction of landscape and plant husbandry. In W. Groenman-van Waateringe and L. H. van Wijngaarden-Bakker (eds.) *Farm Life in a Carolingian Village*. Assen.

Pals, J. P., and Geel, B. van 1976 Rye cultivation and the presence of cornflowers (*Centaurea cyanus* L.). *Berichten van de Rijksdienst voor het Oudheidkundig Bodemonderzoek* 26, 199–204.

Payne, F. G. 1949 The plough in ancient Britain. *The Archaeological Journal* 104, 82–111.

Pearson, G. W. and Stuiver, M. 1986 High-precision calibration of the radio-carbon time scale, 500–2500 BC. *Radiocarbon* 28, No. 2B, 839–862.

Pearson, G. W., Pilcher, J. R., Baillie, M. G. L., Corbett, D. M. and Qua, F. 1986 High-precision C14 measurement of Irish oaks to show the natural C14 variations from AD 1840 – 5210 BC. *Radiocarbon* 28, no. 2B, 911–934.

Pearson, M. C. 1960 Muckle Moss, Northumberland. *Journal of Ecology* 48, 647–666.

Percival, J. 1921 *The Wheat Plant*. London, Duckworth (1974 facsimile edition).

Pfitzenmeyer, C. D. C. 1962 *Arrhenatherum elatius* (L.) Beauv. ex. J. and C. Presl. *Journal of Ecology* 50, 235–245.

Piggott, S. 1958 Native economies and the Roman occupation of the north. I. A. Richmond (ed.) *Roman and Native in North Britain*. London, Nelson, 1–27.

Plicht, J. van der and Mook, W. G. 1987 Automatic radiocarbon calibration: illustrative examples. *Palaeohistoria* 29, 173–182.

Pollard, F. and Cussans, G. W. 1976 The influence of tillage on the weed flora of four sites sown to successive crops of spring barley. *Proceedings of the 1976 British Crop Protection Conference – Weeds*, 1019–1028.

Rackham, D. J. 1978 Skeletal material. In G. Jobey: Iron Age and Romano-British settlements on Kennel Hall Knowe, North Tynedale, Northumberland. *Archaeologia Aeliana* 5th series, 6, 1–28.

Rackham, D. J. 1982a Skeletal material. In G. Jobey: The settlement at Doubstead and Romano-British settlement on the coastal plain between Tyne and Forth. *Archaeologia Aeliana* 5th series, 10, 1–24.

Rackham, D. J. 1982b The faunal remains. In C. C. Haselgrove and V. L. Allon: An Iron Age settlement at West House, Coxhoe, County Durham. *Archaeologia Aeliana* 5th series, 10, 25–51.

Rackham, D. J. 1987 The animal bone. In D. H. Heslop *The Excavation of an Iron Age Settlement at Thorpe Thewles, Cleveland, 1980–1982*. London, Council of British Archaeology, Research Report No. 65, 99–109.

Rees, S. E. 1979 *Agricultural Implements in Prehistoric and Roman Britain*. Oxford, British Archaeological Reports, British Series 69.

Renfrew, J.M. 1973 *Palaeoethnobotany*. London, Methuen.

Reynolds, P. 1981 Deadstock and Livestock. In R. Mercer (ed.) *Farming Practice in British Prehistory*. Edinburgh, Edinburgh University Press, 97–122.

Reynolds, P. J. 1987 *Butser Ancient Farm Yearbook*

1986. Butser Ancient Farm Project Trust.

Reynolds, P. J. 1988 *Butser Ancient Farm Yearbook 1987*. Butser Ancient Farm Project Trust.

Reynolds, P. J. 1989 *Butser Ancient Farm Yearbook 1988*. Butser Ancient Farm Project Trust.

Rivet, A. L. F. 1969 Social and economic aspects. A. F. L. Rivet (ed.) *The Roman Villa in Britain*. London, Routledge and Kegan Paul, 173–216.

Rivet, A. L. F. and Smith, C. 1979 *The Place-Names of Roman Britain*. London, Batsford.

Roberts, B. K., Turner, J. and Ward, P. F. 1973 Recent forest history and land use in Weardale, Northern England. In H. J. B. Birks and R. G. West (eds.) *Quaternary Plant Ecology*. Symposia of the British Ecological Society No. 14. Oxford, Blackwell, 207–221.

Robinson, M. 1989 Seeds and other plant macrofossils. In P. Ashbee, M. Bell and E. Proudfoot, *Wilsford Shaft: excavations 1960–62*. London, English Heritage Archaeological Report No. 11, 78–90.

Runhaar, J., Groen, C. L. G., Meijden, R. van der and Stevens, R. A. M. 1987 Een nieuwe indeling in ecologische groepen binnen de Nederlandse flora. *Gorteria* 13, 276–359.

Sagar, G. R. and Harper, J. L. 1964 *Plantago major* L., *P. media* L., and *P. lanceolata* L. *Journal of Ecology* 52, 189–221.

Salway, P. 1981 *Roman Britain* Oxford, Clarendon Press.

Shennan, I. and Innes, J. B. 1986 Late Devensian and Flandrian environmental changes at The Dod, Borders region. *Scottish Archaeological Review* 4, part 1, 17–26.

Shennan, S. 1988 *Quantifying Archaeology*. Edinburgh, Edinburgh University Press.

Shirlaw, D. W. G. 1966 *An Agricultural Geography of Great Britain*. Oxford, Pergamon Press.

Silverside, A. J. 1977 *A Phytosociological Survey of British Arable Weeds and Related Communities*. University of Durham, PhD Thesis.

Simmonds, N. W. 1945 *Polygonum* L. ex Gaertn., *P. persicaria* L. and *P. lapathifolium*. *Journal of Ecology* 33, 117–139.

Smith, C. 1985 Dod Law 1984. *Archaeological Reports* 8, Durham, Universities of Durham and Newcastle upon Tyne, 26–29.

Smith, C. 1986 Dod Law 1985. *Archaeological Reports* 9, Durham, Universities of Durham and Newcastle upon Tyne, 35–36.

Smith, C. 1990 Excavations at Dod Law West Hillfort, Northumberland. *Northern Archaeology* 9, 1–55.

Smith, L. P. 1984 *The Agricultural Climate of England and Wales*. London, Her Majesty's Stationary Office.

Sobey, D. G. 1981 *Stellaria media* (L.) Vill. *Journal of Ecology* 69, 311–335.

Stamp, L. D. and Beaver, S. H. 1971 *The British Isles. A Geographical and Economic Study*. London, Longman (sixth edition).

Straker, V. 1984 First and second century carbonized cereal grain from Roman London. In W. van Zeist and W. A. Casparie (eds.) *Plants and Ancient Man*. Rotterdam, Balkema, 323–329.

Stuiver, M. and Kra, R. 1986 Proceedings of the 12th International Radiocarbon Conference, Trondheim, Norway. *Radiocarbon* 28, No. 2B, 805–1030.

Stuiver, M. and Pearson, G. W. 1986 High-precision calibration of the radiocarbon time scale, AD 1950–500 BC. *Radiocarbon* 28, No. 2B, 805–838.

Thran, P. and Broekhuizen, S. 1965 *Agro-Climatic Atlas of Europe*. Wageningen, Pudoc, and Amsterdam, Elsevier.

Tomlinson, P. forthcoming The charred cereal deposit from Area I, 86.53, context 841. In M. Gelling (ed.) Peel Castle Excavations 1982–1987: Final Report (compiled by D. Freke), Liverpool, Liverpool University Press.

Topping, P. 1989a The context of cord rig cultivation in later prehistoric Northumberland. In M. Bowden, D. Mackay and P. Topping (eds.) *From Cornwall to Caithness. Some Aspects of British Field Archaeology*. Oxford, British Archaeological Reports, British Series 209, 145–157.

Topping, P. 1989b Early cultivation in Northumberland and the Borders. *Proceedings of the Prehistoric Society* 55, 161–179.

Turnbull, P. and Fitts, L. forthcoming The excavation at Rock Castle, North Yorkshire.

Turner, J. 1979 The environment of northeast England during Roman times as known by pollen analysis. *Journal of Archaeological Science* 6, 285–290.

Turner, J. 1981a The vegetation. In M. Jones and G. Dimbleby (eds.) *The Environment of Man: the Iron Age to the Anglo-Saxon Period*. Oxford, British Archaeological Reports, British Series 87, 67–73.

Turner, J. 1981b The Iron Age. In I. G. Simmons and M. J. Tooley (eds.) *The Environment in British Prehistory*. London, Duckworth, 250–281.

Turner, J. 1983 Some pollen evidence for the environment of northern Britain, 1000 BC to 1000 AD. In J. C. Chapman and H. C. Mytum (eds.) *Settlement in North Britain 1000 BC – AD 1000*. Oxford, British Archaeological Reports, British Series 118, 1–27.

Turner, J., Hewetson, V. P., Hibbert, F. A., Lowry, K. H. and Chambers, C. 1973 The history of the vegetation and flora of Widdybank Fell and the Cow Green reservoir basin, Upper Teesdale. *Philosophical Transactions of the Royal Society of London* Series B, 265, 327–408.

Turner, J. and Kershaw, A. P. 1973 A late- and post-glacial pollen diagram from Cranberry Bog, near Beamish, County Durham. *New Phytologist* 72, 915–928.

Tüxen, R. 1950 Grundriss einer Systematik der nitrophilen Unkrautgesellschaften in der Eurosiberischen Region Europas. *Mitteilungen der Floristisch-Soziologischen Arbeitsgemeinschaft* 2, 94–175.

Veen, M. van der 1982 Botanical remains from Hallshill Farm, Northumberland. Preliminary report. *Ancient Monuments Laboratory Report* No. 3745.

Veen, M. van der 1984 Sampling for seeds. In W. van Zeist and W. A. Casparie (eds.) *Plants and Ancient Man*. Rotterdam, Balkema, 193–199.

Veen, M. van der 1985a Carbonized seeds, sample size and on-site sampling. In N. R. J. Fieller, D. D. Gilbertson and N. G. A. Ralph (eds.) *Palaeoenvironmental Investigations*. Oxford, British Archaeological Reports, International Series 258, 165–174.

Veen, M. van der 1985b Evidence for crop plants from north-east England: an overview with discussion of new results. In N. R. J. Fieller, D. D. Gilbertson and N. G. A. Ralph (eds.) *Palaeobiological Investigations*. Oxford, British Archaeological Reports, International Series 266, 197–219.

Veen, M. van der 1985c Carbonized plant remains. In G. J. Barclay: Excavations at Upper Suisgill, Sutherland.

Proceedings of the Society of Antiquaries of Scotland 115, 159–198.

Veen, M. van der 1985d The plant remains. In R. Miket: Ritual enclosures at Whitton Hill, Northumberland. *Proceedings of the Prehistoric Society* 51, 137–148.

Veen, M. van der 1987a The plant remains. In D. H. Heslop, *The Excavation of an Iron Age Settlement at Thorpe Thewles, Cleveland, 1980–1982*. London, Council for British Archaeology, Research Report No. 65, 93–99.

Veen, M. van der 1987b Plant remains. In I. and G. Jobey: Prehistoric, Romano-British and later remains on Murton High Crags, Northumberland. *Archaeologia Aeliana* 5th series, 15, 151–198.

Veen, M. van der 1988a Carbonized grain from a Roman granary at South Shields, North-East England. H. Küster (ed.) *Der Prähistorische Mensch und Seine Umwelt*. Forschungen und Berichte zur Vor- und Frühgeschichte in Baden-Württemberg, Band 31, Stuttgart, Theiss, 353–365.

Veen, M. van der 1988b The plant remains. In B. E. Vyner: The hill-fort at Eston Nab, Eston, Cleveland. *The Archaeological Journal* 145, 60–98.

Veen, M. van der 1988c The plant remains. In D. Coggins and L. J. Gidney, A late prehistoric site at Dubby Syke, Upper Teesdale, Co. Durham. *Durham Archaeological Journal* 4, 1–12.

Veen, M. van der 1989a A wheat experiment in Britain. *Archäobotanik. Dissertationes Botanicae* 133, 51–56.

Veen, M. van der 1989b National wheat experiment: interim report 1987/8. *Circaea* 6, No. 1, 71–76.

Veen, M. van der 1990 The plant remains. In C. Smith: The Excavations at Dod Law West Hillfort, Northumberland. *Northern Archaeology* 9, 33–38.

Veen, M. van der 1991 Consumption or production: Agriculture in the Cambridgeshire Fens? In J. M. Renfrew (ed.) *New Light on Early Farming*. Edinburgh, Edinburgh University Press, 349–361.

Veen, M. van der forthcoming a The plant remains. In T. Gates: The Excavations at Hallshill Farm, Northumberland.

Veen, M. van der forthcoming b The plant remains from Chesterhouse, Northumberland. *Archaeologia Aeliana*.

Veen, M. van der forthcoming c The plant remains. In P. Turnbull and L. Fitts: The Excavations at Rock Castle, North Yorkshire.

Veen, M. van der forthcoming d The plant remains. In P. Bidwell: The Roman Fort at South Shields. The Excavations of the Headquarters and the South West Gate.

Veen, M. van der forthcoming e The results of the national wheat experiment, 1987–1990. *Circaea*.

Veen, M. van der forthcoming f The plant remains from Newstead, Scotland. In R. F. J. Jones, The excavations at Newstead.

Veen, M. van der and Fieller, N. 1982 Sampling Seeds. *Journal of Archaeological Science* 9, 287–298.

Veen, M. van der and Haselgrove, C. C. 1983 Evidence for pre-Roman crops from Coxhoe, Co. Durham. *Archaeologia Aeliana* 5th series, 11, 23–25.

Vyner, B. E. 1988 The hill-fort at Eston Nab, Eston, Cleveland. *The Archaeological Journal* 145, 60–98.

Walters, , M. 1949 *Eleocharis palustris* (L.) R. Br. em R. and S. *Journal of Ecology* 37, 194–202.

Warington, K. 1924 The influence of manuring on the weed flora of arable land. *Journal of Ecology* 12, 111–126.

Wasylikowa, K. 1978 Plant remains from early and late

Medieval time found on the Wawel Hill in Cracow. *Acta Palaeobotanica* 19, part 2, 115–200.

Wasylikowa, K. 1981 The role of fossil weeds for the study of former agriculture. *Zeitschrift für Archäologie* 15, 11–23.

Waterbolk, H. T. 1971 Working with radiocarbon dates. *Proceedings of the Prehistoric Society* 37, 15–33.

Waterbolk, H. T. 1983 Ten guidelines for the archaeological interpretation of radiocarbon dates. *Pact* 8, II.2, 57–70.

Welch, D. 1966 *Juncus squarrosus* L. *Journal of Ecology* 54, 535–548.

Westhoff, V. and Held, A. J. den 1975 *Plantengemeenschappen in Nederland*. Zutphen, Thieme (second edition).

Westhoff, V. and Maarel, E. van der 1973 The Braun-Blanquet Approach. In R. H. Whittaker (ed.) *Ordination and Classification of Communities*. Handbook of Vegetation Science, part V. The Hague, Junk Publishers, 617–726.

Wheeler, M. 1954 *The Stanwick Fortifications*. Oxford, Society of Antiquaries.

Whittaker, R. H. 1973 Approaches to classifying vegetation. R. H. Whittaker(ed.) *Ordination and Classification of Communities*. Handbook of Vegetation Science, part V. The Hague, Junk Publishers, 323–354.

Whittle, A. 1989 Early Linearbandkeramik (LBK) sites in Central Europe. *Archaeometry* 31, 2, 224–226.

Willerding, U. 1978 Paläo-Ethnobotanische Befunde an mittelalterliche Pflanzenreste aus Süd-Niedersachsen, Nord-Hessen und dem östlichen Westfalen. *Berichte der Deutsche Botanische Gesellschaft* 91, 129–160.

Willerding, U. 1979 Zum Ackerbau in der jüngeren vorrömischen Eisenzeit. *Archaeophysika* 8, 309–330.

Willerding, U. 1980 Anbaufrüchte der Eisenzeit und des frühen Mittelalters, ihre Anbauformen, Standortsverhältnisse und Erntemethoden. In: *Untersuchungen zur eisenzeitlichen und frühmittelalterlichen Flur in Mitteleuropa und ihre Nutzung*. Tl. II. (= Abhandlungen der Akademie der Wissenschaften in Göttingen, phil.-hist. Klasse, 3. Folge, Nr. 116). Göttingen, 126–191.

Willerding, U. 1981 Ur- und frühgeschichtliche sowie mittelalterliche Unkrautfunde in Mitteleuropa. *Zeitschrift für Pflanzenkrankheiten und Pflanzenschutz*, Sonderheft 9, 65–74.

Willerding, U. 1983a Paläo-Ethnobotanik und Ökologie. *Verhandlungen der Gesellschaft für Ökologie* 11 (Festschrift Ellenberg), 489–503.

Willerding, U. 1983b Zum ältesten Ackerbau in Niedersachsen. *Archäologische Mitteilungen aus Nordwestdeutschland* 1, 179–219.

Willerding, U. 1984 Paläo-Ethnobotanische Analyse der Getreideschicht. In I. Gabriel, *Starigard/Oldenburg. Hauptburg der Slawen in Wagrien*. Neumünster, 66–74.

Willerding, U. 1988 Zur Entwicklung von Ackerunkrautgesellschaften im Zeitraum vom Neolithikum bis in die Neuzeit. In H. Küster (ed.) *Der Prähistorische Mensch und Seine Umwelt*. Forschungen und Berichte zur Vor- und Frühgeschichte in Baden-Württemberg, Band 31, Stuttgart, Theiss, 31–41.

Williams, J. T. 1963 *Chenopodium album* L. *Journal of Ecology* 51, 711–725.

Williams, W. T. and Gillard, P. 1971 Pattern analysis of a grazing experiment. *Australian Journal of Agricultural*

Research 22, 245–260.

Willis, A. J. 1973 *Introduction to Plant Ecology*. London, George Allan and Unwin.

Wilson, D. 1983 Pollen analysis and settlement archaeology of the first millennium BC from North-East England. In J. C. Chapman and H. C. Mytum (eds.) *Settlement in North Britain 1000 BC – AD 1000*. Oxford, British Archaeological Reports, British Series 118, 29–53.

Young, R. 1987a *Lithics and Subsistence in North-Eastern England*. Aspects of the Prehistoric Archaeology of the Wear Valley, Co. Durham, from the Mesolithic to the Bronze Age. Oxford, British Archaeological Reports, British Series 161.

Young, R. 1987b Space, pattern and time: some prehistoric and early historic settlement problems in Northern Britain. *Scottish Archaeological Review* 4, part 2, 108–115.

Younger, D. A. 1986 The small mammals from a Roman granary at Arbeia Fort, South Shields. Unpublished typescript. To be published in Bidwell, forthcoming.

Zeist, W. van 1968 Prehistoric and early historic food plants in the Netherlands. *Palaeohistoria* 14, 41–173.

Zeist, W. van 1974 Palaeobotanical studies of settlement sites in the coastal area of The Netherlands. *Palaeohistoria* 16, 223–371.

Zeist, W. van 1981 Plant remains from Iron Age Noordbarge, Province of Drenthe, The Netherlands. *Palaeohistoria* 23, 169–193.

Zeist, W. van 1988 Botanical evidence of relations between the sand and clay districts of the north of The Netherlands in medieval times. In H. Küster (ed.) *Der Prähistorische Mensch und Seine Umwelt*. Forschungen und Berichte zur Vor- und Frühgeschichte in Baden-Württemberg, Band 31, Stuttgart, Theiss 381–387.

Zeist, W. van, and Neef, R. 1983 Plantenresten uit vroeg-historisch Leeuwarden. In W. van Zeist, R. Neef, D. C. Brinkhuizen and S. Jager, *Planten-, Vis-, en Vogelresten uit vroeg-historisch Leeuwarden*. Leeuwarden, Commissie Archeologisch Stadskernonderzoek Leeuwarden, 6–11.

Zeist, W. van, and Palfenier-Vegter, R. M. 1979 Agriculture in Medieval Gasselte. *Palaeohistoria* 21, 267–299.

Zeist, W. van and Palfenier-Vegter, R. M. 1981 Seeds and fruits from the Swifterbant S3 site. *Palaeohistoria* 23, 105–168.

Zeist, W. van, Roller, G. J. de, Palfenier-Vegter, R. M., Harsema, O. H. and During, H. 1986 Plant remains from medieval sites in Drenthe. *Helinium* 26, 226–274.

Zeist, W. van, Cappers, R., Neef, R. and During, H. 1987 A palaeobotanical investigation of medieval occupation deposits in Leeuwarden, The Netherlands. *Proceedings of the Koninklijke Nederlandse Akademie van Wetenschappen* Series B, 90/4, 371–426.

Zeuner, F. E. 1963 *A History of Domesticated Animals*. London, Hutchinson.